POLYIMIDES

Synthesis, Characterization, and Applications

Volume 1

POLYIMIDES

Synthesis, Characterization, and Applications

Volume 1

Edited by

K. L. Mittal

IBM Corporation
Hopewell Junction, New York

Plenum Press • New York and London

7227-0433

CHEMISTRY

Library of Congress Cataloging in Publication Data

Technical Conference on Polyimides (1st: 1982: Ellenville, N.Y.)
  Polyimides: synthesis, characterization, and applications.

  Proceedings of the First Technical Conference on Polyimides, held under the
auspices of the Mid-Hudson Section of the Society of Plastics Engineers,
November 10–12, 1982, in Ellenville, New York

  Includes bibliographical references and indexes.
  1. Polyimides—Congresses. I. Mittal, K. L., 1945–   . II. Title.
TP1180.P66T43   1982   668.9   84-3316
ISBN 0-306-41670-0 (v. 1)
ISBN 0-306-41673-5 (v. 2)

Proceedings of the First Technical Conference on Polyimides, held under the
auspices of the Mid-Hudson Section of the Society of Plastics Engineers,
November 10–12, 1982, in Ellenville, New York

©1984 Plenum Press, New York
A Division of Plenum Publishing Corporation
233 Spring Street, New York, N.Y. 10013

All rights reserved

No part of this book may be reproduced, stored in a retrieval system, or transmitted
in any form or by any means, electronic, mechanical, photocopying, microfilming,
recording, or otherwise, without written permission from the Publisher

Printed in the United States of America

TP1180
P66 T43
1982
v. 1
CHEM

PREFACE

This and its companion Volume 2 chronicle the proceedings of the First Technical Conference on Polyimides: Synthesis, Characterization and Applications held under the auspices of the Mid-Hudson Section of the Society of Plastics Engineers at Ellenville, New York, November 10-12, 1982.

In the last decade or so there has been an accelerated interest in the use of polyimides for a variety of applications in a number of widely differing technologies. The applications of polyimides range from aerospace to microelectronics to medical field, and this is attributed to the fact that polyimides offer certain desirable traits, _inter alia_, high temperature stability. Polyimides are used as organic insulators, as adhesives, as coatings, in composites, just to name a few of their uses. Even a casual search of the literature will underscore the importance of this class of materials and the high tempo of R&D activity taking place in the area of polyimides.

So it was deemed that a conference on polyimides was both timely and needed. This conference was designed to provide a forum for discussion of various ramifications of polyimides, to bring together scientists and technologists interested in all aspects of polyimides and thus to provide an opportunity for cross-pollination of ideas, and to highlight areas which needed further and intensified R&D efforts. If the comments from the attendees are a barometer of the success of a conference, then this event was highly successful and fulfilled amply its stated objectives.

The technical program consisted of 69 papers (45 oral presentations and 24 posters) by about 130 authors from many corners of the globe. The purpose of a conference or symposium is to present the state of knowledge of the topic under consideration and it can best be accomplished by a blend of invited overviews and original unpublished research contributions. This is exactly what was done in this conference. The program contained a number of invited overviews on certain subtopics, and these were augmented by contributed original research papers. The invited speakers were selected so as to represent differing disciplines and interests and they hailed from academic, governmental and industrial research laboratories.

As for the present proceedings volumes, the papers have been rearranged (from the order they were presented) so as to fit them in a more logical fashion. Incidentally, these volumes also contain nine papers which were not included in the formal printed program. Also it should be recorded here that, for a variety of reasons, a few papers which were presented are not included in these volumes. It must be emphasized here that these proceedings volumes are not simply a collection of papers, but the peer review was an integral part of the total editing process. All papers were critically reviewed by qualified reviewers as the comments of peers are a desideratum to maintain high standard of publications. Also it should be recorded that although no formal discussion is included in these proceedings, but there were many enlightening, not exothermic, and lively discussions both formally in the auditorium and in more relaxed places. Particular mention should be made here regarding the poster presentations. The poster session was highly successful as can be gauged from the comments of those who participated in this format.

Coming back to the proceedings volumes, these contain a total of 71 papers (1182 pages) by 164 authors from nine countries. The text is divided into five sections as follows: Synthesis and Properties; Properties and Characterization; Mechanical Properties; Microelectronic Applications; and Aerospace and other Applications. Sections I and II constitute Volume 1, and Sections III-V grace the pages of Volume 2. The topics covered include: synthesis, properties and characterization of a variety of polyimides; metal-containing polyimides; cure kinetics of polyimides; structure-property relationships; polyimide adhesion; mechanical properties of polyimides both in bulk state as well as in thin film form; applications of polyimides in microelectronics; photosensitive polyimides; polyimides as adhesives; aerospace applications of polyimides; polyimide blends; electrophoretic deposition of polyimides; and applications of polyimides in electrochemical and medical fields.

Even a cursory glance at the Table of Contents will convince that there is a great deal of R&D activity in the area of this wonderful class of materials, and all signals indicate that this tempo is going to continue. It is hoped that these proceedings volumes which represent the repository of latest knowledge about polyimides will be useful to both the seasoned researcher (as a source of latest information) and to the neophyte as a fountain of new ideas. As we learn more about this unique class of materials, more pleasant applications will emerge.

Apropos, a comment should be made here regarding the nomenclature being used in the field of polyimides. Dr. Paul Frayer (Rogers Corp.) pointed out that there was a great deal of inconsistency in the nomenclature and different authors were using quite different terms for chemically identical species. He further suggested that a consistent method of nomenclature was needed and urged the attendees to come up with a consensus on this issue and to take the requisite action in this regard in a future conference on polyimides.

Acknowledgements. First of all I am thankful to the Mid-Hudson Section of the Society of Plastics Engineers for sponsoring this event, to Dr. H.R. Anderson, Jr. (IBM Corporation) for permitting me to organize the technical program and to Steve Milkovich (IBM Corporation) for his understanding and cooperation during the tenure of editing. My special thanks are due to the Organizing Committee members (C. Araps, P. Buchwalter, G. Czornyj, M. Gupta, J. Schiller and M. Turetzky) for their invaluable help, and in particular to the General Chairman (Julius Schiller) for his continued interest in and overall responsibility and organization of this conference. All members of the Organizing Committee worked hard in many capacities and unflinchingly devoted their time to make a conference of this magnitude a grand success. Special appreciation is extended to P. Hood, M. Htoo, R. Martinez, R. Nufer and B. Washo for their help and cooperation in more ways than one.

On a more personal note, I am thankful to my wife, Usha, for tolerating, without much complaint, the frequent privations of an editor's wife and for helping me in many ways. I am appreciative of my darling children (Anita, Rajesh, Nisha and Seema) for not only letting me use those hours which rightfully belonged to them but also for rendering home environment conducive to work. Special thanks are due to Phil Alvarez (Plenum Publishing Corporation) for his continued interest in this project, and to Barbara Mutino (Office Communications) for meeting promptly various typing deadlines. The time and effort of the unheralded heroes (reviewers) is gratefully acknowledged. Last, but not least, the enthusiasm, cooperation and contributions of the authors are certainly appreciated without which we could not have these volumes.

K. L. Mittal
IBM Corporation
Hopewell Junction, New York 12533

CONTENTS

PART I: SYNTHESIS AND PROPERTIES

CONTENTS OF VOLUME 2

PART V: AEROSPACE AND OTHER APPLICATIONS

PART I. SYNTHESIS AND PROPERTIES

SYNTHESIS AND CHARACTERIZATION OF ORGANO-SOLUBLE, THERMALLY-CURABLE, PHENYLATED AROMATIC POLYIMIDES

F.W. Harris*, S.O. Norris, L.H. Lanier,
B.A. Reinhardt, R.D. Case, S. Varaprath,
S.M. Padaki, M. Torres and W.A. Feld

Department of Chemistry
Wright State University
Dayton, Ohio 45435

Phenylated aromatic polyimides have been prepared in two steps by the Diels-Alder polymerization of bis-cyclopentadienones with dimaleimides followed by dehydrogenation and in one step by the polymerization of phenylated bis(phthalic anhydrides) with diamines. Although the polymers prepared by both routes were soluble in chlorinated-hydrocarbon solvents, the latter afforded considerably higher molecular weight materials. Polyimides obtained in this manner had intrinsic viscosities as high as 4.0 dl/g and glass transition temperatures that ranged from 238 to 466°C. Thermogravimetric analysis thermograms of the polymers in air and nitrogen atmospheres showed no weight loss until near 530°C. Two new ethynyl-substituted diamines, 4-(2,4-diaminophenoxy)phenylacetylene (1) and 3,5-diaminodiphenylacetylene (2), have been synthesized and polymerized with a phenylated bis(phthalic anhydride). The polymer prepared with diamine 1 underwent cross-linking at 250°C without the evolution of volatile by-products, while the polymer prepared with diamine 2 underwent an analogous cure at 350°C. A phenylated bis(phthalic anhydride) has also been copolymerized with pyromellitic dianhydride (PMDA) and 4,4'-diaminodiphenyl ether to afford soluble copolymers that contain as high as 50 mole % PMDA.

* Present address: Institute of Polymer Science, University of Akron, Akron, OH 44325

3

INTRODUCTION

Although aromatic polyimides retain usable properties at 300°C for months and withstand exposures of a few minutes to temperatures well over 500°C, their applications have been limited[1-3]. This is due to their insolubility which, combined with their high transition temperatures, makes them extremely difficult to fabricate. In fact, these polymers are processed in the form of their polyamic acid precursors, which are then thermally or chemically converted to the imide structure. Processing is still difficult, however, because the precursors are thermally and hydrolytically unstable and because water is evolved during the cure process.

A major goal of this laboratory for the past 10 years has been the preparation of aromatic polyimides that can be easily fabricated. Thus, considerable research has been carried out aimed at the synthesis of polyimides that are (1) soluble in common organic solvents, (2) melt-processable, and (3) thermally curable without the evolution of volatile by-products. Our approach to obtaining solubility in these systems has involved the introduction of pendent phenyl groups along the polyimide backbone. (Phenylated polyphenylenes and phenylated polyquinoxalines[5,6] are soluble in common organic solvents, in contrast to the insolubility of the parent polymers.) The melt-processability of the polyimide systems has been enhanced by lowering their flow temperatures through the incorporation of aryl-ether and meta-phenylene linkages in the backbone. Pendent ethynyl and phenylethynyl groups have also been investigated as potential crosslinking sites. (Ethynyl-terminated polyimide oligomers undergo thermal crosslinking near 250°C without the evolution of volatile by-products[7].) The following is a review of our work in this area.

RESULTS AND DISCUSSION

Diels-Alder Polymerization of Phenylated Biscyclopentadienones and Dimaleimides[8,9].

Our initial approach to obtaining phenylated polyimides involved the Diels-Alder polymerization of phenylated biscyclopentadienones with dimaleimides. The reactions of 3,3'-(p-phenylene)bis(2,4,5-triphenylcyclopentadienone) (1a) and 3,3'-(oxydi-p-phenylene)bis(2,4,5-triphenylcyclopentadienone) (1b) with o-,m-, and p-phenylenedimaleimide (2a-c) afforded quantitative yields of the corresponding polydihydrophthalimides. The polymers attained their maximum intrinsic viscosities in 1 to 3 hr in refluxing α-chloronaphthalene or in 18 to 24 hr in refluxing 1,2,4-trichlorobenzene. In each case, additional heating at reflux resulted in a decrease in viscosity. Since the polymerization reactions are not reversible, decomposition must have occurred, possibly via the dihydrophthalimide linkage.

$$O \quad C_6H_5 \quad C_6H_5 \quad O$$

$$C_6H_5 \quad \text{—Ar—} \quad C_6H_5 \quad + \quad N\text{—}Ar'\text{—}N$$

$$C_6H_5 \quad C_6H_5$$

$\underline{1a}$ Ar = $\underline{p}$-C_6H_4 $\underline{2a}$ Ar$'$= $\underline{o}$-C_6H_4

$\underline{1b}$ Ar = —⟨ ⟩—O—⟨ ⟩—   −CO $\underline{2b}$ Ar$'$= $\underline{m}$-C_6H_4

$\underline{2c}$ Ar$'$= $\underline{p}$-C_6H_4

$$C_6H_5 \quad C_6H_5 \quad C_6H_5 \quad C_6H_5$$

$$\left(N \quad \text{— Ar —} \quad N\text{—}Ar'\right)$$

$$C_6H_5 \quad C_6H_5$$

$\underline{3}$

The polydihydrophthalimides were soluble in DMF and chlorin-ated-hydrocarbon solvents. Their intrinsic viscosities in DMF at 30°C ranged from 0.3 to 1.0 dl/g. Flexible light-yellow films could be cast from chloroform solutions. Thermogravimetric analysis (TGA) thermograms of all the polymers showed an initial break in air near 300°C followed by gradual weight loss to 530°C where complete de-composition occurred (Figure 1). The small weight loss near 300°C was due to dehydrogenation accompanied by slight decomposition.

The polydihydrophthalimides were dehydrogenated by stirring in refluxing nitrobenzene for 12 hr to afford the corresponding aro-matic polyphthalimides. The conversions to the aromatic polyimides, however, were evidently not quantitative, as the TGA thermograms of these polymers still showed a slight weight loss near 300°C. Unfor-tunately, heating the polydihydrophthalimides in nitrobenzene also resulted in a sharp decrease in their intrinsic viscosities. Quan-titative dehydrogenation of the polydihydrophthalimides was accom-plished by heating finely powdered samples under vacuum or in a nitrogen atmosphere to 350°C. The TGA thermograms of the poly-phthalimides prepared in this manner showed no weight loss in air

5

until near 530°C, where complete decomposition occurred (Figure 1). Thermal dehydrogenation, however, also resulted in a sharp decrease in intrinsic viscosity.

The low-molecular-weight polyimides were soluble in DMF, but formed brittle, dark-yellow films. The infrared spectrum of each polymer showed bands characteristic of aromatic polyimides and could be compared directly to spectra of similar model compounds.

Polymerization of Phenylated Bis(phthalic anhydrides) with Diamines.[8,10,11]

Since dehydrogenation of the polydihydrophthalimides was accompanied by decomposition, an alternate synthetic route to polyphthalimides involving the polymerization of phenylated dianhydrides was devised. The reactions of the biscyclopentadienones 1a-c with maleic anhydride followed by dehydrogenation with bromine provided the bis(phthalic anhydrides) 4a-c.

The dianhydrides were then polymerized with a series of diamines in m-cresol containing isoquinoline. The diamines used were m- and p-phenylenediamine, 4,4'-diaminodiphenyl ether, 4,4'-diaminodiphenylmethane, 4,4'-diaminobiphenyl, and 1,3-bis(3-aminophenoxy)benzene. The polymerization mixtures were stirred at ambient temperature for 1 hr and then heated at 202–203°C for 3 hr. Under these conditions, the intermediate polyamic acid was converted to the corresponding polyimide. The water that evolved from imidization was removed by distillation. The reaction mixtures were added to absolute ethanol to precipitate the white polyimides in nearly quantitative yields. The polymers were then heated under vacuum at 250°C for 4 hr to insure complete imidization. Infrared analysis indicated that 1 to 5% of the imide rings were not closed prior to the heat treatment.

$$\underline{4a\text{-}c} \ + \ H_2N\text{-}Ar'\text{-}NH_2 \ \longrightarrow$$

5

The phenylated polyimides were soluble in chlorinated hydro-carbons and had intrinsic viscosities as high as 4.0 dl/g (Table I). Films cast from chloroform were slightly yellow, transparent, and highly flexible. The infrared spectra of the films showed absorption bands at 1780 and 1730 cm^{-1}, characteristic of imides and could be compared directly to the spectra of model compounds.

The glass transition temperatures (Tg's) of the polymers ranged from 238 to 466°C. The incorporation of a flexible ether linkage in the dianhydride monomer resulted in a lower polymer Tg than the incorporation of the same unit in the diamine (compare polymers 5c and 5f). In fact, this observation was made during the course of the work and led to the synthesis of dianhydride 4c. This monomer was used in the preparation of polymer 5k, which displayed the lowest Tg. (Thermomechanical analysis of polymer 5k employing a penetration mode (10-g load) showed initial penetration at 160°C with maximum penetration occurring near 238°C.) TGA thermograms of the phenylated phthalimides showed no weight loss in air or nitro-gen atmospheres until near 530°C (Figure 1). In nitrogen, after an initial 25-30% weight loss near 530°C, the polymers essentially maintained their weight to 900°C. The TGA thermograms were similar to those obtained for unsubstituted polypyromellitimides[3]. Iso-thermal aging studies of the polymers showed no weight loss after 72 hr in air at 350°C or in nitrogen at 400°C. The polymers did undergo decomposition in air at 400°C (Figure 2).

Synthesis and Cure of Phenylated Polyimides Containing Pendent Ethynyl Substituents.[12-15]

In order to prepare thermally-crosslinkable systems, two new ethynyl-substituted diamines, 4-(2,4-diaminophenoxy)phenylacetylene (6) and 3,5-diaminodiphenylacetylene (7) were synthesized by the following routes:

Table I. Soluble, Phenylated Polyimides.

Polymer	Ar	Ar'	$[\eta]^a$	$T_g{}^b (°C)$
5a			0.35	466
5b			0.86	416
5c			1.11	413
5d			1.22	280
5e			4.03	399
5f			0.40	371
5g			2.80	360
5h			1.06	357
5i			1.38	356
5j			0.75	261
5k			1.00	238

(a) Intrinsic viscosity obtained at 30°C in sym-tetrachloroethane.

(b) Average values of the highest penetration rate on TMA at ΔT= 20°C/min (10 g load).

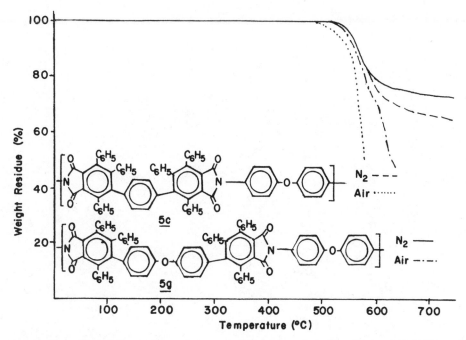

Figure 1. Thermogravimetric analysis of polymers 5c and 5g (heating rate 5° C/min).

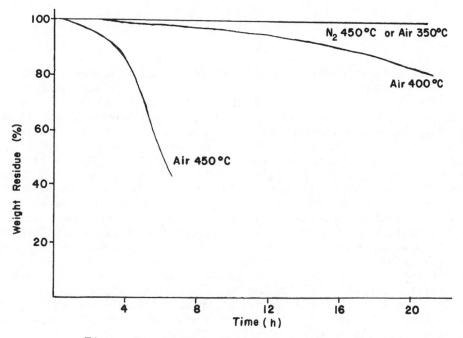

Figure 2. Isothermal aging of polymer 5g.

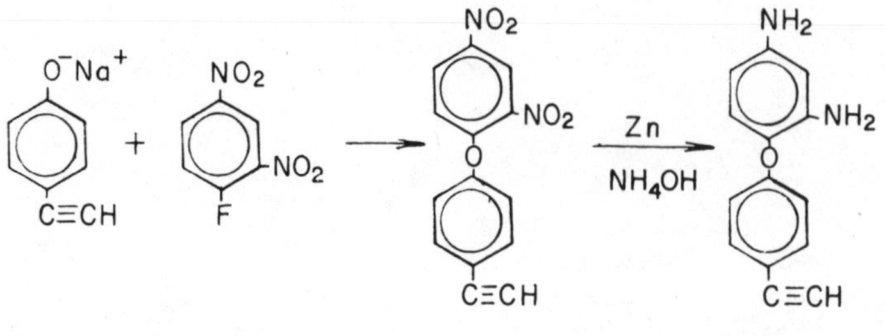

Scheme 1 (structures):

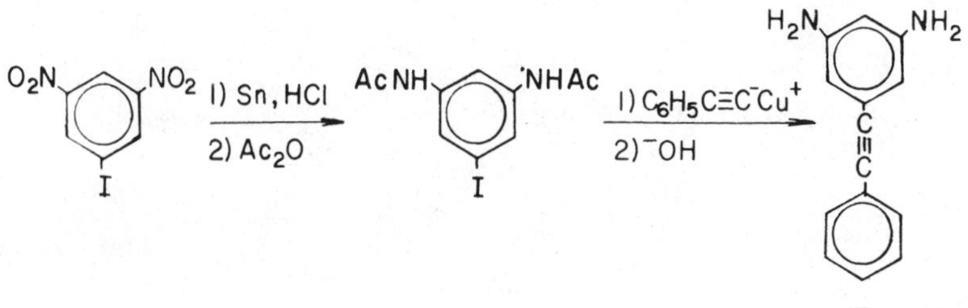

O⁻Na⁺ — benzene ring with C≡CH (para) + NO₂ / NO₂ substituted fluorobenzene (F) → diphenyl ether with two NO₂ groups and C≡CH, then Zn / NH₄OH → diamino diphenyl ether with C≡CH

$$\text{(top ring: NO}_2\text{, NO}_2\text{)} \quad O \quad \text{(bottom ring: C}\equiv\text{CH)}$$

$$\xrightarrow[\text{NH}_4\text{OH}]{\text{Zn}}$$

compound **6** (NH₂, NH₂ diphenyl ether with C≡CH)

Scheme 2 (structures):

O₂N / NO₂ substituted iodobenzene (I) → 1) Sn, HCl 2) Ac₂O → AcNH / NHAc substituted iodobenzene (I) → 1) C₆H₅C≡C⁻Cu⁺ 2) ⁻OH →

compound **7** (H₂N, NH₂ ring with C≡C–C₆H₅)

The polymerization of 6 with 4b in refluxing m-cresol resulted in an insoluble crosslinked polymer. Hence, the polymerization was carried out at a lower temperature (140–150°C) in an m-cresol/toluene mixture containing isoquinoline. The water that evolved during the polymerization was removed as an azeotrope by distillation. Polymer 8, however, had an inherent viscosity of only 0.2 (0.250 g/dl in sym-tetrachloroethane at 30°C) and formed brittle films from chloroform solutions. The low molecular weight was probably due to the sterically hindered 2-amino group in 6, which was expected to be less reactive in the polymerization. The differential scanning calorimetry (DSC) thermogram of 8 showed a strong exotherm near 250°C, which is characteristic of the thermal polymerization of ethynyl groups. The TGA thermogram of the uncured polymer showed an initial weight loss near 400°C. Samples could be crosslinked by heating at 250°C for 3 hr. Completely insoluble material could also be obtained by heating the polymer for 4 hr at 225°C, for 6 hr at 210°C, or for 24 hr at 190°C. The DSC thermogram of the cured polymer showed no exotherm, and the TGA thermogram showed no weight loss in air before 400°C.

10

The polymerization of diamine 7 with 4b was then carried out under the conditions described for polymer 5. Polyimide 9, which was obtained in a nearly quantitative yield, was soluble in aromatic and chlorinated-hydrocarbon solvents and had an inherent viscosity of 0.67 (0.250 g/dl in sym-tetrachloroethane at 30°C). TMA with a penetration mode showed a Tg near 355°C. The TGA thermogram indicated that the polymer was stable to near 500°C in nitrogen and 450°C in air. The DSC thermogram showed a broad exotherm near 450°C, which was attributed to the thermal polymerization of the pendent phenylethynyl groups. Crosslinking could be carried out by heating the polymer under vacuum at 350°C for 4 hr. However, flexible films cured in this manner became quite brittle, which is indicative of a high crosslink density.

In order to lower the crosslink density in the cured resins, two copolymerizations of 7 and 4,4'-diaminodiphenyl ether (10) with dianhydride 4b were carried out. The molar ratios of 7 to 10 employed were 10:90 and 20:80. The copolymers obtained (11a,b) had inherent viscosities of 0.82 and 0.69 (0.250 g/dl in sym-tetrachloroethane at 30°C), respectively. The copolymer's properties were very similar to those of polymer 9. Both also underwent crosslinking when heated at 350°C for 4 hr. The reduction in crosslink density was witnessed by the fact that thin, flexible films of each material remained flexible after curing.

Copolymerization of Phenylated Bis(phthalic anhydrides) and Pyromellitic Dianhydride (PMDA) with Diamines.[16]

In an attempt to reduce the cost of phenylated systems, the copolymerization of 4b and PMDA and diamines was investigated. Thus, various amounts of 4b and PMDA were copolymerized with 10 in m-cresol containing isoquinoline. The molar ratios of the two dianhydrides employed varied from 75:25 to 25:75. Polymers prepared

11

7 + 4b →

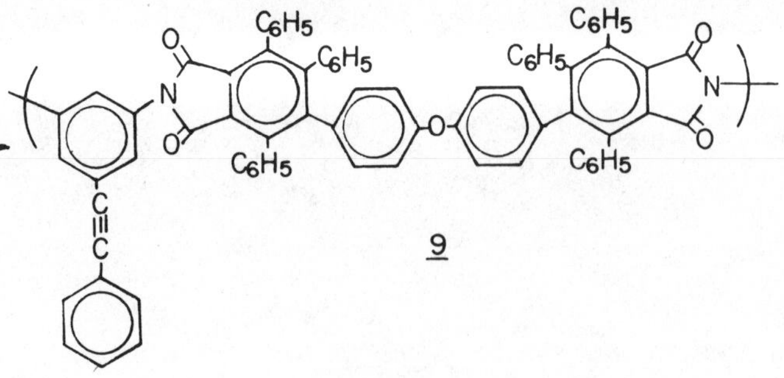

9

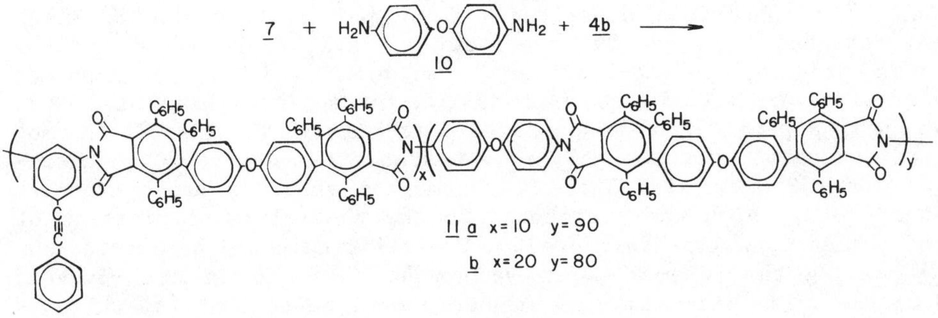

7 + H₂N—⬡—O—⬡—NH₂ + 4b ⟶

10

11 a x=10 y=90

b x=20 y=80

with up to 50 mole % PMDA remained soluble in chlorinated-hydro-carbon solvents. Polymers containing more than this amount of the symmetrical dianhydride, however, were insoluble in common organic solvents.

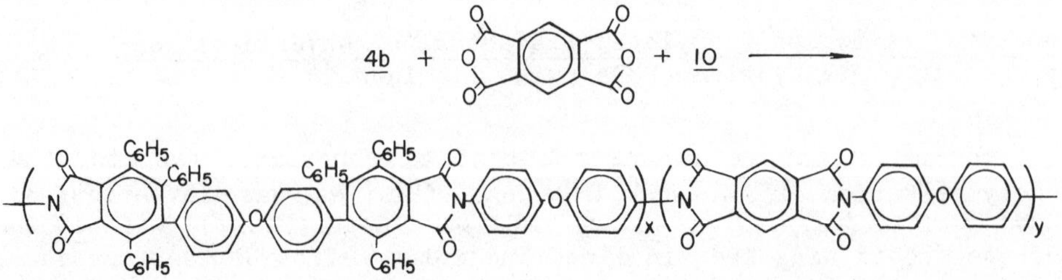

4b + [PMDA dianhydride] + 10 ⟶

CONCLUSIONS

High-molecular-weight polyimides that are soluble in common organic solvents can be prepared by the polymerization of phenylated bis(phthalic anhydrides) with aromatic diamines. These phenylated polymers display excellent thermal stability and can be cast into thin flexible films. Their Tg's can be varied from 238 to 466°C by using different diamine and dianhydride combinations. Thermal crosslinkability can be built into these systems through the use of diamines containing ethynyl and phenylethynyl moieties. Material costs can be reduced by replacing up to 50 mole percent of the phenylated dianhydride with PMDA.

Future work will concentrate on further lowering the Tg's of these systems so they can be melt-processed at temperatures considerably below their cure temperatures. Less expensive routes to phenylated dianhydrides will also be investigated.

ACKNOWLEDGEMENT

Partial support of this research by the Research Corporation and by the Air Force Materials Laboratory, Non-Metals Division, Polymer Branch, WPAFB, OH, is gratefully acknowledged.

REFERENCES

1. C. E. Sroog, A. L. Endrey, S. V. Abramo, C. E. Berr, W. M. Edwards and K. L. Olivier, J. Polym. Sci., A, 3, 1373 (1965).
2. J. I. Jones, F. W. Ochynski and F. A. Rackley, Chem. Ind. (London), 1686 (1962).
3. C. E. Sroog, in "Encyclopedia of Polymer Science and Technology", Vol. 2, H. F. Mark, N. G. Gaylord and N. M. Bikales, Editors, p. 247, Wiley, New York, 1969.
4. H. Mukamal, F. W. Harris and J. K. Stille, J. Polym. Sci., A-1, 5, 2721 (1967).
5. W. Wrasidlo and J. M. Augl, J. Polym. Sci., A-1, 7, 3393 (1969).
6. P. M. Hergenrother and H. H. Levine, J. Polym. Sci., A-1, 5, 1453 (1967).
7. A. L. Landis, N. Bilow, R. H. Boschan and R. E. Lawrence, Am. Chem. Soc. Polym. Div., Polymer Preprints, 15 (2), 537 (1974).
8. F. W. Harris, S. O. Norris, L. H. Lanier and W. A. Feld, Am. Chem. Soc. Div. Org. Coat. Plast., Preprints, 33 (1), 160 (1973).
9. F. W. Harris and S. O. Norris, J. Polym. Sci., A-1, 11, 2143 (1973).
10. F. W. Harris, W. A. Feld and L. H. Lanier, in "Applied Polymer Symposium No. 26", N. Platzer, Editor, pp. 421-428, Wiley, New York, 1975.

11. F. W. Harris, B. A. Reinhardt, R. D. Case and W. A. Feld, Am. Chem. Soc. Polym. Div., Polymer Preprints, 19 (1), 556 (1978).

12. S. Varaprath, "Synthesis of Phenylated Polyimides Containing Pendent Phenylethynyl Groups", M. S. Thesis, Wright State University, Dayton, Ohio, 1977.

13. S. Padaki, "Synthesis of a Phenylated Polyimide Containing Pendent Ethynyl Groups", M. S. Thesis, Wright State University, Dayton, Ohio, 1977.

14. F. W. Harris, S. M. Padaki and S. Varaprath, Am. Chem. Soc. Polym. Div., Polymer Preprints, 22 (1), 215 (1981).

15. F. W. Harris, S. M. Padaki and S. Varaprath, (1983), J. Polym. Sci., submitted for publication.

16. F. W. Harris and M. Torres, (1983), J. Polym. Sci., submitted for publication.

POLYIMIDES FROM WATER SOLUBLE PRECURSORS

R. A. Pike

United Technologies Research Center
Silver Lane
East Hartford, Connecticut   06108

The development of water soluble polyimide resin precursors is described in terms of the synthesis and the chemistry of the resulting polymers. Selected properties of films and composites fabricated from the water soluble polyamic acids are presented and compared with organic solvent based resins where applicable. The advantages and differences found in processing the water soluble polymers are described.

15

INTRODUCTION

The impact of rising organic solvent costs and environmental restrictions has in the past several years resulted in a concerted effort to develop water dispersible and water soluble polymer systems without sacrifice of product performance. This trend has been particularly pronounced in the coatings industry with the introduction of a variety of water soluble alkyds, acrylics, polyester and epoxy resin systems.[1,2,3,4] Other water soluble systems include a photocurable elastomer, cathodic electrocoating compositions and Nylon type polyamides.[5,6,7] The increased emphasis on high temperature applications, dictated primarily by aerospace requirements, has produced several commercially available polymer systems, the most prominent of which are the polyimides. Modifications include silicone, amide and ester combinations which cover a wide variety of applications. Some of the latter types have been converted to water dispersible systems.[8,9] The so-called condensation and Mannich addition type polyimide resins require use of expensive basic solvents such as N-methyl pyrrolidone, N, N-dimethylacetamide or diglyme for synthesis and application. The PMR addition type polyimides have the distinct advantage, in terms of solvent requirements, of using alcohol which results in considerable cost savings over the condensation polymers. However, the addition type resins have been restricted primarily to composite matrix and some adhesive applications. As a result, the effort to convert polyimides to water based solvent systems has concentrated, to date, on the condensation type polymers having potential applications as electrical coatings and films, as well as composite matrices and adhesives.

The following sections of this paper review the development of water soluble polyimides and describe some of the chemistry associated with their synthesis, solution properties and applications.

EXPERIMENTAL

The general procedure for synthesis of the various resin systems is adequately described in the appropriate references. The modified aliphatic-aromatic water soluble Essex type polymers were synthesized using methylene dianiline obtained from commercial sources and used as-received. The acid, 1,2,3,4-butane tetracarboxylic acid (BTC) was obtained from Mitsui Chemical or synthesized in-house by nitric acid oxidation of tetrahydrophthalic acid. BTC is currently available from Mallinckrodt.

GPC analysis of BTC based resins was carried out at Essex Magnet Wire Division, Ft. Wayne, Ind., using a Waters instrument with Styragel columns and dimethylformamide solvent.

Dielectric analysis was performed as previously described.[16] Resin enamel was applied to two 4 in. x 4 in. squares of 181 style glass cloth and "B" staged in an air oven at 120°C - 10 minutes. The two prepreg cloth strips were inserted into a press with the appropriate electrodes and a thermocouple. The change in dissipation factor vs temperature was then monitored over the desired heat cycle.

Reduced viscosity measurements were run at 25°C $\pm$ 0.5° using an Ostwald-Fenske viscometer.

POLYMER TYPES

Two approaches have been used in the synthesis of water soluble polyimides; (a) polymerization in organic solvent followed by isolation of carboxylic-acid-containing species and water solublization by salt formation with amines, and (b) direct synthesis in a water/co-solvent/alkaline system to produce water solublized polymer directly. The types of polymers which have been synthesized using these two routes are listed in Table I.

Table I. Polymer Types.

<u>DeSoto, Inc.</u> - A bis-maleimide reacted with a primary diamine and an anhydride to produce residual acid groups.

<u>Nitto Electric, Japan</u> - A polyamide amic acid formed by reaction of BTC and methylene dianiline in a glycol/water solvent.

<u>UTC, Essex Magnet Wire Div</u>. - A polyamide amic acid formed by reaction of BTC, methylene dianiline, and other primary diamines in a water/co-solvent system.

<u>LARC-TPI, NASA-Langley</u> - The amic acid formed by reaction of benzophenone tetracarboxylic dianhydride and m, m'-diaminobenzophenone in diglyme solvent.

One of the earliest reported synthesis of water dispersible

polyimides involved use of a Mannich type addition of a bis-maleimide, made with a primary diamine such as polyoxypropylene diamine and maleic anhydride, with methylene dianiline. The resulting polyimide was then reacted with an anhydride using the secondary amino groups to introduce carboxyl groups along the polymer chain.[10] Precipitation in water produced a powder which was dispersible in water using triethylamine. The reaction sequence is shown in Figure 1. The polymers, by electrodeposition, gave impact resistant, flexible coatings on aluminum. The base polymers

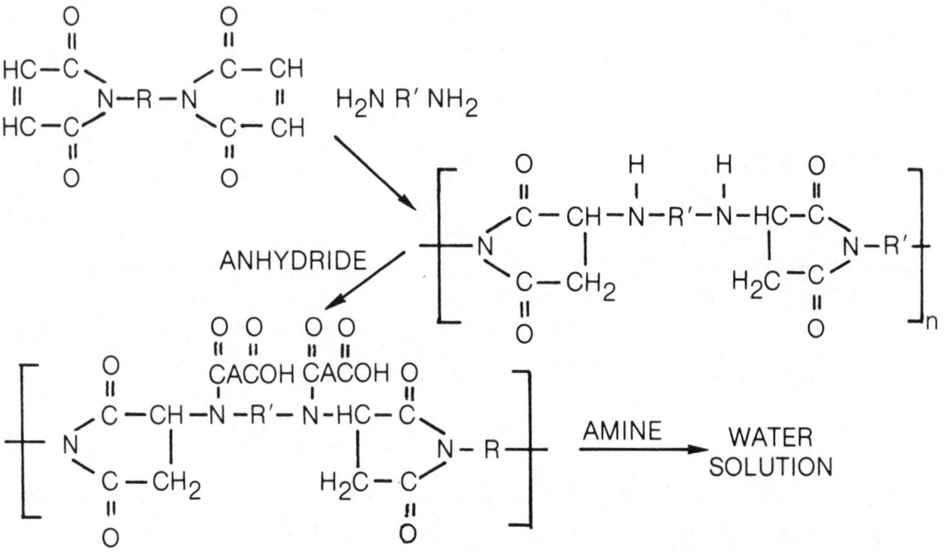

Figure 1. De Soto Inc. Water Soluble Polyimide.

were also modified by reaction with sulfhydryl-terminated low molecular weight rubber polymers to provide increased flexibility without sacrificing water dispersibility. The distinctive feature of this synthetic approach was that the polyimide entity was formed prior to introduction of the carboxyl groups which provided water solubility.

The first description of a completely water soluble polyimide resin having commercial application involved the use of the conventional amic acid condensation approach.[11] The described resin, made by the reaction of 1,2,3,4-butane tetracarboxylic acid(BTC) with an aromatic diamine such as methylene dianiline, in a water soluble solvent and water gave on addition of ammonium hydroxide the ammonium salt of the amic acid. The preferred co-solvents were identified as glycol type compounds. The resulting amic acid

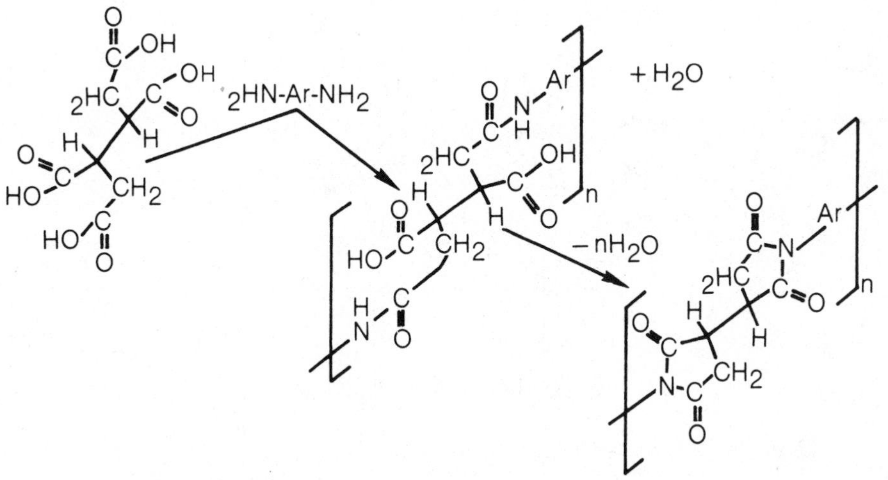

Figure 2. Polymerization and curing sequence of BTC based polyimide.

polymers, designed primarily as wire coating enamels, were readily converted to the imide form using conventional thermal cure cycles. The sequence of reactions involved is shown in Figure 2. This same polymer composition has been synthesized in organic solvent by reaction of the dianhydride of BTC with aromatic diamines.[12] Results of studies, described below, indicated that the polymer structures of the water vs organic solvent synthesized resins were different.

19

A cooperative development effort conducted by the United Technologies Corporation Essex Group, Magnet Wire Division and United Technologies Research Center has resulted in modifications of the BTC – aromatic diamine water soluble polyimide resins which improved flexibility and thermal resistance compared to the glycol-water synthesized resins.[13] The major changes in the synthesis involved elimination of the glycol type co-solvents and use of additional amines in conjunction with the methylene dianiline. Water soluble resins based on BTC by this alternate approach have potential application as composite matrix and adhesive base resins as well as films, coatings, and fiber sizing resins.

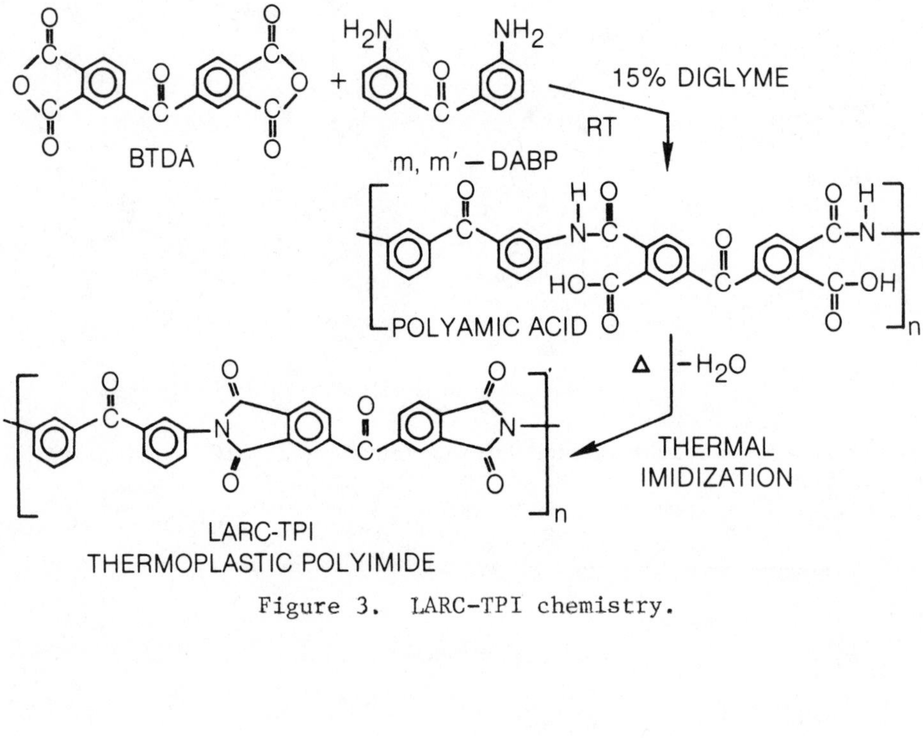

Figure 3. LARC-TPI chemistry.

The recent development of thermoplastic polyimide polymers reported by NASA-Langley Research Center[14] presented an alternate approach for the elimination of organic solvents which involved polymerization of a dianhydride and diamine in diglyme solvent to produce a polyamic acid as illustrated in Figure 3. On precipitation in water, the resulting amic acid powder was water dispersible using ammonium hydroxide or water soluble amines. No additional water soluble co-solvent was required to achieve water solubility.

UTRC is currently being funded by NASA to characterize and evaluate the NASA-TPI water soluble resin systems in graphite reinforced composite, and adhesive applications.[15]

STRUCTURE OF BTC BASED POLYMERS

Studies were carried out to determine if there were any structural differences between the water soluble BTC/MDA polymer synthesized in water-glycol and that made in an organic solvent such as N-methylpyrrolidone (NMP).[12] Dielectric analysis, which can be used to detect solvent and reaction product losses well as gel points[16], showed that the glycol-water synthesized polyimide gave, during cure, evidence of glycol elimination after the gel point of the resin as illustrated in Figure 4. The dissipation curve for the organic solvent resins showed no evidence of glycol elimination during cure and gave essentially a smooth curve. The

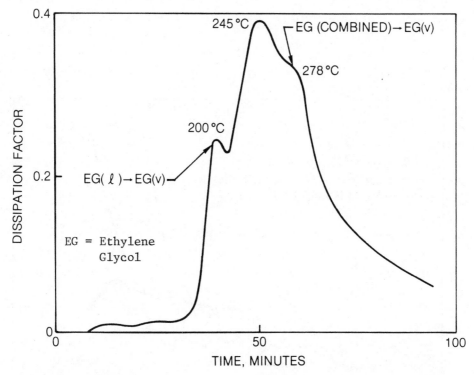

Figure 4. Dissipation factor vs time; cure profile of BTC/MDA polymer.

molecular distribution curves of the two types of resin, as illustrated in Figure 5, markedly demonstrated the effect of the glycol reactant on molecular weight distribution. Infrared analysis of the various GPC fractions of the glycol-water resin showed the presence of glycol ester groups in the polymer chain. In contrast, the water soluble BTC based polyimide synthesized in water without the presence of glycol gave a distribution curve similar to the organic solvent version but of somewhat lower molecular weight.

The infrared spectra of cured films from the two types of polyimide also revealed a difference in the water-glycol resin. As shown in Figure 6, both resins contained imide as indicated by the 1770 cm^{-1} peak. However, the peak at 1605 cm^{-1} indicated the presence of aromatic amide linkages in the polymer structure. Assignment of this peak to amide has been previously reported.[17] In addition, spectra of polyimide-amide resins synthesized from methylene di-p-phenylene-diisocyanate and trimellitic anhydride showed a 1:1 ratio of peak intensity of the 1770/1605 cm^{-1} absorp-

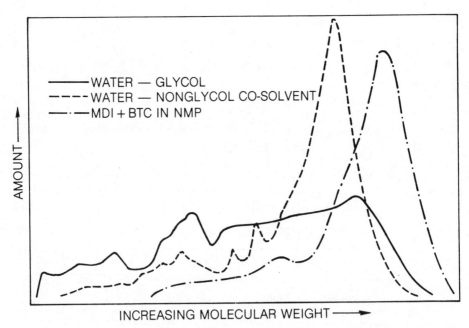

Figure 5. Solvent effect on BTC/MDA polymerization.

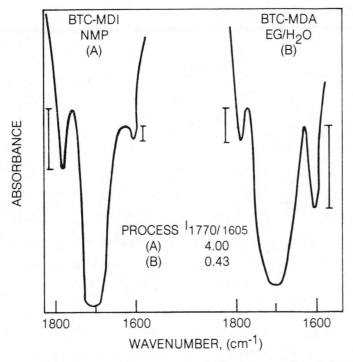

Figure 6. Solvent effects on the imide/amide ratio of BTC polyimide resins.

tion bands. Thus, the presence of both combined glycol and amide linkages in the glycol-water resin suggested that the structure of the Nitto polymer was best represented by the schematic shown in Figure 7. There appears to be, based on the absorption peak ratios, a majority of amide linkages. The spectra of the organic solvent synthesized resin also showed the presence of a small amount of amide linkage as does the Essex water soluble polyimide synthesized in a water/non-glycol co-solvent combination.

EFFECT OF ADDITIVES ON THE STRUCTURE OF BTC BASED POLYMERS

Since the presence of amide linkages in the polyimide polymer would tend to produce films or coatings having lower thermal stability (see below), it was of interest to study the effects of varying co-solvent and alkaline catalysts on the imide/amide ratio of synthesized polymer. Based on the reported mechanism for amide formation in water solution,[18] it would be predicted that the

23

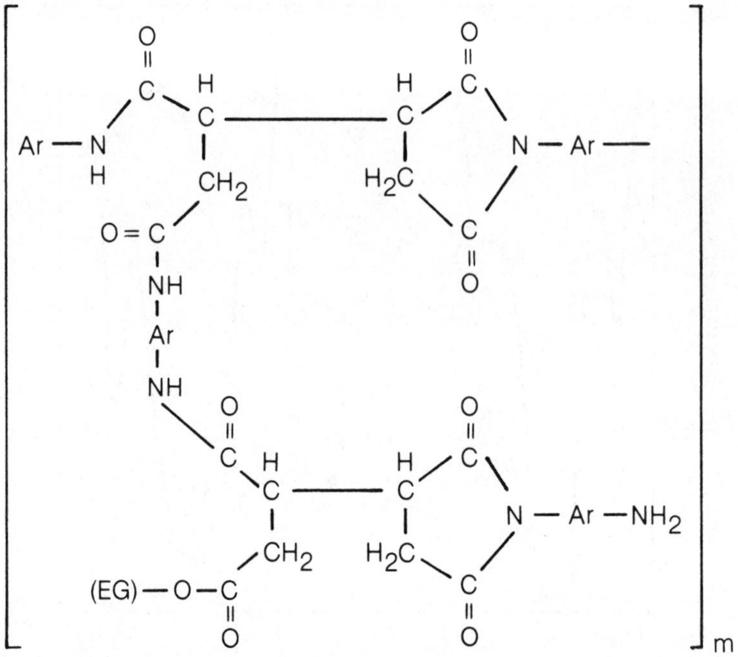

Figure 7.   BTC based glycol/water polyimide structure.

presence of amines would influence the relative amount of amide formed in the BTC/MDA reaction.   To determine the extent of this effect, a series of resins were synthesized using low concentrations (<5w/o) of amines.   The results are illustrated in Figure 8 in terms of imide/amide ratio and pKa of the amine catalyst.   As indicated, only a slight increase in imide content was found using amines having a pKa below 8.   The strongest base, triethylamine, resulted in the largest increase in imide content.

A similar series of resins were synthesized using the triethylamine catalyst with varying hydroxyl containing co-solvents.   The change in imide content with co-solvent, as expressed in terms of activity coefficient, calculated from reported data,[19] is illustrated in Figure 9.   The activity coefficient assimilates such factors as number of C-atoms and hydroxyl content into one expression. It is apparent from the data that a change in the imide/amide ratio occurred as the hydroxyl content of the co-solvent decreased and the

24

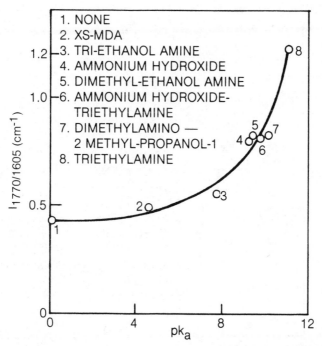

Figure 8. Change in imide/amide ratio, effect of added amine on
the BTC + MDA in ethylene glycol/water reaction.

hydrocarbon content increased. The trend in increasing imide content can also be expressed in terms of increasing base strength of
the co-solvent, i.e., the order of increasing basicity is ethylene
glycol $\tilde{=}$ propylene glycol, water, methylcellosolve, butyl cellosolve, dioxane.[20] This suggests, as discussed below on viscosity
effects, that if solvation by the co-solvent displaces water from
hydrated acid, which favors dimerization of carboxyls, then the
reaction to form amide groups which results ultimately in imide is
favored. This hypothesis seems reasonable since in dimer formation
(hydrogen bonding) by COOH groups, the carbon of the carboxyl
groups would tend towards having similar positive charge thereby
favoring nucleophillic attack by an amine on both COOH groups at
the same time. This could be the cause of forming amide groups on
the 1 and 2 carboxyls of BTC which would prohibit imide formation.
Thus, a decrease in COOH dimerization results in more selectivity
in the reaction sequence. If this hypothesis has merit, then a
correlation between the IR imide/amide ratio and γA/γw (activity
coefficient of addend/activity coefficient of water) should

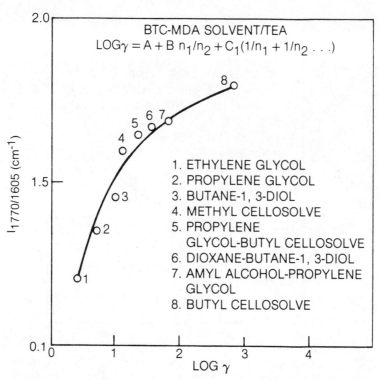

Figure 9. Change in imide/amide ratio, effect of co-solvent on the
BTC + MDA triethylamine catalyzed reaction.

exist as discussed below. Using data from the above indicated
reference,[19] the ratio of the activity coefficients of four of the
co-solvents vs water was calculated (sufficient data are not available
for the remaining systems) and plotted as shown in Figure 10. There
appears to be a direct correlation between IR ratio and the activity
coefficient of the co-solvent. Thus, the selectivity of the polymerization reaction in terms of co-solvent appears to be related to the
viscosity effect on the finished polymer in terms of added solvent;
the controlling factor being the hydration or solvation of the
carboxyl groups involved. As indicated in Figure 9, with
the alcohol-ether type co-solvents an IR ratio of more than
2.0 is unlikely. This is due in part to decreasing solubility of
the polymerizing system as the hydroxyl content of the co-solvent
decreases. To overcome this effect, use of a solvent which solublizes
the resin, such as NMP, is required. An increase in imide content,
equivalent to that obtained in the organic solvent system, can be
achieved by use of the appropriate co-solvents and catalysts as
illustrated by the data listed in Table II.

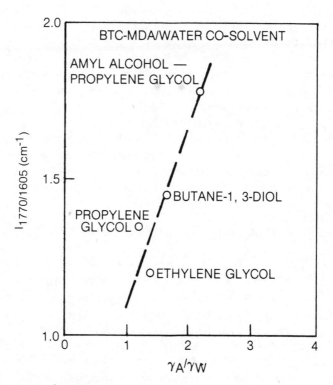

Figure 10. Imide/amide ratio, change with activity coefficient of co-solvent.

Table II. Imide/Amide Ratio Change on Polymerization-Effect of Co-Solvent on BTC-MDA Reaction.[a]

Solvent	Imide/Amide Ratio ($1770/1605$ cm^{-1})
NMP (no water)	1.9
NMP/butylcellosolve (1/1)[b]	1.5
NMP/butylcellosolve (1/3)+TEA[c]	2.8
NMP/butylcellosolve (1/1)+TEA[c]	4.3
NMP[BTC+MDI - no water]	4.0

a = finished resin, 40% solids; b= co-solvents 9 wt. percent (w/o) based on finished resin; c = triethylamine ~3.4 w/o based on reactants.

With the proper ratio of butylcellosolve and NMP under alkaline conditions, the imide/amide ratio was increased to an equivalent value obtained in the organic solvent system. The infrared spectra of the two cured resin films were essentially identical. It is apparent that by selection of co-solvent and alkaline catalyst, the structure of the polymer synthesized in water from BTC and MDA (or other diamines) can be controlled to produce optimum film properties depending upon the intended application. A study which described the effect of water co-solvents on system stability, surface tension reduction, and evaporation rate has been reported.[20] These factors must also be controlled and understood in the application of water soluble polyimide systems.

VISCOSITY EFFECTS IN WATER SOLUBLE POLYIMIDES

Reduced viscosity measurements have been used to determine the coil size of a polymer molecule. The coil size is related to the degree of intramolecular bonding. Low viscosity solutions are an indication of small coil volumes and a high degree of intramolecular bonding; high viscosity indicates an increase in coil size and a decrease in intramolecular bonding. Since the polyamic acid precursors contain carboxyl groups the solution properties should be analogous to those of polymethacrylic acid and polyacrylic acid which have been extensively investigated.[22] To determine if the water solutions of polyimide behave similar to these systems, reduced viscosity curves were obtained on two systems containing different amounts of combined glycol using dioxane as an addend. The resulting curves are shown in Figure 11. Based on the referenced study[22] it can be hypothesized that two phases exist in an aqueous polymer solution; one phase is the aqueous phase containing the addend (dioxane in this case); the other the polymer itself. Equilibrium exists between these two phases. Active sites in the polymer are in three states: 1) intramolecular bonded, 2) hydrated with water or 3) solvated with addend. Such sites are free carboxyl, amide, ester and imide groups. If the solvent mixture is ideal, the activity coefficients and their ratio, R, are unity. Intramolecular bonding will thus increase with addend concentration if $\gamma A/\gamma W <1$, i.e., if the polymer solvates better with water than with addend. Conversely, if $\gamma A/\gamma W >1$, i.e., if the addend adds more readily to the chain than does water, the degree of intramolecular bonding will decrease as more addend is introduced. In such cases, the addend may be regarded as a hydrogen bond breaker.

γW and γA are a measure of the effectiveness of water and added

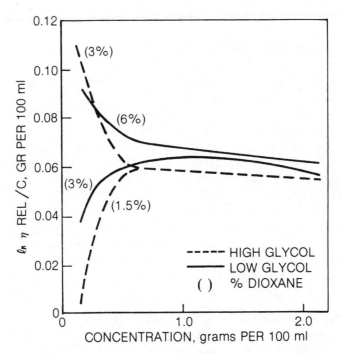

Figure 11. Reduced viscosity of BTC/MDA polymer vs concentration: effect of added dioxane.

molecules, respectively, as they occur in the solvent mixture in reacting with the polymer and measure the total effectiveness of addend in replacing water on the chain. If this ratio is larger than 1 then removal of a molecule of water from the system and its replacement by one of addend will have the effect of increasing the solvation tendency, and intramolecular bonds will be broken. Thus, an addend which is effective in reducing the degree of intramolecular bonding in aqueous solution has at least $\gamma A/\gamma W > 1$.

The results in Figure 11 demonstrated the effectiveness of dioxane as a hydrogen bond breaker (decrease in intramolecular bonding) in the two resins. With the low retained glycol resin, a concentration of 3% dioxane was not sufficient to eliminate all intramolecular bonding (due to hydration) in the polymer chain. Thus, with decreasing polymer concentration below 1.0g/100ml, the viscosity decreased. However, 6% dioxane caused the viscosity to increase with decreasing polymer concentration indicating the dioxane had

solvated all active sites, hydration had decreased and the degree
of intramolecular bonding had decreased. A similar effect was found
with the high retained glycol resin except that only 3% dioxane was
required to show a viscosity increase. The lower amount of dioxane
required to produce this effect reflected the fact that a greater
amount of ester linkages were present in the resin. The free
hydroxyl group in a monoester linkage would act like an alcohol in
solvating active sites, reducing the amount of hydration and as a
result decrease the degree of intramolecular bonding. Thus, less
dioxane would be required to obtain the viscosity increase. It was
of interest that with PMAA (polymethacrylic acid) a similar change
in viscosity (that is, increase in viscosity with decreasing concentration) occurred using ethyl alcohol (15%) added to a water solution
of PMAA. Below 15% the solution viscosity decreased with decreasing
concentration.[22] Information on these effects which occur in water
solutions of polyamic acids is of importance in film or coating
applications using either electrolytic deposition or thermal cure
processes since the adhesion as well as stresses developed during
cure will be influenced by the conformation of the applied polymer.

This same type of addend or co-solvent influence on the degree
of hydration or carboxylic acid dimerization can be used to explain
the effect of the various co-solvent alcohols described above on
the course of the condensation of BTC-MDA to yield either imide or
amide linkages in the synthesized polymer.

The water solutions of the LARC-TPI resin made by dissolving
precipitated powder, from the diglyme solution, into water-ammonium
hydroxide yielded the reduced viscosity curves shown in Figure 12.
The response of the two resins studied was the same, i.e., increasing
viscosity with decreasing concentration. The intrinsic viscosities
in DMF of the two diglyme resin solutions were 0.4 (low) and 2.1
(high). That no additional solvent was required to obtain the
observed effect may be due to low levels of hydrogen bonding in the
LARC-TPI related to the structural differences between this resin
and the BTC based polymers. An alternate explanation could be that
small amounts of diglyme remaining in the powder after processing
would act like dioxane in decreasing the degree of hydration and
intramolecular bonding which resulted in the increased viscosity
at low concentration levels. Reduced viscosity curves of water
solutions of LARC-TPI powder and alkanol amines showed the same
general trend of increasing viscosity with decreasing concentration.

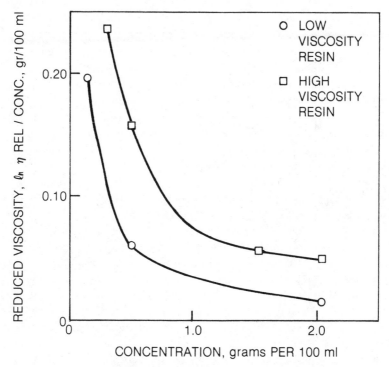

Figure 12.   Viscosity vs concentration of LARC-TPI resin in
11/1 H_2O/ammonium hydroxide solvent.

FILM PROPERTIES OF WATER SOLUBLE POLYIMIDES

The stress-strain characteristics of the water soluble films
evaluated to-date are shown in Figure 13 and illustrate the differences between the LARC-TPI type polyimide and the BTC based resins.
The former resin produced a higher strength (equivalent to Kapton[®])
and had a higher modulus of elasticity as would be predicted. The
BTC based resin films can be tailored to produce films of varying
modulus and elongation depending upon the type of primary diamines
used along with MDA in the resin synthesis. Thus, high elongation
can be achieved with a slight sacrifice in tensile strength by
varying the structure of the amine components. The ultimate tensile
strengths of the BTC-based resin films tended to be lower than
LARC-TPI or the Kapton[®] films.

The loss in modulus and strength properties with increasing
temperature for the Essex BTC based resin films is illustrated in

31

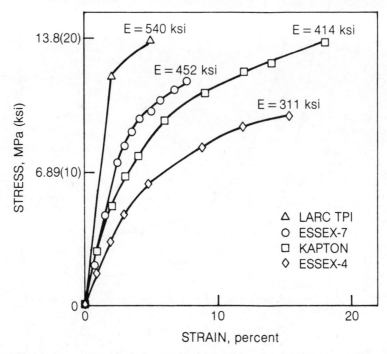

Figure 13. Stress-strain behavior of polyimide films.

Figure 14. As shown, both modulus and strength decreased with increasing temperature. Strength retention with temperature for the LARC-TPI resin films cast from diglyme solution have been reported.[23]

Comparative thermal properties as measured by TGA of the water soluble polyimide systems studied to date are shown in Figure 15 using Kapton® polyimide as a control. The two BTC-based systems showed essentially the same thermal response with increasing temperature. The LARC-TPI resin was intermediate between the aliphatic acid modified resins and the Kapton® film. Isothermal weight loss at 200°C, Figure 16, showed no major differences between the two aliphatic-aromatic resins. At 250°C, however, a distinct difference in response between the Essex and Nitto type resin films was found. The lower thermal capability of the Nitto type BTC based resin was attributed to the presence of residual glycol as well as a higher amide content as discussed above.

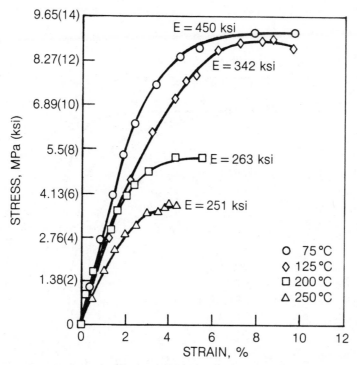

Figure 14. Stress-strain behavior of BTC-based polyimide film,
effect of temperature.

COMPOSITE PROPERTIES OF WATER SOLUBLE POLYIMIDES

Fiber reinforced composites are potentially a large application
area for water based polyimide resin systems. The adverse effects
on composite properties attributed to residual organic solvent
(NMP or diglyme) would be eliminated by conversion to water systems.
The composite data listed in Table III for the BTC-based Essex-7
resin and Table IV for LARC-TPI illustrate that water solutions of
these resins give acceptable composite properties. An important
difference between the organic and water systems, i.e., degree of
fiber wetting, is illustrated using the Essex resin system. Composite
fabrication with the as made resin resulted in poor wetting of the
graphite fiber and inferior composite properties. Addition of two
types of nonionic surfactants gave room temperature composite
properties equivalent to a standard epoxy system. Elevated temperature response of the BTC based resins has not as yet been determined.
The LARC-TPI water soluble resin resulted in composite properties
equivalent to those reported for comparable laminates made using
the diglyme solvent system.[14]

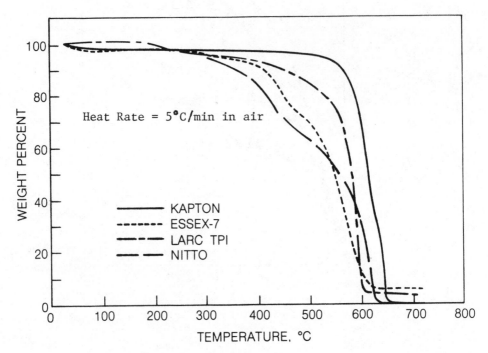

Figure 15. Thermogravimetric analysis of thin polyimide films.

Table III. Composite Properties of Essex Water Soluble Polyimide[a]

Additives	Flex mod GPa(10^6 psi)	Flex stg[b] MPa(Ksi)	SBS MPa(Ksi)[b]	Fiber v/o
None[c]	128(18.6)	1145(166.0)	70.3(10.2)	64.0
DuPont FSN[d]	121.4(17.6)	1400(203.0)	86.0(12.5)	54.0
U.C. NP-27[e]	106.9(15.5)	1460(212.0)	83.4(12.1)	45.3
Epoxy	117(17.0)	1380(200.0)	89.6(13.0)	60.0

a) Celion 6000 graphite fiber, epoxy sized; b)flexual strength S/D= 20/1, short beam shear S/D=4/1; c) surface tension = 62 dyne/cm; d) surface tension = 25 dyne/cm; e) surface tension = 35 dyne/cm; f) v/o = volume percent.

Table IV. Composite Properties of Water Soluble LARC-TPI[a]

Property	Temperature, °C		
	RT	200°	250°
Flexural Strength, MPa(Ksi)[b]	1310(191)	444(64.4)	150(21.8)
Flexural Modulus GPa(psi x 10[6])	121.4(17.6)	94.4(13.7)	32(4.65)
Short Beam Shear, MPa(Ksi)[c]	66.6(9.7)	43.4(6.3)[d]	16.2(2.35)

a) Celion 6000 graphite fiber; b)   three point, S/D=20/1, v/o fiber = 61; c)   S/D=4/1, v/o fiber = 50 volume percent.

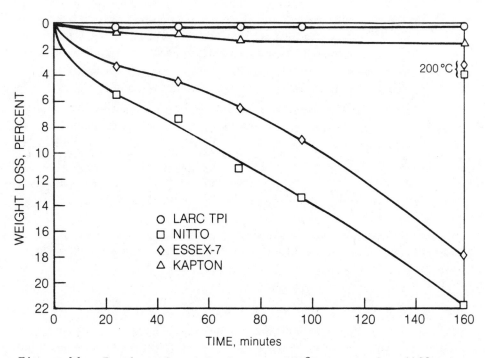

Figure 16.   Isothermal weight loss at 250°C in air for different polyimides.

ADDITIONAL WATER SOLUBLE POLYIMIDE SYSTEMS

The technique used in converting the LARC-TPI resin from an organic to a water based system can be modified for use with other polyimide resins. DuPont NR-150B2 and Pyre-ML, NASA-Lewis PMR-15, and an Essex amide-imide polymer have been reformulated into water solvent form. Evaluation of these systems in composite, film and adhesive applications has been initiated and the results will be reported when complete.

ADVANTAGES AND DISADVANTAGES OF WATER BASED POLYIMIDES

Based on the limited evaluation of the herein described polyimide water solutions, the comparative status with organic solvent systems is summarized below:

Advantages: Environmentally acceptable.

Reduced solvent cost (recoverable).

Resin storable as powder if synthesized in organic solvent.

Structure and properties controllable by additives in water solution synthesized resins.

Properties equivalent or superior to organic solvent systems.

Potential Disadvantages:

Water contamination – effects unknown.

Wetting problems – effects of surfactants on ultimate properties unknown.

Solution stability (storage life) – appears to be acceptable to date for some BTC/MDA and the LARC-TPI formulations.

The probability of eliminating residual traces of DMF, NMP or diglyme from cured polyimide resins may result in improved aging properties particularly at elevated temperatures. Cost savings resulting from the polyimides initially synthesized in organic solvent will be dependent upon the ability to recycle the solvent after resin precipitation has occurred. The commercial availability of polyamic acid powders is attractive not only from a resin stability standpoint, but also presents a shipping and storage volume advantage.

The introduction of water soluble high temperature polyimide resin systems into the composite, wire coating, film, fiber and fiber sizing as well as adhesive application areas presents a unique opportunity not only from an environmental standpoint, but also in presenting the challenge of new chemistry and processing techniques to achieve the optimum advantages of these systems.

ACKNOWLEDGEMENTS

The technical assistance and guidance of Dr. T. St. Clair of NASA–Langley, Mr. L. Payette and staff of Essex Group, UTC, and Dr. M. A. DeCrescente, Dr. F. P. Lamm and Mrs. J. P. Pinto of UTRC is gratefully acknowledged.

REFERENCES

1. M. E. Woods, Mod. Paint and Coatings, 65, No.9, 21 (1975).
2. Amoco Chemicals Corp., Bulletin TMA-120. Naperville, Ill., 1981.
3. W. J. Van Westrenen, J. Oil Color Chem. Assoc., 62, 246 (1979).
4. E. G. Bozzi, Mod. Paint and Coatings, No.11, 57 (1979).
5. G. D. Jones, C. W. Hoornstra, D. E. Leonard and H. B. Smith; J. Appl. Polymer Sci., 23, 115 (1979).
6. V. D. McGinnis, US Patent 4,140,816 Feb. 20 (1979) SCM Corp.
7. N. Ogata and Y. Hosoda, J. Polymer Sci: Polymer Chem. Ed., 16, 1159 (1978).
8. C. W. McGregor, J. Karkoski and J.D. Shurboff, US Patent 4,290,929 Sept. 22, 1982, Essex Group Inc., United Technologies Corp.
9. K. Schmitt, F. Gude and S. Brundt, US Patent 3,882,085 May 6, 1975, Veta Chemie Aktiegesellschaft.
10. G. G. Vincent and T. E. Anderson, US Patent 3,652,511 March 28, 1972, DeSoto, Inc.
11. M. Kojima, Y. Noda, Y. Suzuki and T. Okamoto, US Patent 3,925,313 Dec. 9, 1975, Nitto Electrical Ind. Co., Japan.
12. D. F. Loncrini and J. M. Witzel, J. Polymer Sci., A-1, 7, 2185 (1969).
13. UTC Essex Group, Magnet Wire Division, FT Wayne, Indiana – Patents applied for.
14. A. K. St. Clair and T. L. St. Clair, in Proc 26th Nat. SAMPE Sym. Los Angeles, CA, April 28, 1981.
15. R. A. Pike-NASA Contract NAS1-16841-United Technologies Research Center, East Hartford, CT. 4th Bi-Monthly Report, August 12, 1982.
16. R. A. Pike, F. P. Lamm and J. P. Pinto, J. Adhesion 12,143 (1981); ibid, 13,229 (1982).
17. D. Kruh and R.J. Jablonski, J. Polymer, Sci., Polymer Chem Ed., 17, 1945 (1979).
18. H. Morawetz and P.S. Otaki, J. Am. Chem. Soc., 85,463 (1963).
19. G. J. Pierotti, C. H. Deal and E. L. Derr, Ind. Eng. Chem., 51, 95 (1959).

20. C. M. Hansen, Ind. Eng. Chem., Prod. Res. Dev., 16,266 (1977).
21. I. Gyenes, "Titration in Non-Aqueous Solvents", D.Van Nostrand
    Co., New York, 1967.
22. J. Eliassaf and A. Silberberg, J. Polymer Sci., XLI, 33 (1959).
23. V. L. Bell, L. Kilzer, E. M. Hett and G. E. Stokes, J. Appl.
    Polymer Sci., 26, 3805 (1981).

ADVANCES IN ACETYLENE SUBSTITUTED POLYIMIDES

A.L. Landis and A.B. Naselow

Hughes Aircraft Company
El Segundo, California 90245

An acetylene-substituted isomer of a polyimide was prepared and thermally isomerized into acetylene-substituted polyimide. Subsequent cure through the acetylene end groups yielded resin with mechanical and thermal properties similar to "Thermid 600" the first of the acetylenic polyimides to be marketed. Advantages of the isomer designated "IP-600" over the polyimide HR-600 are its lower melting range, high solubility in common solvents, and longer gel time. These advantages lead to markedly improved processing characteristics.

BACKGROUND

Thermosetting acetylene-substituted polyimides developed during the early to mid-seventies have been described in various prior publications as well as in the patent literature[1-11]. These materials, of which over fourteen examples have been described, cure into void free, or near void-free resins when compression molded at temperatures above their melting ranges, generally 220-260°C. Their glass transition temperatures (Tg's) determine their ultimate utility, but values of 350°C and above can be achieved, depending upon specific structure and upon the ultimate postcure conditions used in their processing. Air postcures at a temperature of 370°C for a minimum of about 14 hours appear to give the highest Tg's. One of these resins has been produced commercially by Gulf Oil Chemicals Company under license from Hughes Aircraft Company under the trade name Thermid 600, and several modifications of this material are available. Glass and carbon fiber-reinforced composites

derived from Thermid 600 (Figure 1) have been studied extensively
and laminates have been produced which retain up to 80 percent of
their flexural strength when exposed to air at 316°C for 1000 hours
(see Ref. 11, p. 146).

In spite of its excellent heat resistance and good mechanical
properties, for certain applications such as large area laminates,
Thermid 600 suffers from two major deficiences, namely,its high
melting temperature and its high rate of cure. As a consequence
it has a gel time of less than three minutes at its customary cure
temperature (250°C) and this factor has restricted its use in the
above specified application.

Th solution to this problem was discovered in the early
eighties when it was found that an isomeric form of the Thermid
600 prepolymer could be made which had a much lower melting temperature (160±15°C vs. 202+5°C), a much longer gel time (up to 30
minutes), and good solubility in low boiling common solvents such
as tetrahydrofuran, dioxane, acetone/toluene mixtures, etc. The
longer gel time is a direct consequence of the lower prepolymer
melting range, since acceptable melt flow characteristics can be
achieved at temperatures at which the acetylene groups homopoly-
merize more slowly than they do at the temperatures required to
melt the imide isomers. As it is cured, the isomeric material
isomerizes into the normal imide structure and thus its cured resin
properties are at least equivalent to those previously reported for
"Thermid 600". The new isomeric materials, designated HR-60XP, are
the subject of this paper.

Acetylene-Substituted Polyimide Isomers

Differential scanning calorimetric analysis shows that the
polymerization exotherm of HR-600P has its maximum at 235 to 240°C,
and this value is essentially the same as that for Thermid 600.
The comparison is shown in Figure 2. These values are in the same
range as most of the other acetylene-substituted oligomers made to
date. It will be noted, however, that as the molecular weight per
acetylene group increases, there is a tendency for the temperature
of the exotherm maxima to increase also and thus values as high as
280°C have been observed on higher molecular weight homologues.

Because of its lower cure temperature requirements, and its
better melt flow characteristics, HR-600P can very effectively be
processed in a vacuum bag in a conventional autoclave. Comparisons
of the cure cycles of HR-600P, LaRC-160 (a development of T. St.
Clair of NASA Langley Research Center) and NR-150-B2 (from E.I.
DuPont Company) are shown in Figure 3.

40

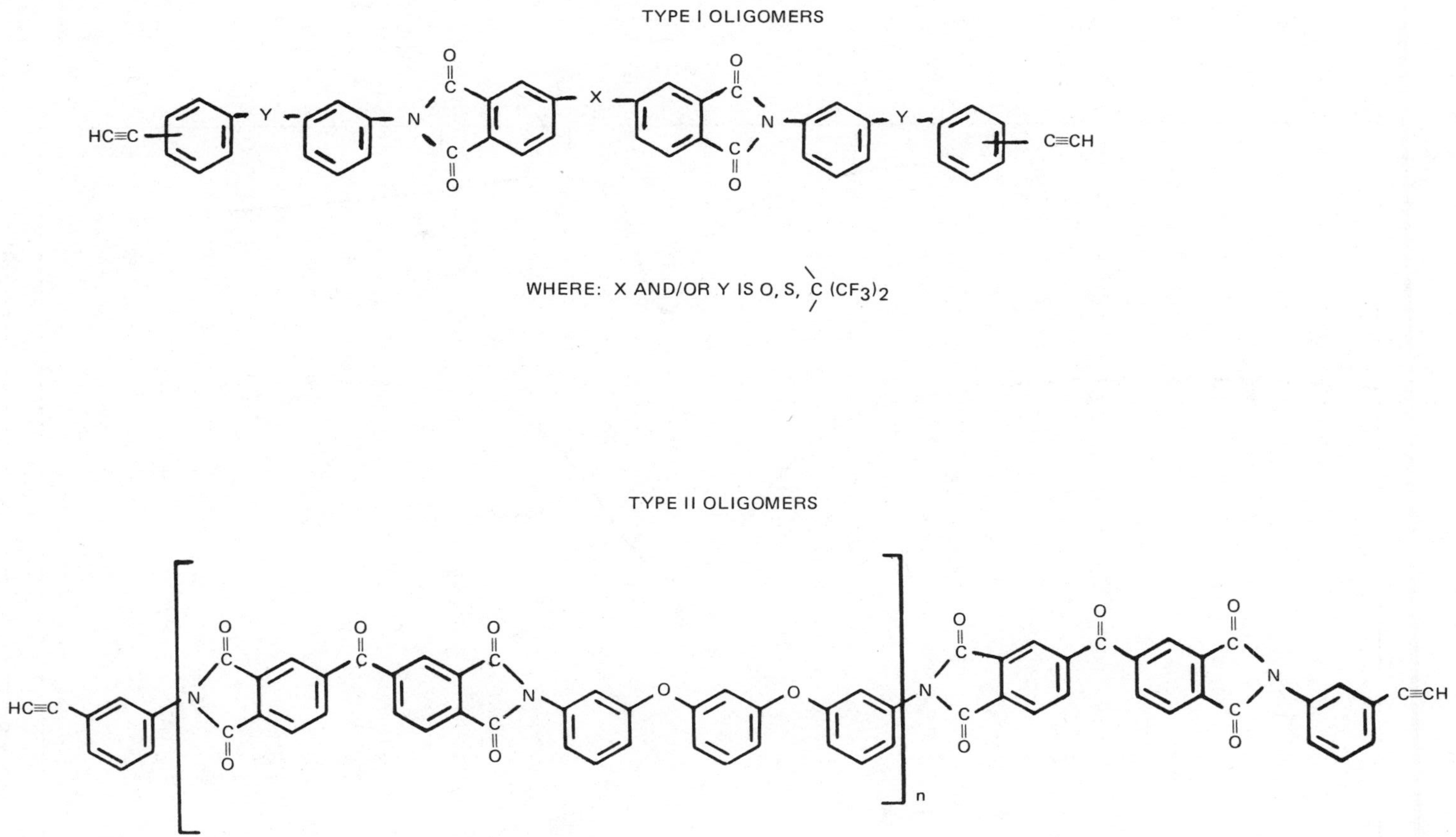

Figure 1. Acetylene-substituted polyimides. Specific designations: Specific designations: Thermid 600 when n = 1; HR-60X when n = X

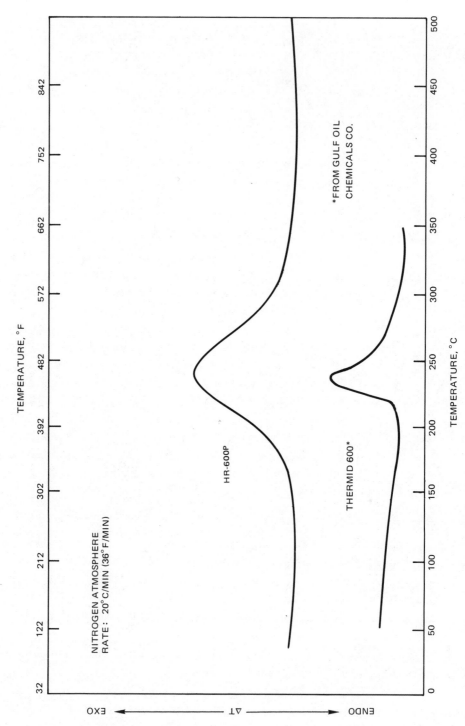

Figure 2. Differential scanning calorimetry curves for Thermid 600 and HR-600P resins.

42

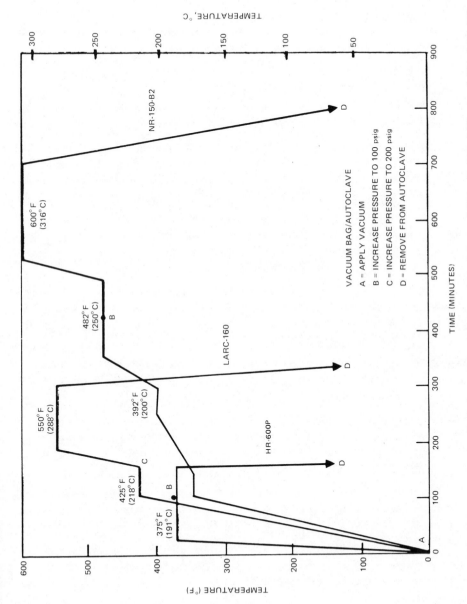

Figure 3. Comparison of processing parameters of HR-600P, LaRC-160, and NR-150-B2.

Unidirectional eight ply graphite fiber-reinforced laminates were prepared from HTS graphite fiber and HR-600P in a vacuum bag autoclave and subsequently flexural strength test specimens and short beam shear strength specimens were machined from the laminate. Specimens were then thermally aged in air at 288°C (550°F) for a period of 750 hours and the mechanical properties were monitored periodically. Results of these tests showed that at 288°C the flexural strength showed no change for 550 hours and only a 10 percent loss after 750 hours. In contrast, the ambient temperature flexural strength of thermally aged samples decreased from 130,000 psi to 70,000 psi in 750 hours with a uniform rate of drop during this period (see Figure 4). The test results also are presented in tabular form in Table I.

Elevated temperature Short Beam Shear Strength tests (at 288°C) showed a continual, but very slow, drop as a function of thermal aging, falling from 5,000 psi to 4,500 psi over the 750 hour period. This also represents a 10 percent reduction. In contrast, the ambient temperature strength fell from 7,500 psi to 4,500 psi or a 40 percent drop over a 750 hour period. Results of this study are shown graphically in Figure 5 and in tabular form in Table II. Gel points of HR-600P and Thermid 600 were also determined on a Rheometrics Dynamic Spectrometer. In this comparative study the rheometric behavior of the Thermid 600 was investigated at 250°C, whereas the HR-600P resin was studied at both 190°C (375°F) and 210°C (420°F). The higher temperature used for the Thermid 600 was necessary because the resin does not have adequate flow at lower temperatures and thus cannot be processed at lower temperatures. It is evident from the results of this study that Thermid 600 had a gel time of only 1 minute under these conditions, whereas HR-600P had gel times of 5-1/2 minutes and 15-16 minutes at 210°C and 190°C, respectively. Since both of these latter temperatures are suitable for processing HR-600P, the observed gel times are quite realistic. Results of the study are shown graphically in Figure 6 and the investigation illustrates quite conclusively the marked advantage of the "P" resin over the conventional "Thermid" resin. In these graphs, G' and G" are the elastic and loss component, respectively, of the Dynamic Modulus. The point at which the two curves cross is interpreted as being the gel point.

Glass transition temperatures of HR-600P vary with postcure conditions and this is the same observation as made with the original Thermid 600 resin. Tg's as high as 350°C (662°F) have been achieved, using a 371°C (700°F), 15-hour postcure in Argon. Although 371°C air postcures have commonly been recommended for the earlier Thermid 600 resin when maximum Tg's were desired, the achievement of such high values in Argon is new in our experience.

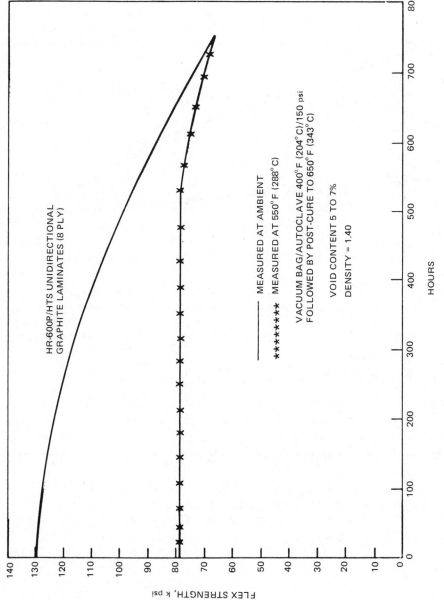

Figure 4. Flexural strength as a function of thermal aging at 550°F in air.

45

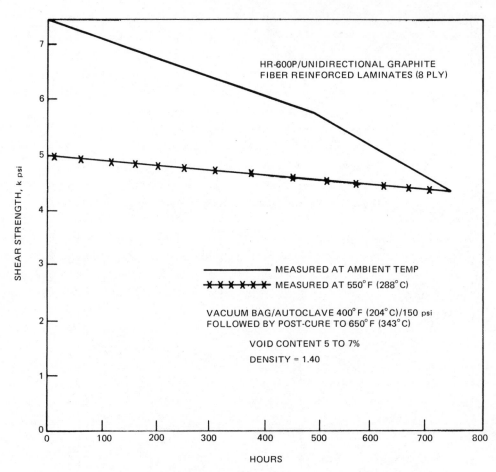

Figure 5. Short beam shear strength as a function of thermal aging at 550°F in air.

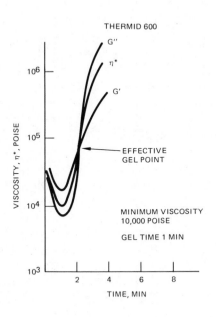

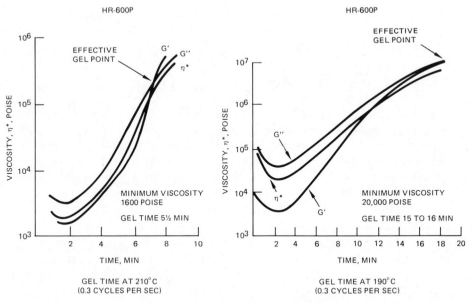

Figure 6. Gel times of Thermid 600 and HR-600P.

Table I. Flexural Strength and Modulus After 550°F (288°C)
Heat Aging.

Graphite/Polyimide (HTS/HR-600P) Unidirectional Laminates (8 Ply)					
Hours at 550°F (288°C)	% Wt. Loss	% of Original Value Ambient Test		% of Original Value 550°F (288°C) Test	
		Strength	Modulus	Strength	Modulus
0	0	100	100	100	100
250	1.25	90	93	100	105
500	2.41	82	90	100	114
750	4.59	51	86	84	117

Table II. Short Beam Shear Strength After 550°F (288°C) Heat
Aging.

Graphite/Polyimide (HTS/HR-600P) Unidirectional Laminates (8 Ply)			
Hours at 550°F (288°C)	% Wt. Loss	% of Original Value Ambient Test	% of Original Value 550°F (288°C) Test
0	0	100	100
250	2.17	88	86
500	3.87	79	100
750	6.12	59	87

SUMMARY

An acetylene-substituted isomeric analogue of the acetylene-substituted polyimide designated "Thermid 600" was found to have a lower melting range, higher solubility in common solvents, and longer gel time than the original Thermid. Its longer gel time (15 to 16 minutes at 190°C) is attributed to its superior melt characteristics and lower melt temperature. In contrast "Thermid 600" required processing at or above 250°C under which conditions its gel time is less than 3 minutes. The isomerized isomeric analogue has a Tg of 350°C when cured to 371°C for 15 hours and has mechanical properties similar to those of "Thermid 600".

REFERENCES

1. A. L. Landis, N. Bilow, R. H. Boschan, R. E. Lawrence and T. J. Aponyi, Polymer Preprints, 15, No. 2, 533 (1974).
2. N. Bilow, A. L. Landis and T. J. Aponyi, Symposium Preprints, SAMPE National Meeting, San Diego, CA, April 28-30, 1975.
3. N. Bilow and A. L.Landis, National Technical Conference, SAMPE, Vol. 8, p. 94, Seattle, WA, October 12-14, 1976.
4. N. Bilow, 23rd National Technical Conference, SAMPE, Anaheim, CA, May 2-4, 1978.
5. N. Bilow and A. L. Landis, Polymer Preprints, 19, No. 2, 23 (1978).
6. N. Bilow, A. L.Landis and L. J. Miller, U.S. Patent 3,845,018 (1974).
7. N. Bilow, U.S. Patents 4,108,836 (1978) and 4,098,767 (1978).
8. N. Bilow and A. L. Landis, U. S. Patents 3,879,439 (1975) and 4,276,407 (1981).
9. N. Bilow and B. G. Kimmel, U. S. Patent 4,100,138 (1978).
10. N. Bilow, in "Resins for Aerospace", C. A. May,Editor, ACS Symposium Series No. 132, Ch. 13, Washington, DC, 1980.
11. A. Landis and A. B. Naselow, Preprints, 14th National Technical Conference, SAMPE, Vol. 29, pp. 139-150, October 11-15, 1982.

SILICON-MODIFIED POLYIMIDES: SYNTHESIS AND PROPERTIES

G.N. Babu

Polymer Research Laboratory, Department of Chemistry

Indian Institute of Technology, Powai,Bombay-400076,India

Three new silicon-containing dianhydrides, 4,4-bis(3,4-dicarboxyphenyldimethylsilyl)biphenyl 12, 1,3-bis(3,4-dicarboxyphenyl)-1,1,3,3,-tetramethyldisiloxane dianhydride, 18, and 1,5-bis(3,4-dicarboxyphenyl)-1,13,3,5,5-hexamethyltrisiloxane dianhydride, 20 were synthesized by multistep synthesis. Dianhydrides, containing di-methylsiloxy groups, were polymerized with m,m and p,p-diaminodiphenylmethane to poly(amic acids) which were converted to polyimides. The objective of the present investigation was to study the effects of increasing the distance between and varying the isomeric positions of diamines/dianhydrides on glass transition temperatures of polymers.

INTRODUCTION

Polyimides, among the class of heterocyclic polymers, have remarkable high temperature resistance. They are, however, extremely intractable, being insoluble in most organic solvents. The excellent thermal, thermooxidative and chemical stability of the wholly aromatic polyimides are derived from the structural features of these polymers. A number of methods have been used to overcome intractability while maintaining reasonably high temperature chara-cteristics. These include incorporation of flexible, non-symmetri-cal and thermally stable linkages in the backbone. For example, polyimides obtained from 3,3',4,4'-benzophenonetetracarboxylic dianhydride (BTDA) soften at lower temperatures than do polyimides from pyromellitic dianhydride[1]. Similarly, incorporation of per-fluoroalkyl units in the dianhydride results in polymers with in-creased solubility[2,3]. Also, the use of diamines cotaining flexible spacers like methylene, oxygen, sulfur and sulfone provides polyimides with lower softening points than does m- and p-phenylene. Further improvemetns in the solubility of the polyimides have been demon-strated for various isomeric positions of the phenyl ring[4,5]. Harris and coworkers[6] reported the synthesis of a range of phenyla-ted polymers. This approach leads to a dilution of the phthalimide groups in the polymer chain. Rudakov et al[7] were the first to study systematically the relationship between the effect of increase in p-phenoxy moieties and softening points of the resulting polyimides. Several researchers have exploited this approach for various other systems[8-10]. Processing characteristics of the polyimides can also be realized by copolymerizing different dianhydrides[11] or different diamines[12-15].

Excluding silicones, very little has been reported on the pre-paration and evaluation of the thermally stable polymers contain-ing the silyl moiety in the polymer backbone. Synthesis and proper-ties of siloxane based heterocyclic polymers have been reported[16-19]. Kovacs et al.,[20,21] reported the synthesis and properties of siloxane based 1,3,4-oxadiazole, amide and benzimidazole polymers. These were soluble in organic solvents and their solution cast films were flexible and exhibited good adhesion to glass and metal surfaces. On the contrary aromatic polyoxadiazoles containing no silicon were thermally stable, infusible, insoluble materials which could not be processed.

RESULTS AND DICUSSION

Synthesis of Silyl Monomers

Bis(3,4-dicarboxyphenyl)dimethylsilane dianhydride 5 and 1,4-bis(3,4-dicarboxyphenyldimethylsilyl)benzene dianhydride 10 were synthesized according to published procedure[23]. The syntheses of these compounds are outlined in Schemes I and II.

Synthesis of 4,4'-bis(3,4-dicarboxyphenyldimethylsilyl)di-phenyl dianhydride 13 was carried out according to Scheme III. The tetramethyl derivative 11 was obtained by two different synthetic routes. Reaction of compound 7 in the presence of bis(1,5-cyclo-octadiene)nickel (0) gave tetra methyl compound 11 in 28% yield. Alternatively, compound 11 was synthesized starting from 4,4'-dibromobiphenyl. The synthetic scheme is similar to Scheme II.

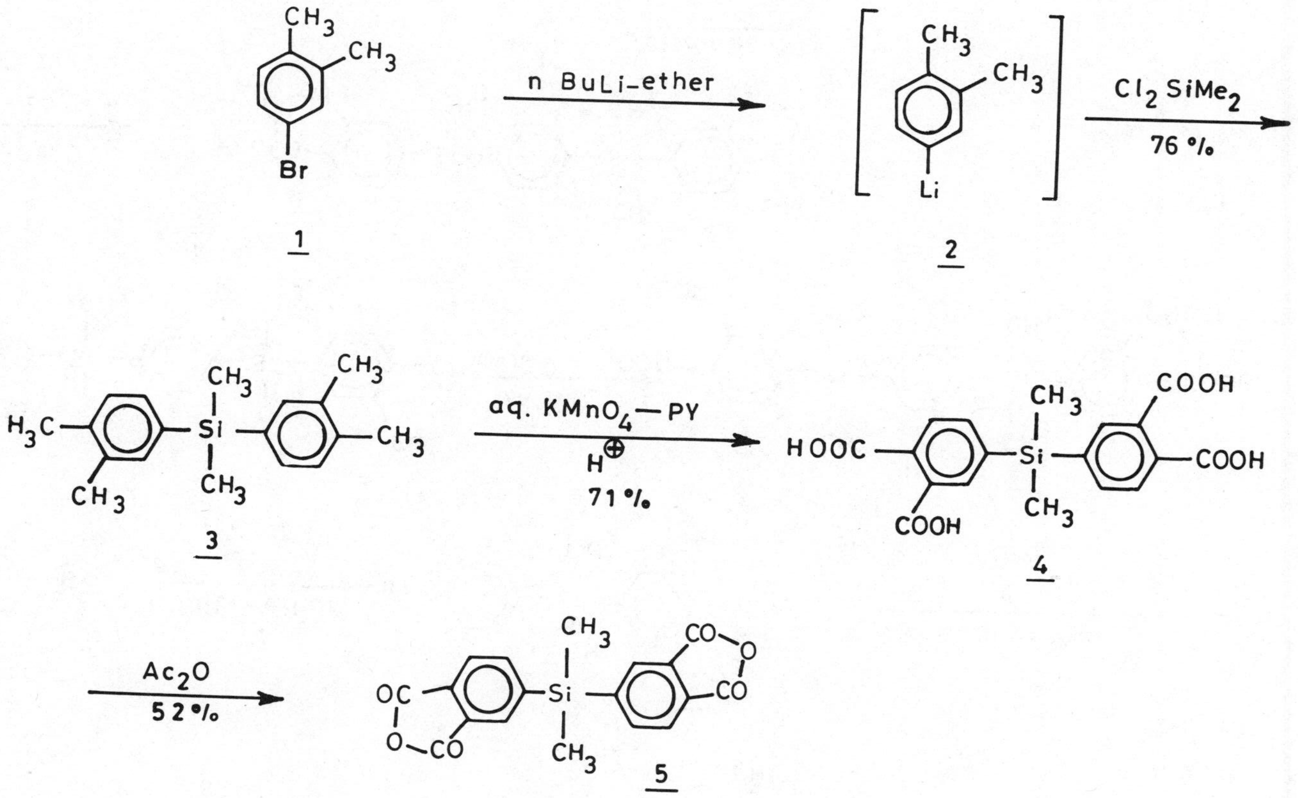

Scheme I

53

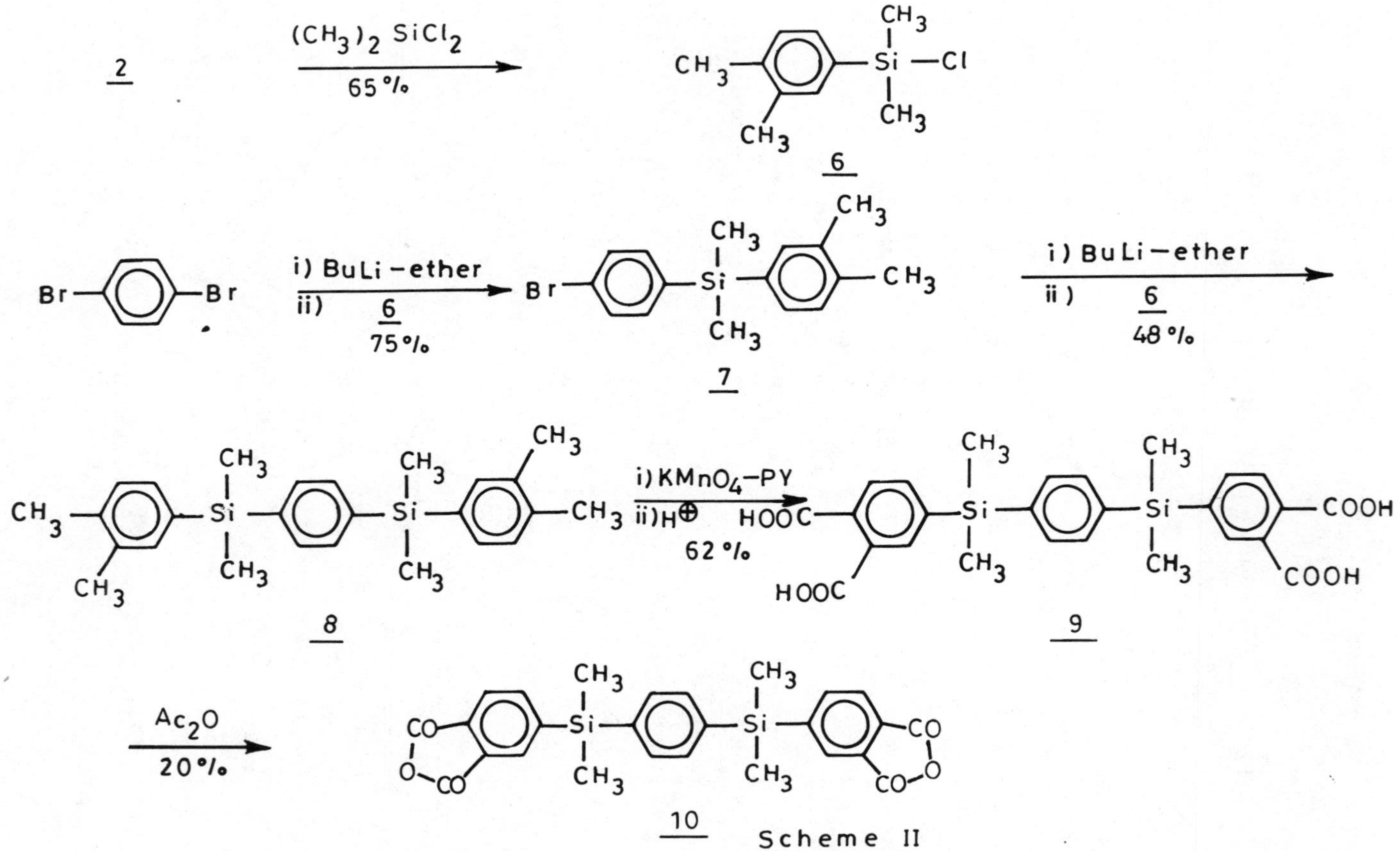
2
(CH3)2 SiCl2
65 %
CH3
Si
Cl
CH3
CH3
CH3
6
Br
Br
i) BuLi-ether
ii) 6
75 %
Br
CH3
Si
CH3
CH3
CH3
7
i) BuLi-ether
ii) 6
48 %
CH3
CH3
Si
CH3
CH3
Si
CH3
CH3
CH3
CH3
8
i) KMnO4-PY
ii) H⊕
62 %
HOOC
HOOC
CH3
Si
CH3
CH3
Si
CH3
COOH
COOH
9
Ac2O
20 %
CO
O
CO
CH3
Si
CH3
CH3
Si
CH3
CO
O
CO
10

Scheme II

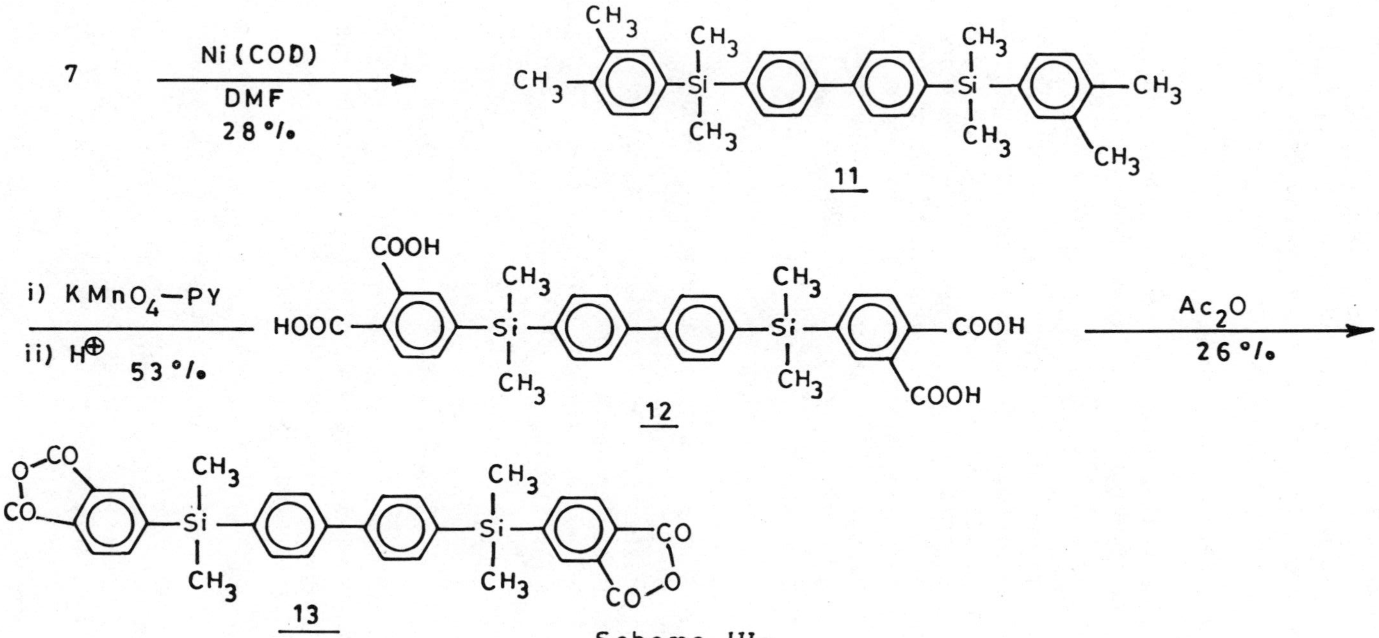

7 $\xrightarrow[\text{28\%}]{\begin{array}{c}\text{Ni(COD)}\\\hline\text{DMF}\end{array}}$ 11 $\xrightarrow[\text{26\%}]{\text{Ac}_2\text{O}}$

11 $\xrightarrow[\text{ii) H}^{\oplus}\ \ 53\%]{\text{i) KMnO}_4-\text{PY}}$ 12 $\longrightarrow$ 13

Scheme IIIa

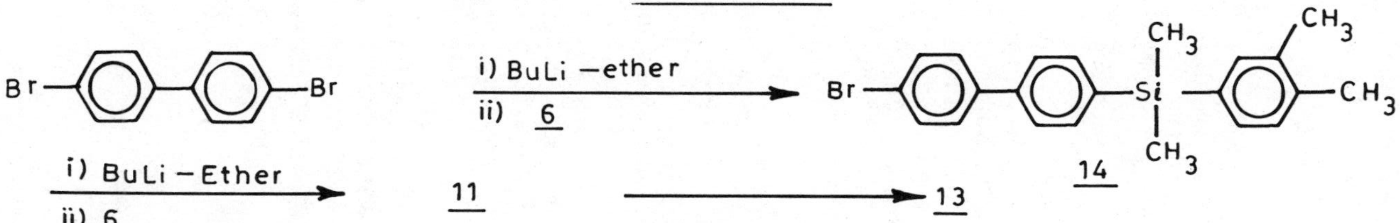

6 $\xrightarrow[\text{ii) 6}\ \ 36\%]{\text{i) BuLi}-\text{Ether}}$ 11

13 $\xrightarrow[\text{ii) 11}]{\begin{array}{c}\text{i) BuLi}-\text{ether}\\\hline\text{6}\end{array}}$ 14

Scheme IIIb

55

1,3-bis(3,4-dicarboxyphenyl)-1,1,3,3-tetramethyldisiloxane dianhydride 18 and 1,5-bis(3,4-dicarboxyphenyl) 1,1,3,3,5,5-hexamethyltrisiloxane dianhydride 20 were synthesized according to Schemes V and VI, respectively. The monomers 15 and 16 were pre- ed as shown in Scheme IV. Dimethyldichlorosilane was hydrolyzed with a deficiency of water to give a mixture of cyclic and linear siloxanes which includes disiloxane dichloride 15 and trisiloxane dichloride 16.

Synthesis of Polyimides

Polyimides were prepared by reacting silyl dianhydride with diamines. The two step polymerization is shown in Scheme VII involving the ring opening polyaddition and subsequent cyclodehydra- tion reaction. The ring opening polyaddition proceeds readily in DMF at room temperature forming polyamic acid; the results are summerized in Table 1. A perusal of the data reveals that poly- amic acids obtained from 3,3' diaminodiphenylmethane gives higher viscosities than those obtained from 4,4' diaminodiphenylmethane 23a under similar conditions. This may be attributed to the symmetry and the basicity of the diamines. These polyamic acids were subjected to thermal cyclodehydration to form the corresponding polyimides. The imidization was performed at 100, 200 and 300° C each for one hour under a stream of nitrogen.

Infrared Spectra

The infrared spectra of all polyamic acids showed broad absorption bands around 1610, 1650 and 3400 cm^{-1}. The absorption band at 1650 cm^{-1} is characteristic of secondary amide groups. In polyimides these bands disappear and new absorption peaks appear at about 1775, 1710, 1080, 800 and 710 cm^{-1}. The bands at 1775 and 1710 cm^{-1} are related to the stretching vibrations of two carbonyls that are weakly coupled. The bands at 1080, 800 and 710 cm^{-1} are associated with vibrations of cyclic structures.

Solubility

Siloxane containing polyimides dissolve in polar aprotic sol- vents such as DMF, DMA and DMSO. However, polymers obtained from 10 and 13 are not soluble. Further, the solubility of polyimides containing unsymmetrical diamines is more than for symmetrical diamines.

Glass Transition Temperature

The T_g was considered to be the single thermal transition most suitable for reflecting potential changes in processability of polyimides resulting from polymer structure changes. The determi-

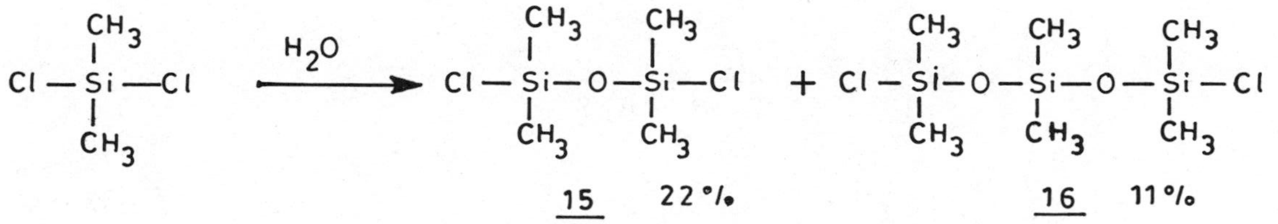

Scheme IV

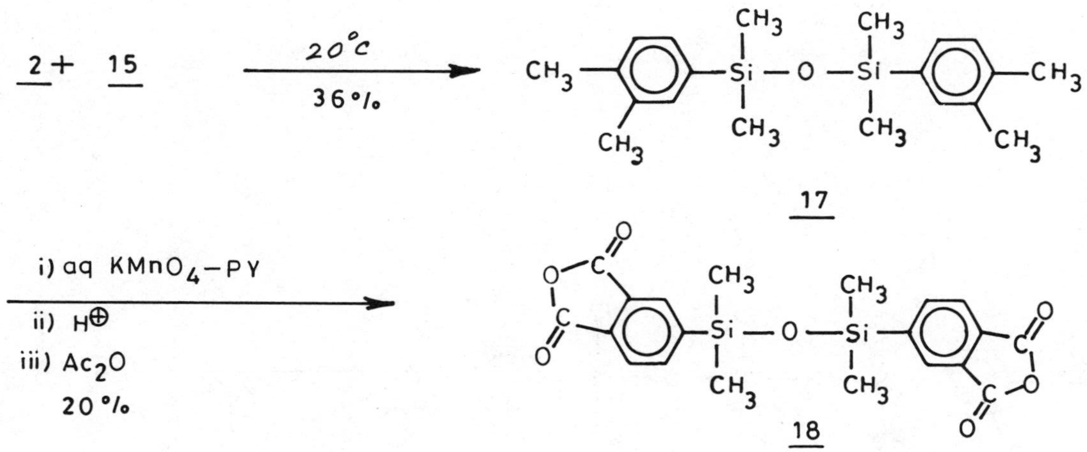

H₂O →

15   22%

16   11%

Scheme V

2 + 15

20°C
36%

17

i) aq KMnO₄–PY
ii) H⊕
iii) Ac₂O
20%

18

Scheme V

57

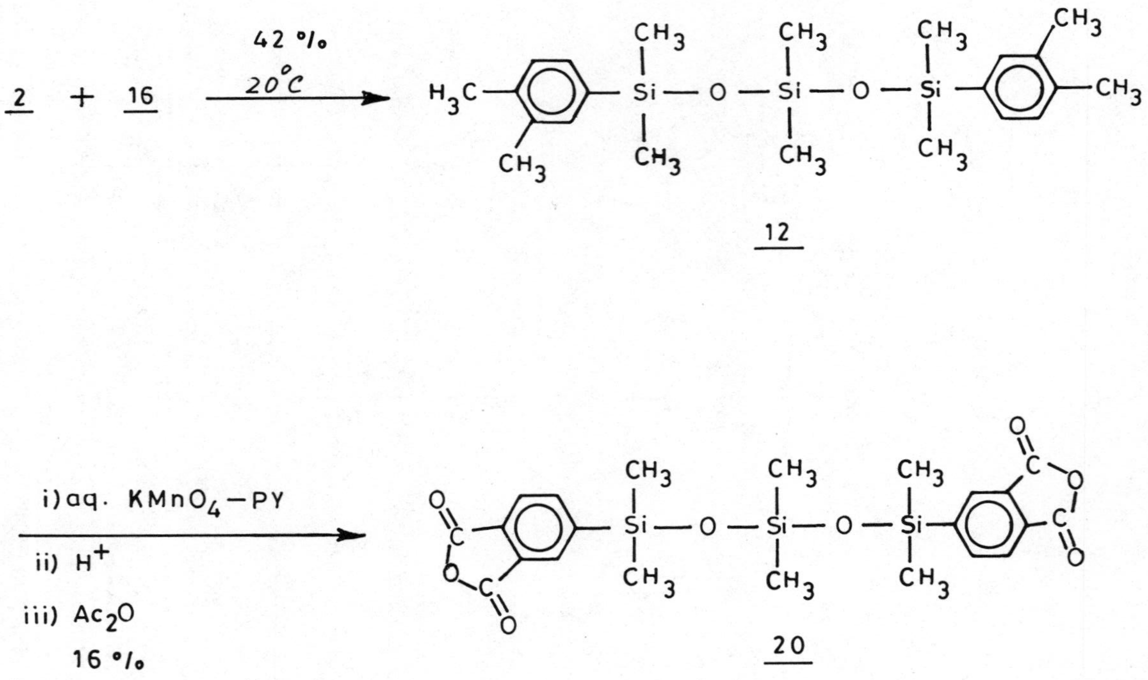

Scheme VI

nation of T_g was accomplished by means of thermomechanical analysis (TMA). The glass transition temperatures of various polyimides are listed in Table 1. The polyimides from m,m'-diamine isomers have lower T_g as compared to p,p'-diamine isomers. Insertion of siloxane units in the dianhydride lower the T_g of the polymers to a significant extent (See Figure 1).

Thermal Analysis

The heat stabilities were assessed by thermogravimetric analyses and isothermal heating experiments in air. All polymers showed an initial gain in weight. This has been attributed to the oxidation of the methylene groups to carbonyl. Similar results were obtained in our earlier studies[22]. TGA studies reveal that polyimides are more stable than m,m'diamines (Table II).

EXPERIMENTAL SECTION
Monomers

Bis(3,4-dimethylphenyl)dimethylsilane (13), chlorodimethyl (3,4-dimethylphenyl)silane (6), dimethyl (3,4-dimethylphenyl) (4-bromophenyl)silane (7), 1,4-bis(3,4-dimethylphenyldimethylsilyl) benzene (8) and tetra carboxylic acids from tetramethyl derivatives were made according to the published procedure[23]. Efforts were, however, made to improve the yields wherever possible.

Bis(1,5-cyclooctadiene)nickel (0) : Bis(1,5-cyclooctadiene)nickel (0), Ni(COD) was prepared according to the published procedure[26] and was used without further purification.

1,3-Dichlorotetramethyldisiloxane (15)and 1,5-dichlorohexamethyl trisiloxane 16[24] : Freshly distilled dimethyldichlorosilane (129g, 1.0 mol) in ether (100 ml) was added dropwise at room temperature to a stirred mixture of water (18g, 1.0 mol) and ether (500 ml) over a period of one hour. The two phase mixture was separated and the ether phase was washed with 5% aqueous sodium carbonate. The dried ether layer was concentrated on a rotary evaporator and the residue was fractionally distilled, yielding 1,3-dichloro-tetramethyldisiloxane (22%), bp 136-138° C (literature value bp. 41/20mm) and 1,5-dichlorohexamethyltrisiloxane (11%), bp 174-177° C (literature value bp. 79° C/20mm). The infrared spectrum showed bands at 2950,2850,2800,1422,1290,1260 cm^{-1} (Si-Me), 1180, 1050-1090 cm^{-1} (broad, Si-0-Si)

Dimethyl(3,4-dimethylphenyl)(4-bromobiphenyl)silane (14) : Freshly distilled pp'dibromobiphenyl (156g, 0.5 mol) was placed in a dry, nitrogen purged, 2-liter three necked flask fitted with mechanical stirrer, dry ice condenser and a mercury overhead pressure valve Anhydrous ether (450 ml) was added to the flask and it was cooled to -78° C. n-Butyllithium (0.50 ml) of 2.38 M in hexane was added dropwise and the resulting mixture was stirred for ca. 10 hr. Following the usual work-up, distillation of crude afforded 65% of 14.

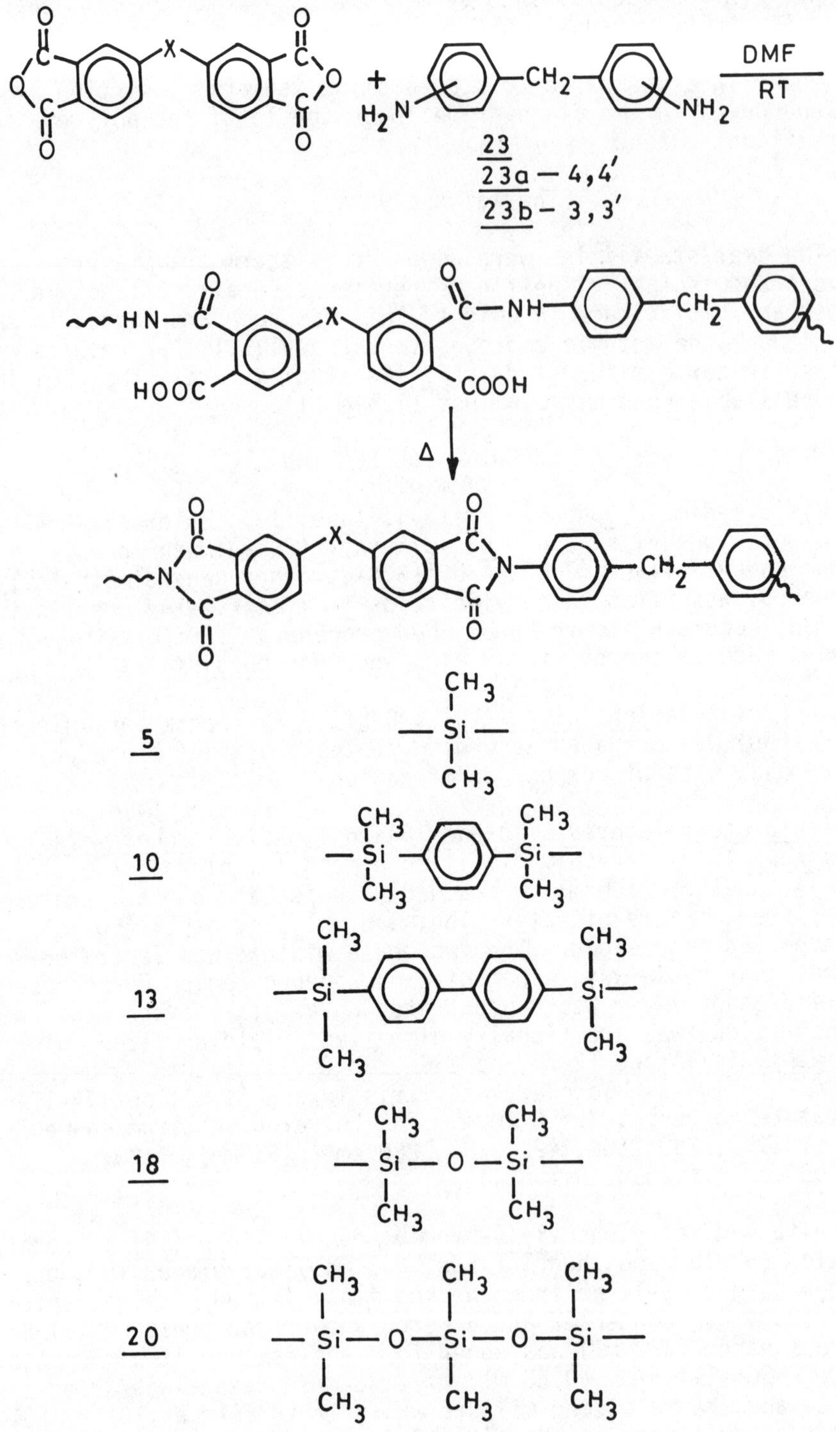

Scheme VII

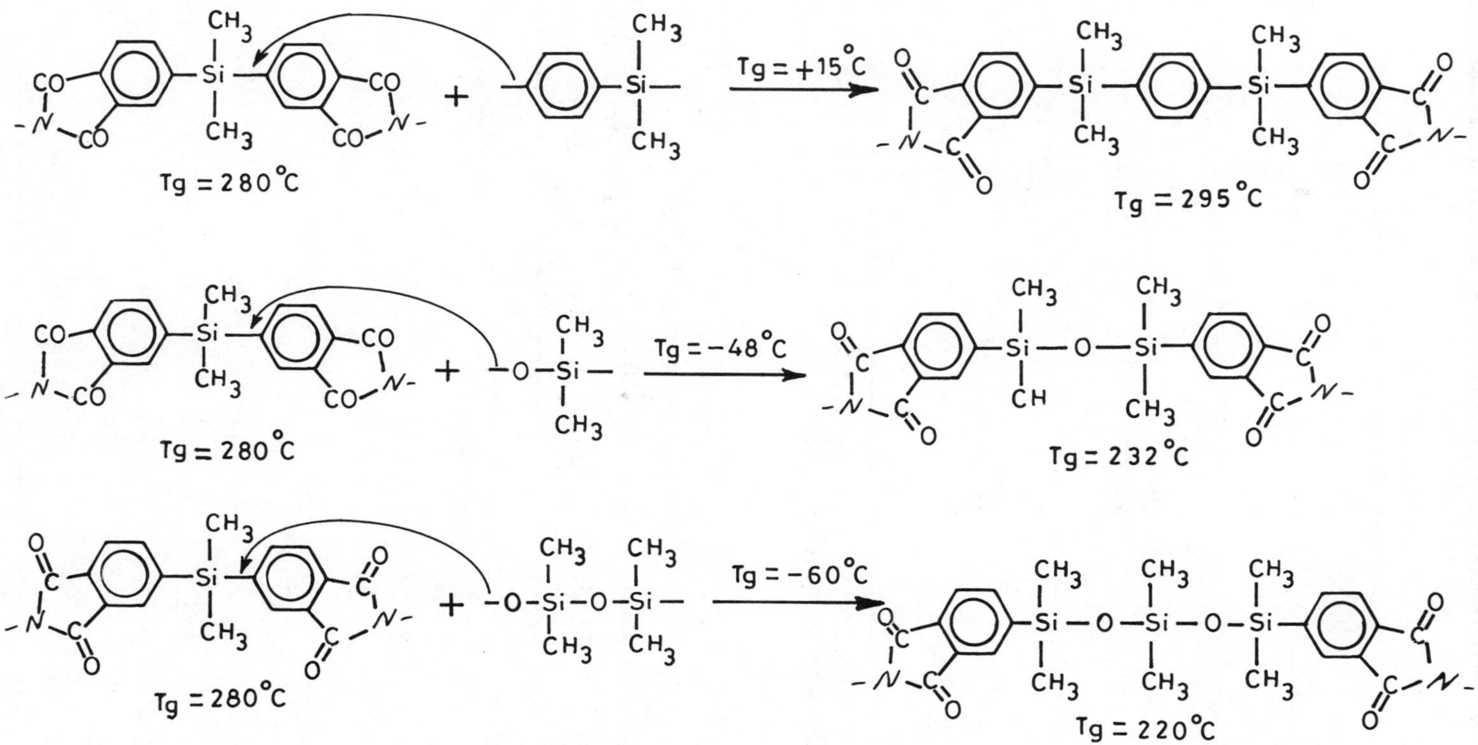

Figure 1. Change in glass transition temperature by insertion of dimethylsiloxy units into polyimides from SIDA and 4,4'-diaminodiphenyl methane.

Table I. Properties of Polymers from Silyldianhydrides and Isomeric Diaminodiphenylmethane.

Diamine	Dianhydride	a inh, dl/g	T_g^b, °C
23a	5	0.62	280
23a	10	0.63	295
23a	13	0.84	-
23a	18	0.32	232
23a	20	0.26	220
23b	5	0.81	238
23b	10	0.74	246
23b	13	0.89	285
23b	18	0.38	217
23b	20	0.31	208

a, intrinsic viscosity of polyamic acid; 0.5% in DMF at 30°C.
b, from TMA measurements.

Table II. Thermal Analysis of Polyimides.

Diamine	Dianhydride	Mode	TGA and Isothermal aging for 100 hr. in air weight loss % at		
			250°C	300°C	500°C
23a	5	TGA	3	1	53
		IA	12	-	-
23b	5	TGA	8	19	69
		IA	32	-	-
23a	10	TGA	2	3	36
		IA	26	-	-
23b	10	TGA	6	8	52
		IA	28	-	-
23a	18	TGA	10	12	39
		IA	18	-	-

4,4'-Bis(3,4-dimethylphenyldimethylsilyl)biphenyl(11): To a mixture of 14 (49.25g, 0.125 mol) in 220 ml of anhydrous ether at 0° C was added dropwise n-butyllithium (0.125 mol) of 2.4 M in hexane. Yield of crude was 36%. The crude was recrystallized from hexane to afford pure 11. mp. 121-124° C.

4,4'-Bis(3,4-dimethylphenyldimethylsilyl)biphenyl(11): To a slurry of 5.42g (0.02 mol) of bis(1,5-cyclooctadiene)nickel (0) in 150 ml of dry deoxygenated DMF was added under nitrogen atmosphere 6.36g (0.02 mol) of 7. The resulting mixture was poured into 200 ml of a 2 % HCl solution. The aqueous mixture was extracted with dichloro-methane and the solvent was removed under vacuum to afford white crystals in 28 % yield mp. 122-126° C. IR (KBr) 1260 cm^{-1} (Si-Me).

1,3-Bis(3,4-dimethylphenyl)-1,1,3,3-tetramethyldisiloxane(17): Freshly distilled 4-bromo-o-xylene (185g, 1 mol) was placed in a dry, nitrogen purged, 2-liter three necked flask fitted with a mechanical stirrer, dry ice condenser and a mercury over pressure valve. Anhydrous ether (450 ml) was added to the flask and it was cooled to -5 to 0° C. n-Butyllithium (425 ml, 1mol) of 2.38 M in hexane was added dropwise to the cooled solution while maintaining a positive nitrogen purge. The reaction was stirred at room tempera-ture over a period of 6 hr. Compund (15) was added dropwise at 20°C. The mixture was allowed to stir overnight. The salt, lithium chloride, was filtered and the filtrate was concentrated to remove ether. The yield of the residue was 36 %.

1,5-Bis(3,4-dimethylphenyl)-1,1,3,3,5,5-hexamethyltrisiloxane(19): The tetramethyl derivative was synthesized from compounds 2 and 16 at 20° C in 42 % yield.

4,4-Bis(3,4-dicarboxyphenyldimethylsilyl)biphenyl (12): A mixture of 50 ml pyridine and 25 ml water was placed in a 1-liter resin kettle along with 3.78g of compound 11. Potassium permanganate (15g) was slowly added to the refluxing and well stirred solution; the addition was continued for ca. 1 hr. After refluxing for an addi-tional period of 4 hr., a negative test for permanganate[25] was obtained and the hot mixture was filtered to remove manganese dioxide The MnO$_2$ was washed several times with boiling water, then the filtrate was concentrated and pyridine was removed on a rotary evaporator. The concentrated basic solution was acidified to p^H 1.0 with 3 N HCl to precipitate crude product. After filtering, the product was dissolved in 95% ethanol and placed on a strong acid cation exchange column with 95% ethanol. The tetra acid was obtained in 53% yield. The recrystallized compound did not show a definite melting point and was degraded beyond 400°C. IR(KBr) 1700 cm^{-1} (carbonyl) and 1255, 804 cm^{-1} (Si-Me).

4,4'-Bis(3,4-dicarboxyphenyldimethylsilyl)biphenyl dianhydride (13)

The tetra acid (12) was placed in a flask with a 4 mol excess of acetic anhydride, and acetic acid was distilled from the mixture. The hot mixture was poured into a crystalline dish to precipitate dianhydride in 26% yield, m.p. 224-229°C. Anal. Calcd. for $C_{32}H_{26}O_6Si_2$: C, 68.32; H, 4.63. Found: C, 67.84; H, 4.43. IR(KBr) 1845, 1775 cm^{-1} (doublet, anhydride carbonyl), 1260, 804 cm^{-1} (Si-Me).

3-Bis(3,4-dicarboxyphenyl)-1,1,3,3-tetramethyl disiloxane dianhydride (18).

The oxidation of (17) (27.36 g, 0.008 mol) with potassium permanganate (14.7 g, 0.093 mol) in 50 ml pyridine and 25 ml water was conducted according to the procedure outlined for (12). The crude tetra acid was treated with excess acetic anhydride and acetic acid and refluxed to get compound (18) in 20% yield (based on tetra acid). The anhydride was recrystallized from n-hexane m.p. 156-159°C. Anal. Calcd. for $C_{12}H_{18}O_6Si_2$ (18); C, 48.32; H, 5.37. Found: C, 47.93, H, 5.21. IR(KBr) 1843, 1775 cm^{-1} (doublet, anhydride carbonyl); 1260 cm^{-1}, (Si-Me) and 1180, 1055-1110 cm^{-1} broad absorption (Si-O-Si).

1,5-Bis(3,4-dicarboxyphenyl)-1,1,3,3,5,5-hexamethyltrisiloxane dianhydride (20).

The dianhydride was obtained according to the procedure described for (18) in 16% yield. The recrystallized compound (in n-hexane) showed a m.p. in the range 168-172°C. Anal. Calcd. for $C_{14}H_{24}O_7Si_3$ (20): C, 43.30; H, 6.18. Found: C, 43.13; H, 6.09. IR(KBr) 1850, 1775 cm^{-1} (doublet, anhydride carbonyl) 1260, 805 (Si-Me).

Synthesis of Polymers

The poly(amic acid) precursors in the polyimides were prepared by the customary solution polymerization methods in N,N-dimethylformamide (DMF) solvent. The polymerizations were carried out in a three necked flask equipped with dropping funnel, nitrogen inlet and a stirrer. Dianhydrides were added at one time in quantities equimolar to the diamines. Polymerizations were carried out for about 10 hr. The inherent viscosities at 0.5% concentration in DMF at 30°C were determined. The solutions were stored at -10°C until further use.

Films were prepared from the polymers by spreading the poly(amic acid) solutions, usually at 15% solids concentration, onto glass plate. Thin films were made for IR studies. The films for thermomechanical analysis and thermogravimetric analysis were dried under vacuum overnight. The films were further cyclodehydrated to the polyimides in nitrogen at 100, 200 and 300°C for 1 hr at each temperature.

Thermomechanical Analysis (TMA): The glass transition temperatures (T_gs) of polyimide films were determined using E.I. DuPont Model 940 Thermomechanical Analyzer in nitrogen.

Thermogravimetric analysis was carried out with E.I. DuPont Model 940 thermobalance in air at a heating rate of 10°C/min.

CONCLUSIONS

Studies on the combined effects of meta vs. para orientation of the amine groups of aromatic diamines and the dilution of the imide content on the glass transition temperature were conducted to undertake possible ways to improve PI processibility. The use of a methylene group to connect the benzene rings of the diamines and systematic variation of silyl groups in dianhydrides afforded a comparison of the thermal properties of polyimides. A T_g of as low as 220°C could be obtained in polyimides containing hexamethyl trisiloxane groups.

REFERENCES

1. V.L. Bell, L. Kilzer, E.M. Hett and G.M. Stokes, J. Appl. Polym. Sci., 26, 3805 (1981).
2. L.W. Frost and I. Kesse, J. Appl. Polym. Sci., 8, 1039 (1964).
3. J.P. Critchley and M.A. White, J. Polym. Sci., Polym. Chem. Ed., 10, 1809 (1972).
4. V.L. Bell, B.L. Stump and H. Gager, J. Polym. Sci., Polym. Chem. Ed., 14, 2275 (1976).
5. T.L. St. Clair, A.K. St. Clair and E.N. Smith, Polymer Preprints, 17, 359 (1976).
6. F.W. Harris, W.A. Feld and L.H.Lanier in "Applied Polymer Symposium" No 26 (N. Platzer Ed.) John Wiley and Sons, N.Y. (1975).
7. A.P. Rudakov, M.I. Bessonov, Sh. Tuichiev, M.M. Koton, F.S. Florinskii, B.M. Ginzburg and S. Ya. Frenkel, J. Polym. Sci. USSR, 12 , 720 (1970).
8. G.L. Brode, J.H. Kawakami, G.T. Kwiatkowski and A.W. Bedwin J. Polym. Sci., Polym. Chem. Ed., 12, 575 (1974).
9. R.J. Jones, M.K. O'Rell and J.M. Horn, U.S. Pat. 4,111,906 (1978).
10. R.J. Jones, M.K. O'Rell and J.H. Horn, U.S. Pat. 4,196,277; 4,203,922 (1980).
11. F.E. Rogers, U.S. Pat., 3,356,648 (1967).
12. G.N. Babu and S. Samant, Europ. Polym. J., 17, 421 (1981).
13. G.N. Babu and S. Samant, Die Makromol. Sci., 183, 1129(1982).
14. G.N. Babu and S.Samant, J. Macromol. Sci., Chem. A17, 425 (1982).
15. C.P. Pathak, S. Samant, M.V.R. Murthy and G.N. Babu, ACS Org. Coat. and Plastic. Preprints, 46, 154 (1982).

16. J.E. Mulvaney and C.S. Marvel, J.Polym. Sci., $\underline{50}$, 541 (1961).
17. H. Kuckertz, Makromol. Chem., $\underline{98}$, 101 (1966).
18. J.C. Bonnet and E. Marechal, Bull. Soc. Chim. France, 3562 (1972).
19. N.J. Johnston and R.A. Jewell, Div. Org. Coat. & Plast. Chem. Preprints, $\underline{33}$, 169 (1973).
20. H.N. Kovacs, A.D. Delman and B.B. Simms, J. Polym. Sci., A1, $\underline{6}$, 2103 (1968).
21. H.N. Kovacs, A.D. Delman and B.B. Simms, J. Polym. Sci., A1 $\underline{8}$, 869 (1970).
22. S. Samant, Ph.D. Thesis, Indian Institute of Technology, Bombay (1983).
23. J.R. Pratt and S.F. Thames, J. Org. Chem., $\underline{38}$, 4271 (1973).
24. L.N. Breed, M.E. Whithead and R.L. Elliot, Inorg. Chem. $\underline{6}$, 1254 (1970).
25. L.F. Fieser and M. Fieser, 'Reagents for Organic Synthesis' Vol I p. 943, Wiley N.Y. (1967).
26. M. Butcher, R.J. Mathews and S. Middleton, Aust. J. Chem., $\underline{26}$, 2067 (1973).

MODIFIED POLYIMIDES BY SILICONE BLOCK INCORPORATION

A. Berger

M&T Chemicals Inc.
Rahway, New Jersey 07065

Copolymerization of a variety of dianhydrides with molar quantities of organic diamines and/or amino-functional di- and/or polysiloxanes is discussed.

Whereas conventional polycondensation reactions are effected in high boiling dipolar aprotic solvents such as DMF, NMP, DMAc, etc. or very corrosive solvents such as cresylic acids, etc., the PSI polymerization can be run in all of the above solvents and additionally in chlorinated solvents, hydrocarbon solvents, and polyglyme solvents. After being converted to their finalized form they still maintain solubility in the above solvents.

Accordingly, the resulting properties arising from a co-polymerization reaction of the corresponding monomers will be strictly a function of the composition of the polymer and will vary in properties between the extremes of silicones and polyimides, depending solely upon the nature of the siloxane diamine and quantity incorporation into the co-polymerization reaction.

Certain modified PSI polymers do exhibit excellent adhesion, weatherability resistance (UV, corona discharge, etc.), good thermal cycling properties, controlled elongation, high solubility, etc. and still have high thermal stability, scratch resistance, tear resistance, etc. associated typically with all aromatic polyimides.

Furthermore, PSI polymers can be melt polymerized and can be produced with a minimal of contamination leading to a variety of applications wherein cleanliness is of utmost importance.

67

INTRODUCTION

Copolymerization of a variety of dianhydrides with molar quantities of organic diamines and/or aminofunctional di- and/ or polysiloxanes is discussed.

Such polymers will vary in properties between the two extremes of silicones and polyimides depending solely on the composition of the polymer. The most desirable properties of silicones can be incorporated into the polyimide polymer still maintaining the salient features of "an all aromatic" polyimide.

The silicone blocks lend themselves to cross-linking mechanism not available to a typical polyimide adding extra dimension to the copolymer system.

Polyimides and related polymers, when synthesized from aromatic reactants are generally insoluble and infusible. This is a result of the rigid, inflexible structures present in these polymers. Although these properties are valuable for high temperature service applications, they make it virtually impossible to process these polymers by conventional methods.

It is possible to improve the processability of these polymers by the incorporation of flexible segments that connect the rigid imide structures. The free volume associated with the flexible segments permits the polymers to be fusible and soluble. Such polymers are easily processed, but this processability is obtained with some sacrifice in physical properties such as heat distortion temperature, thermal stability, hardness and modulus. On the other hand, the impact resistance, weatherability and other characteristics of the polyimide are improved. Depending on the relative amounts of the flexible and rigid (imide, etc.) portions, and on their structures, a considerable variation in physical properties is possible.

The above concept was recognized a long time ago, and a number of silicone modified polyimides were prepared in efforts to improve their tractability.

Accordingly, Holub [1,2] in a series of patents discloses co-polymers derived from mixtures of 1,3 bis (Y-aminopropyl) tetra-methyldisiloxane and several of its dimethylsiloxy equilibrates with aromatic diamines and benzophenone tetracarboxylic acid dianhydrides.

Whereas the number of dimethyl siloxy units were limited to a maximum of about four in the Holub disclosures, Fessler[3] describes copolymeric silicone-polyimides involving the Bis-amino functional equilibrates with benzophenone tetracarboxilic acid dianhydride having dimethylsiloxy unit greater than twenty-five.

68

Other dianhydrides, such as pyromellitic tetra acid dianhydride
were likewise polymerized with the aminofunctional disiloxane.
Both the synthesis of Bis(aminopropyl) tetramethyldisiloxane and
1,3 Bis(m-aminophenyl) tetramethyl disiloxane and its polymerization with pyromellitic dianhydride was discussed by Kuckertz [4],
Greber [5,6] likewise carried out polymerizations of pyromellitic
tetra acid dianhydride and trimellitic anhydride with 1,3 Bis
(aminopropyl) tetramethyldisiloxane.

Heath and Wirth [7] also describe the preparation of poly
(siloxane-ether-imides) based on a polycondensation reaction
involving Bis(aminobutyl) tetramethyl disiloxane and 4,4' Bis
(3,4 dicarboxyphenoxy) diphenyl sulfide tetra acid dianhydride.

The common feature of all of the above work is the incorporation of Bis(aminoalkyl) siloxanes into the polyimide formulation.
In one case[4] there is also mention of a Bis(aminoaryl) siloxane.
However, such incorporations led to brittle, less soluble, low
molecular weight products.

An alternative approach to silicone-modified polyimides was
described by Johnston[8], and Pratt[9]. These authors prepared
siloxane containing dianhydrides and copolymerized these monomers
with organic diamines.

M&T APPROACH

In the M&T approach to silicone-modified polyimides, a
copolymer involving a variety of dianhydrides with molar quantities of organic diamines and/or aromatic aminofunctional di- and/
or polysiloxanes were synthesized in a variety of organic solvents.

The typical polycondensation reaction involves:

"A block co-polymer polyamic acid"

"A block co-polymer polyimide"

The nature of the aminofunctional siloxanes is the subject matter of two U.S. patents[10,11].

Unlike the conventional dipolar aprotic solvents typically utilized for polymerization reactions involving polyimide formation, i.e., DMAc, NMP, DMF, etc., incorporation of the subject aminofunctional siloxane enabled polymerization to take place in chlorinated solvents, hydrocarbon solvents, glyme solvents, etc. depending primarily on the final composition of the product. This factor was significant since the inability to remove the last traces of dipolar aprotic solvents leads to destruction of coating material at high temperature. Uniquely to this system was the capability of running polymerization reactions in the melt wherein product recovery could easily be achieved solvent free.

Because of the thermoplasticity of the resulting final product, such polymers can be redissolved in a variety of solvents and applied to surfaces via spraying techniques, spinning techniques, etc. and cured via solvent evaporation. There is no need to invoke a programmed thermal cycle process as is conventional for coatings in their amic acid form. Accordingly, thicker films can be deposited, cured at lower temperatures, and failure due to incomplete imidization be eliminated.

Another feature inherent in silicone modified polyimides is the ability to alter the nature of the siloxane diamine by equilibration reactions.

70

Equilibration

Typically a monofunctional silane is reacted with a difunctional silane and an equilibration catalyst leading to an equilibrated product.

Typically

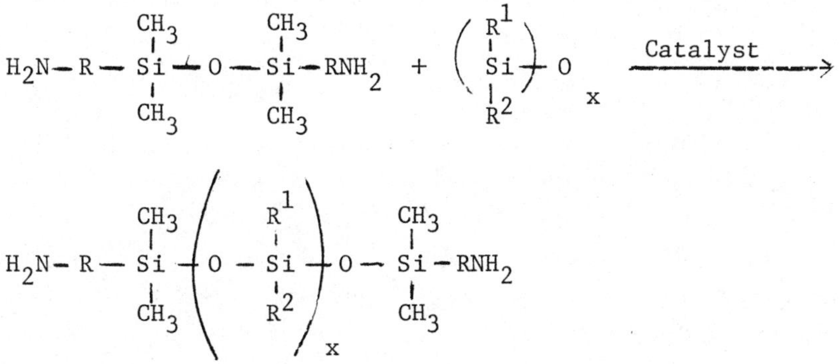

$$H_2N-R-\underset{\underset{CH_3}{|}}{\overset{\overset{CH_3}{|}}{Si}}-O-\underset{\underset{CH_3}{|}}{\overset{\overset{CH_3}{|}}{Si}}-RNH_2 \;+\; \left(\underset{\underset{R^2}{|}}{\overset{\overset{R^1}{|}}{Si}}O\right)_x \xrightarrow{\quad\text{Catalyst}\quad}$$

$$H_2N-R-\underset{\underset{CH_3}{|}}{\overset{\overset{CH_3}{|}}{Si}}\left(O-\underset{\underset{R^2}{|}}{\overset{\overset{R^1}{|}}{Si}}-O\right)_x \underset{\underset{CH_3}{|}}{\overset{\overset{CH_3}{|}}{Si}}-RNH_2$$

R^1 and R^2 may each be alkyl, vinyl, phenyl, functional alkyl, etc. Catalyst is a base as KOH, CsOH, $R_4 \overset{\oplus}{N} \overset{\ominus}{OH}$, $R_4 \overset{\oplus}{P} \overset{\ominus}{O}-Si-(CH_3)_3$, etc.

More generally, the equilibrated polysiloxane diamine looks like

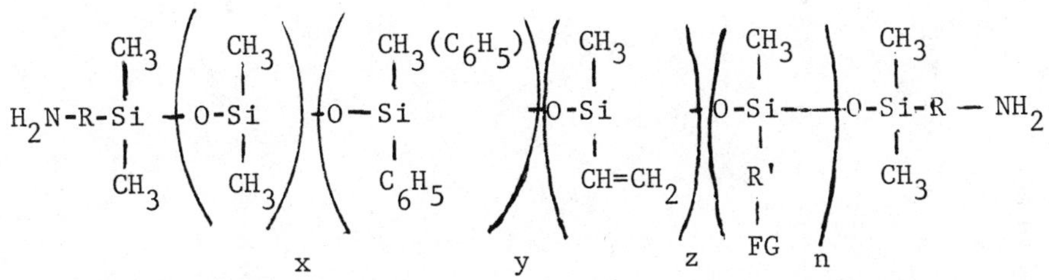

$$H_2N-R-\underset{\underset{CH_3}{|}}{\overset{\overset{CH_3}{|}}{Si}}\left(O-\underset{\underset{CH_3}{|}}{\overset{\overset{CH_3}{|}}{Si}}\right)_x\left(O-\underset{\underset{C_6H_5}{|}}{\overset{\overset{CH_3(C_6H_5)}{|}}{Si}}\right)_y\left(O-\underset{\underset{CH=CH_2}{|}}{\overset{\overset{CH_3}{|}}{Si}}\right)\left(O-\underset{\underset{\underset{FG}{|}}{R'}}{\overset{\overset{CH_3}{|}}{Si}}\right)_z\underset{\underset{CH_3}{|}}{\overset{\overset{CH_3}{|}}{O-Si}}_n-R-NH_2$$

x, y, z and n are each independent of each other and can be zero, or greater.

FG - functional group as $-CN$, $-\overset{\overset{O}{\|}}{C}OR$, $-OH$, etc.

R' = alkylene or arylene

Benefits of Block Copolymer by Silicone Incorporation (Ref. 12-14).

(1) Controlled Tg - lower Tg material, may extend service temperature range of material.

(2) Enhanced adhesion - surface priming unnecessary.

(3) Impact properties of polymers improved by incorporation of soft segments.

(4) Good weatherability - resistance to UV, corona discharge, nuclear bombardment, etc.

(5) Controlled elongation - applications comparable to RTV's with the advantage of no catalyst, no mixing, infinite shelf life.

(6) Improved dielectric properties over silicones, approaching and matching polyimides.

(7) Minimal contamination - high purity monomers and solvents other than dipolar aprotic lead to products having minimal contamination by chloride, sodium, potassium, etc.

(8) Ease of preparation via melt polymerization.

(9) Enhanced solubility - order of 40-60% concentration available - deposition of thick films, minimal handling of and contamination of solvent.

Cross-linking Routes

Cross-linking typically accomplished in polyimide polymers by

(1) Incorporation of nadic end groups.

(2) Incorporation of acetylenic end groups.

(3) Incorporation of Diels Alder components.

(4) Incorporation of alkylene linkages and cross-linking via peroxides, bis-maleimides, etc.

Additionally, when vinylmethylsiloxy units are incorporated into an equilibrated aromatic amino functional siloxane, then a different mode of cross-linking is available having no counter-part in conventional polyimide polyimide polymers.

$$-O-Si(CH_3)(CH_3) - \left(O-Si(CH_3)(CH_3) \right) \xrightarrow{UV} O-Si(CH_3)(CH_2\cdot) - \left(O-Si(CH_3)(CH_3) \right)_x + H\cdot$$

$$-O-Si(CH_3)(CH_2\cdot) - \left(O-Si(CH_3)(CH_3) \right) + \left(O-Si(CH_3)(CH=CH_2) \right)\left(O-Si(CH_3)(CH_3) \right)_x \longrightarrow$$

$$-O-Si(CH_3)(CH_2) - \left(O-Si(CH_3)(CH_3) \right)_x$$

with CH$_2$—·CH—O-Si(CH$_3$)—(O-Si(CH$_3$)(CH$_3$))$_x$

$$\xrightarrow[\text{Abstraction}]{\text{Hydrogen}}$$

Leads to cross-linked polymer

Benefits of Silicone-block Polyimide Copolymers (PSI) as opposed to Silicones

(1) Greater thermal stability
     (reversion problems minimal)

(2) Ease of application

  a) No mixing
  b) No catalyst
  c) No shelf-life problems
  d) Requires no refrigeration

(3) Greater strength of unreinforced films

(4) Less sensitive to degradation by water at high temperature

(5) Greater scratch resistance

(6) Better dielectric properties

(7) Coefficient of expansion can better match materials of construction

(8) Cleanliness of system

CONCLUSION

Polyimides having flexible segments via aminofunctional siloxane incorporation have been synthesized. As anticipated, improved tractability and solubility have resulted.

There are a variety of compositions depending on the nature of the dianhydride, aromatic amine and the nature of the aminofunctional siloxane and/or polysiloxane.

The electrical, mechanical and thermal properties will vary again, depending on the composition of the formulation and lie between silicones on one end and an all-aromatic polyimide composition on the other end.

REFERENCES

1. U. S. Patent 3,325,450.
2. U. S. Patent 3,740,305.
3. U. S. Patent 3,736,290.
4. J. H. Kuckertz, Makromol. Chem., 98, 101 (1966).
5. G. Greber and R. Pense, Angew. Chem., 78, 610 (1966).
6. G. Greber, Angew. Makromol. Chem., 4/5, 212 (1968).
7. D. R. Heath and S. G. Wirth, Fr. Demande 2,236,887.
8. N. J. Johnston, ACS Organic Coating & Plastic Preprints, 33, 169 (1973).
9. J. R. Pratt, Ph.D. Thesis - Univ. of Southern Mississippi, (1974).

10. U.S. Patent 4,139,547 (1983).
11. U.S. Patent 4,395,527 (1983).
12. B. Chowdhury, these proceedings, vol. 1, pp. 401-415.
13. M&T 3500 Siloxane Polyimide Data Sheet, M&T Chemicals, Rahway, N.J., 1983.
14. M&T 2065 Siloxane Polyimide Resin Preliminary Data Sheet, M&T Chemicals, Rahway, N.J., 1983.

V-378A, A NEW MODIFIED BISMALEIMIDE MATRIX RESIN FOR HIGH MODULUS GRAPHITE

S. W. Street

HITCO Materials Division
710 East Dyer Road
Santa Ana, California 92707

V-378A is a modified bismaleimide matrix resin which exhibits significant advantages over previously available bismaleimides and 350°F curing epoxy resins. The Tg of this new resin is in excess of 700°F. It has excellent tack at ambient temperatures and is available at prepreg resin solids contents as low as 30% for "net" resin (i.e. no bleed) cures. High modulus graphite unidirectional tapes based on V-378A exhibit excellent uncured or wet transverse strength. V-378A can be cured under standard low autoclave pressure of 100 psi and temperatures of 350°F similar to current 350°F curing epoxy systems. Unrestrained post cure at temperatures of 475 - 500°F are required to maximize elevated temperature strength retention. Cured V-378A/high modulus graphite composites exhibit strength retention at temperatures up to 600°F, excellent resistance to conditions of high humidity, no microcracking after repeated thermal spikes, and excellent tensile, compressive and shear strength. V-378A offers easy, epoxy-like processing characteristics with no volatile condensation products given off during cure with significantly superior high temperature strength retention after severe humidity aging.

INTRODUCTION

Addition polyimides cure with no evolution of gaseous by-products at relatively low temperatures and may be cured at low pressures to yield composites with excellent hot-wet strength retention. These properties have made them excellent candidates as matrix resins for advanced composites. However, commercially available bismaleimides are solids and difficult to handle in preimpregnated form.

V-378A is an addition polyimide composed of a mixture of bismaleimides and other reactive ingredients formulated to provide good prepreg properties and handling, facile cure and excellent composite mechanical properties.

Epoxy resins, the dominant matrix system for high modulus graphite (HMG) do provide very good composite properties and have been extensively used in the manufacture of sporting goods such as golf clubs, tennis racquets, fishing rods, etc. and aircraft parts including ailerons, rudders, flaps, vertical and horizontal stabilizers, and complete wing sections for commercial and military aircraft. Even the more sophisticated 175°C curing epoxy resins do not provide sufficient hot-wet strength retention for some applications. In addition, these resins are commonly supplied at relatively high resin contents requiring "bleeding" to achieve the desired resin content.

Use of this technique (bleeding) results in non-uniformity of resin content in the composite since more resin is removed next to the bleeder. This non-uniformity of resin content induces stresses and often results in warped parts. Use of a "net" resin concept tends to overcome these problems. V-378A exhibits excellent fiber wet out during the initial prepregging step and can thus be provided as a "net" resin prepreg.

The 175°C curing epoxy resin systems exhibit long gel times at 175°C and very low "watery" viscosities which tend to make it difficult to achieve low void composites. V-378A, in contrast, gels fairly rapidly at 125°C and exhibits moderate to low viscosity during heat up making it much easier to achieve very low void content composite structures.

RESULTS AND DISCUSSION

Lay-up and Cure. V-378A exhibits light to medium tack at a prepreg "net" resin content of 30% which is significantly lower than epoxies typically supplied at 42%. The "net" resin concept is particularly useful for preparation of composites having both thick and thin sections. Due to the low temperature gel and relatively low exotherm, no difficulty has been encountered in curing laminates up to four inches in thickness.

Standard Nylon vacuum bags and sealant tapes (same as used for 175°C epoxy) may be used versus expensive polyimide film bag and silicone sealant tape required for higher temperature curing polyimides.

To cure the V-378A prepreg lay-up, 85 psi autoclave pressure and vacuum are applied at the beginning of the cycle, and the part heated to 60 - 90°C at a rate of 2 - 5°C/minute. The vacuum is released, the bag vented to the atmosphere, pressure increased to 100 psi and heat up continued to 175°C. The part is cured four hours at 175°C, cooled to ambient temperature, removed and post cured, unrestrained, from room temperature to 246°C for four hours. The four hour post cure at 246°C is required to develop good mechanical properties. Table I illustrates a study of ± 45° tensile ultimate and strain using more extensive post cures and indicates the four hour post cure at 246°C to be satisfactory.

Use of a one to two hour hold at 60 - 90°C for very thick parts allows more time before gelation in order for air, trapped during lay up, to be released.

Figure 1 shows the effect on viscosity of heating the resin to 67°, 70° and 90°C with holds of five hours, three and one-half hours and two and one-half hours respectively. A heating rate of 1°C/minute was used.

Figure 2 illustrates the effect of heating rate with lower minimum viscosities obtained at faster heating rates.

Effects of Moisture on Prepreg. The 175°C epoxy curing matrix resins have exhibited much variability in processing due to moisture pickup of the uncured prepreg. This affinity for moisture results in variations in gel time, foaming during cure and porosity in the cured composite.

In contrast, V-378A prepreg exposed to 100% RH and 52°C for one hour, then laid up and cured, exhibited no deleterious effects. Figures 3 and 4 exhibit the effect of a sixteen hour high humidity (50% vs 90% RH) exposure at 24°C on uncured films of V-378A. No porosity or foaming was noted during the Rheometric Dynamic Spectroscopic viscosity determination.

Mechanical and Physical Properties. Figure 5 exhibits a Tg of V-378A (via Rheometric Dynamic Spectroscopy) in excess of 370°C. Translation of properties on high modulus graphite are excellent. Table II lists mechanical properties of V-378A/T-300·6K tape composites. Retention of dry flexure at 310°C is approximately

Table I. Effect of Post Cure on ±45° Tensile V-378A/T-300.6K Unitape Composites.

I. PREPREG PROPERTIES

RESIN CONTENT, % WT	30.0 (NET)
TACK	LIGHT – MEDIUM
RECOMMENDED STORAGE TEMPERATURE	18°C OR BELOW

II. LAYUP AND CURE

EIGHT PLIES V378A, ± 45° (LOT 2W4872) WERE LAID UP ON A FREKOTE 33 RELEASED CAUL PLATE. ONE PLY TX1040 ADDED ON TOP OF LAYUP PLUS RELEASED TOP CAUL PLATE. COROPRENE SIDE DAMS APPLIED WITH 3 MIL TEDLAR ("L" SLITS AT EACH CORNER) TAPED TO TOP OF CORPORENE DAMS. TWO PLIES 1581 BREATHER OVER ENTIRE LAYUP TO VENT HOLES

APPLY VACUUM PLUS 85 PSI AND HEAT FROM ROOM TEMPERATURE TO 82°C AT 3 ± 2°C MIN, HOLD 30' AT 82°C, THEN VENT BAG TO ATMOSPHERE AND INCREASE PRESSURE TO 100 PSI. HEAT FROM 232°C TO 177°C. HOLD FOUR HOURS AT 232°C. COOL, REMOVE AND POST CURE AS NOTED BELOW:

POST CURE	4 HR AT 246°C	8 HR AT 246°C	12 HR AT 246°C	24 HR AT 246°C	4 HR AT 246°C +1 HR AT 288°C
± 45° TENSILE ULT, KSI AT R.T.	21.49	22.97	23.41	21.16	19.60
	23.71	22.46	22.18	22.31	20.15
	23.77	22.37	23.04	21.74	19.62
AVERAGE	23.0	22.6	22.9	21.7	19.8
± 45° TENSILE STRAIN, μ IN./IN. AT R.T.	16740	17400	16920	15300	17000
	16600	16960	16600	11400	16440
	16700	17660	15000	15600	13500
AVERAGE	16680	17340	16170	14100	15650
± 45° TENSILE ULT, KSI AT 350°F	18.64	15.10	17.38	16.85	16.80
	17.23	15.81	14.12	16.87	–
	16.55	17.53	17.79	16.77	16.48
AVERAGE	17.5	16.1	16.4	16.8	16.6
± 45° TENSILE STRAIN, μ IN./IN.	24600+	28000+	25600+	24400+	25000+
	28600+	27340+	–	21400+	25800+
	29400+	24000+	25000+	26600+	–
AVERAGE	27500+	26400+	25300+	24100+	25400+

LAMINATE DENSITY, GR/CC	1.56
FIBER VOLUME, %	64.6
VOIDS, %	0.1
LAMINATE RESIN SOLIDS, % WT	27.2

Table II. Typical Mechanical and Physical Propertis of Cured
V-378A/T-300.6K, HMG/PI Tape Composites.

TEST	"AS-IS"	4 WK, 71°C 98% RH	8 WK, 71°C 98% RH
0° FLEXURE, ULT, KSI, R.T.	265.0	263.0 (1.73)**	254.0 (1.78)**
0° FLEXURE, ULT, KSI, 177°C	197.5	157.0	135.0
0° FLEXURE, ULT, KSI, 232°C	179.0	—	—
0° FLEXURE, ULT, KSI, 288°C	122.0	—	—
0° FLEXURE, ULT, KSI 316°C	107.0	—	—
0° FLEXURE MODULUS, MSI, R.T.	19.8	20.7	18.4
0° FLEXURE MODULUS, MSI, 177°C*	20.7	18.3	15.3
0° FLEXURE MODULUS, MSI, 232°C	19.2	—	—
0° FLEXURE MODULUS, MSI, 288°C	18.4	—	—
0° FLEXURE MODULUS, MSI, 316°C	17.6	—	—
0° HORIZONTAL SHEAR, KSI, R.T.	18.3	16.1 (1.85)**	13.7 (1.91)**
0° HORIZONTAL SHEAR, KSI, 177°C	10.9	8.0	7.4
0° HORIZONTAL SHEAR, KSI, 232°C	9.2	—	—
0° HORIZONTAL SHEAR, KSI, 288°C	6.4	—	—
0° HORIZONTAL SHEAR, KSI, 316°C	5.8	—	—
90° TENSILE, ULT, KSI, R.T.	9.0	4.3 (1.74)**	4.6 (1.87)**
90° TENSILE, ULT, KSI, 177°C	5.9	1.40	1.58
90° TENSILE MODULUS, MSI, R.T.	1.40	1.60	1.40
90° TENSILE MODULUS, MSI, 177°C	1.00	0.70	0.85
90° TENSILE STRAIN, μ IN./IN., R.T.	6,570	2,400	3,100
90° TENSILE STRAIN, μ IN./IN., 177°C	5,900	2,200	1,800
± 45° TENSILE ULT, KSI, R.T.	23.0	22.6 (1.60)**	22.3 (1.59)**
± 45° TENSILE ULT, KSI, 177°C	15.7	15.8	15.5
± 45° TENSILE MODULUS, MSI, R.T.	2.43	3.00	2.50
± 45° TENSILE MODULUS, MSI, 177°C	1.63	1.35	1.34
± 45° TENSILE STRAIN, μ IN./IN. R.T.	22,500	22,600	22,300
± 45° TENSILE STRAIN, μ IN./IN. 177°C	29,280	30,000	29,000
0° TENSILE ULT, KSI, R.T.	228.9	—	—
0° TENSILE MODULUS, MSI, R.T.	21.8	—	—
0° TENSILE STRAIN, μ IN./IN. R.T.	10,470	—	—

*5-MIN SOAK AT TEST TEMPERATURE
**% WEIGHT GAIN OF MOISTURE LISTED IN PARENTHESES

	0° TENSILE	0° FLEXURE	± 45° TENSILE	90° TENSILE
COMPOSITE, SPECIFIC GRAVITY, GR/CC	1.60	1.60	1.57	1.60
COMPOSITE, VOID CONTENT, %	0.5	−0.9	0.9	−0.9
COMPOSITE, FIBER VOLUME, %	68.3	64.8	63.6	64.8
COMPOSITE, RESIN SOLIDS, % WT	25.2	28.6	29.0	28.6

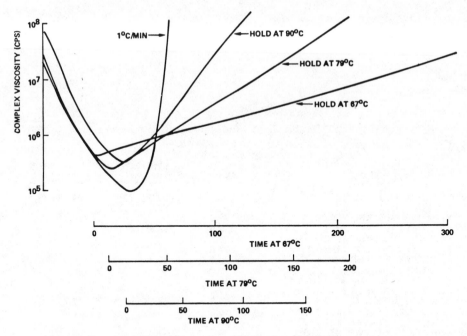

Figure 1. V-378A (lot WR 6080) isothermal cure curves.

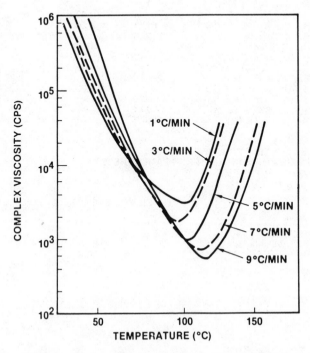

Figure 2. Effect of heating rate and temperature on complex
viscosity of V-378A.

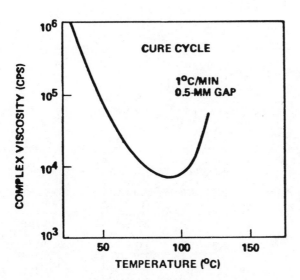

Figure 3. Complex viscosity of V-378A on exposure to 50% RH at 24°C (for 18 hrs.).

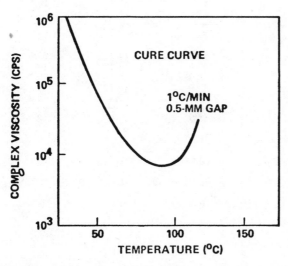

Figure 4. Complex viscosity of V-378A on exposure to 90% RH at 24°C (for 16 hrs.).

Table III. Effect of V-378A Prepreg Ageing at 23°C Prior to Cure.

DAYS AGING	0 FRESH	7 DAYS	14 DAYS	21 DAYS
FLEX/MODULUS AT R.T., KSI/MSI	289.1/18.9	266.9/17.6	302.8/19.1	285.1/18.5
	280.2/19.0	268.2/19.3	287.8/18.9	260.8/19.2
	290.3/18.9	256.2/17.1	273.8/17.3	304.7/20.3
	263.0/18.3	273.4/16.7	295.5/18.6	270.0/18.8
	—	262.9/18.4	274.1/18.4	285.2/19.0
AVERAGE	280.7/18.8	265.5/17.8	286.8/18.5	281.2/19.2
FLEX/MODULUS AT 177°C, KSI/MSI	228.4/17.1	218.6/17.8	208.3/19.4	198.4/18.3
	243.4/19.6	218.1/17.3	209.8/18.4	204.8/17.5
	225.0/17.7	224.2/18.1	206.2/18.3	205.0/17.3
	231.0/18.6	222.2/19.4	205.6/18.6	202.4/17.6
	—	216.3/18.5	216.1/17.2	194.8/18.2
AVERAGE	232.0/18.3	220.9/18.2	209.2/18.4	201.1/17.8
SBS AT R.T., KSI	17.1	18.2	17.5	17.1
	17.8	18.4	18.0	17.0
	18.5	18.4	17.8	—
	16.9	18.2	18.1	15.8
	17.9	17.2	18.1	15.9
AVERAGE	17.7	18.1	17.9	16.5
SBS AT 177°C, KSI	11.5	11.2	10.6	10.0
	11.5	11.1	10.4	10.3
	11.5	10.7	10.7	10.6
	11.9	10.4	10.8	10.3
	12.1	11.4	10.3	10.3
AVERAGE	11.7	11.0	10.6	10.3
DENSITY GR/CC	1.59	1.57	1.57	1.58
RESIN SOLIDS, %	28.1	31.1	30.5	28.0
FIBER VOLUME, %	65.1	61.5	62.0	64.4
VOIDS, %	0.40	0.10	0.50	0.90

40% of the ambient temperature value. Wet flexure retention (98% RH 71°C, 30 days) at 175°C was approximately 60%. The ± 45° tensile ultimate and strain values at ambient and 175°C of 23 ksi/22,500 μ in./in. and 15.7 ksi/29,000 μ in./in. respectively are also impressive. It is of interest to note the increase in composite strain at 175°C rather than a decrease usually noted in 175°C epoxies. No significant degradation in ambient or 175°C ultimate or strain values of the ± 45° tensile are exhibited after wet conditioning. Epoxies exhibit a significant drop in 175°C wet properties.

All wet elevated temperature testing was conducted using a five minute "soak" time, for the specimen to get to temperature in the preheated test chamber, in order to minimize drying of the specimen. V-378A composites lose moisture (and regain) at a much more rapid rate than 175°C epoxies.

Transverse or 90° tensile ultimate and strain values of V-378A at ambient of 9.2 ksi/7,700 μ in./in. and 6.1 ksi/6,400 μ in./in. at 175°C are also higher than most 175°C epoxy systems.

Effects of Prepreg Aging at 75°F. Table III exhibits flexure and shear tested at ambient and 177°C from prepreg which was laid up fresh and aged at 0, 7, 14 and 21 days intervals prior to cure. Ambient temperature tack retention, however, is much shorter than for typical 177°C curing epoxies. A vacuum bag debulk step is suggested at 2 - 3 hour intervals for large parts. Fresh prepreg should be laid up promptly and excess prepreg stored at -18°C. Storage stability of the prepreg is in excess of six months at -18°C.

General Mechanical Properties. Extensive mechanical properties on V-378A/T-300·6K unitape composites were determined by the University of Dayton Research Institute and reported under AF contract F 33615-78-C-5172. These data include 0°, ± 45° and 90° tensile, 0° and 90° compressive, 0° and 90° flexure and shear at -55°C, 22°C, 177°C and 232°C as well as static and creep testing.

Two major aircraft companies have found V-378A to exhibit open hole tensile values 50 - 70% higher than 177°C curing epoxies.

Table IV contains properties of V-378A/woven graphite fabric composites with flexure and shear tests up to 371°C.

Properties of V-378A/7781 E-glass fabric are listed in Table V and also exhibit excellent strength retention at temperatures up to 371°C.

Table IV. Properties of V-378A/T-300.3K, 8HS Polyimide/Bidirectional
Fabric.

I. PREPEG PROPERTIES

WET RESIN CONTENT, % WT	32.0
VOLATILES, % WT, 10' AT 135°C	7.1
FLOW, % — 15 PSI, 135°C, 15'	9.0

II. LAYUP AND CURE (NET RESIN, NO BLEEDER)

EIGHT PLIES V-378A/T-300·3K, 8HS FABRIC ON FREKOTE 33 RELEASED CAUL PLATE,
ONE PLY TX1040 ON TOP LAYUP PLUS RELEASED TOP CAUL PLATE AND TEDLAR TOP
RELEASE FILM WITH "L" SLITS AT CORNERS TAPED TO COROPRENE SIDE DAMS, TWO
PLIES 7581 BREATHER OVER ENTIRE LAYUP PLUS VENT HOLES. APPLY NYLON
VACUUM BAG WITH SEALANT

APPLY VACUUM PLUS 85 PSI AND HEAT FROM ROOM TEMPERATURE TO 82°C AT
3 ± 2°C MIN. VENT BAG TO ATMOSPHERE AT 82°C AND HOLD 30' AT 82°C. INCREASE
PRESSURE TO 100 PSI, THEN CONTINUE HEATING TO 177°C. CURE FOUR HOURS AT
177°C AND COOL TO 60°C OR BELOW BEFORE REMOVING. POSTCURE FOUR HOURS
AT 232°C UNRESTRAINED

III. MECHANICAL PROPERTIES

	RESULTS	
TEST	AS-IS	WET*
FLEX/MODULUS AT R.T., KSI/MSI	114.9/9.2	
FLEX/MODULUS AT 177°C, KSI/MSI	105.0/8.8	
FLEX/MODULUS AT 232°C, KSI/MSI	112.4/9.0	
FLEX/MODULUS AT 316°C, KSI/MSI	65.1/8.8	
FLEX/MODULUS AT 371°C, KSI/MSI	29.8/7.3	
SHORT BEAM SHEAR AT R.T., KSI	9.0	
SHORT BEAM SHEAR AT 177°C, KSI	7.7	
SHORT BEAM SHEAR AT 232°C, KSI	6.2	
SHORT BEAM SHEAR AT 316°C, KSI	4.5	
SHORT BEAM SHEAR AT 371°C, KSI	2.8	
TENSILE/MODULUS AT R.T., KSI/MSI	73.6/10.3	
TENSILE/MODULUS AT 177°C, KSI/MSI	80.3/10.4	
COMPRESSIVE/MODULUS AT R.T.	74.4/9.1	69.8/8.6
COMPRESSIVE/MODULUS AT 177°C**	67.9/8.4	49.8/9.1
LAMINATE RESIN SOLIDS, % WT	30.9	
LAMINATE DENSITY, GR/CC	1.55	
FIBER VOLUME, %	60.8	
VOID CONTENT, %	1.6	

*WET = 30 DAYS, 100% RH, 160°F
**4' SOAK AT TEST TEMPERATURE

Table V.   Properties of V-378A/7581 CS272, Polyimide/Bidirectional
            E-Glass Fabric.

I. PREPREG PROPERTIES

RESIN CONTENT, % WT   30%

II. LAYUP AND CURE (NET RESIN, NO BLEEDER)

TWELVE PLIES V-378A/7581 LOT NO. 2W4822 ON FREKOTE 33 RELEASED CAUL PLATE.
ONE PLY TX1040 ON TOP LAYUP PLUS RELEASED TOP CAUL PLATE AND TEDLAR TOP
RELEASE FILM WITH "L" SLITS AT CORNERS TAPED TO COROPRENE SIDE DAMS, TWO
PLIES 7581 BREATHER OVER ENTIRE LAYUP PLUS VENT HOLES. APPLY NYLON
VACUUM BAG WITH SEALANT

APPLY VACUUM PLUS 85 PSI, HEAT FROM R.T. TO 177°C AT 3 ± 2°C/MIN, HOLD 30'
AT 82°C THEN VENT BAG TO ATM AT 82°C AND INCREASE PRESSURE TO 100 PSI,
THEN CONTINUE HEAT TO 177°C. CURE FOUR HOURS AT 177°C AND COOL TO 60°C
OR BELOW BEFORE REMOVING. POSTCURE FOUR HOURS AT 246°C

III. MECHANICAL PROPERTIES

	RESULTS	
WARP ONLY – 4 EACH	AS-IS	*WET
FLEX/MODULUS AT R.T., KSI/MSI	100/3.8	
FLEX/MODULUS AT 177°C, KSI/MSI	89/4.0	
FLEX/MODULUS AT 232°C, KSI/MSI	84/4.0	
FLEX/MODULUS AT 316°C, KSI/MSI	75/3.9	
FLEX/MODULUS AT 371°C, KSI/MSI	48/3.5	
COMPRESSIVE/MODULUS AT R.T., KSI/MSI (ASTM D695 – DOGBONES)	78/5.3	72/4.0
COMPRESSIVE/MODULUS AT 177°C, KSI/MSI	62/4.5	46/4.0
COMPRESSIVE/MODULUS AT 232°C, KSI/MSI	57/4.3	
TENSILE/MODULUS AT R.T., KSI/MSI	69/4.7	
TENSILE/MODULUS AT 177°C, KSI/MSI	64/4.1	
TENSILE/MODULUS AT 232°C, KSI/MSI	64/4.0	
LAMINATE RESIN SOLIDS, % WT	26.8	
LAMINATE DENSITY, GR/CC	2.0	
LAMINATE FIBER VOLUME, %	57.7	
LAMINATE VOIDS, %	0.1	

*WET = 30 DAYS AT 71°C, 95-100% RH

Table VI lists properties of V-378A/6781 S-glass and illustrates the higher strengths attainable with S-glass.

Some properties of HI-TEX graphite fiber with V-378A, HITCO'S recent entry in the high strength, high modulus field, are listed in Table VII.

Elevated Temperature Stability of Cured Composites. Flexure and shear of 0.080" thick V-378A/T-300 tape composites after aging six months at 177°C and nine months at 232°C in circulating air ovens are listed in Table VIII. Composite weight loss after six months at 177°C is less than 0.6%. Flexure at ambient and 177°C appear to have increased slightly. Six month aging at 232°C exhibited a weight loss of 2.3% and good retention of flexure. Nine months aging produced composite weight loss of 3.4%. Retention of shear and flexure was quite high both at ambient and 232°C. Degradation appeared to be greatest on the surface.

Two new sets of panels were prepared and one set coated with 1 mil of Skybond 703 (Monsanto), a condensation type polyimide, as a protective coating. The control panel exhibited a weight loss of 4%, while the Skybond 703 coated panel lost 2.3% weight after aging one year at 232°C. Test of flexure and shear indicated about 15% better retention of 450°F flexure and slightly better shear on the coated panel after the one year at 232°C.

Smoke Density. Figure 6 illustrates the very low smoke density exhibited by V-378A composites. After a twenty minute burn, the smoke density (NBS Smoke Chamber) is about 1.5.

APPLICATIONS

V-378A is being evaluated in a number of applications for commercial and military aircraft as well as industrial applications. One of the most impressive of these is for manufacture of the complete wing skins and ribs for the new F-16XL cranked arrow fighter plane. The first prototype of this new concept aircraft flew on July 3, 1982 and is under intensive evaluation by General Dynamics.

Figure 7 is a view of a V-378A/T-300.6K composite wing skin during layup. Figure 8 shows several skins in various stages of layup. Figure 9 is a completed skin removed from the tool and after post cure. This part had less than 0.030" warp after post cure. Extensive C-scans indicated essentially no voids. Figure 10 illustrates the completed wing structure with internal ribs mechanically fastened to the upper and lower wing skins. The root thickness of each skin is approximately 0.4" tapering to approximately 0.060" at the tip. The completed wing attached to the fuselage is

Table VI. Properties of V-378A/6781 CS272, Polyimide/Bidirectional S-Glass Fabric.

I. PREPREG PROPERTIES

RESIN CONTENT, % WT   30%

II. LAYUP AND CURE (NET RESIN, NO BLEEDER)

TWELVE PLIES V-378A/6781 LOT NO. D07207 ON FREKOTE 33 RELEASED CAUL PLATE, ONE PLY TX1040 ON TOP LAYUP PLUS RELEASED TOP CAUL PLATE AND TEDLAR TOP RELEASE FILM WITH "L" SLITS AT CORNERS TAPED TO COROPRENE SIDE DAMS, TWO PLIES 7781 BREATHER OVER ENTIRE LAYUP PLUS VENT HOLES. APPLY NYLON VACUUM BAG WITH SEALANT

APPLY VACUUM PLUS 85 PSI, HEAT FROM R.T. TO 177°C AT 3 ± 2°C/MIN, HOLD 30' AT 82°C THEN VENT BAG TO ATM AT 82°C AND INCREASE PRESSURE TO 100 PSI, THEN CONTINUE HEAT TO 177°C. CURE FOUR HOURS AT 177°C AND COOL TO 60°C OR BELOW BEFORE REMOVING. POSTCURE FOUR HOURS AT 246°C

III. MECHANICAL PROPERTIES

	RESULTS	
WARP ONLY — 4 EACH	AS-IS	*WET
FLEX/MODULUS AT R.T., KSI/MSI	117/4.6	
FLEX/MODULUS AT 177°C, KSI/MSI	102/4.6	
FLEX/MODULUS AT 232°C, KSI/MSI	89/4.3	
FLEX/MODULUS AT 316°C, KSI/MSI	67/4.1	
FLEX/MODULUS AT 371°C, KSI/MSI	43/3.6	
COMPRESSIVE/MODULUS AT R.T., KSI/MSI (ASTM D695 — DOGBONES)	76/5.1	74/5.1
COMPRESSIVE/MODULUS AT 177°C, KSI/MSI	65/4.8	50/4.5
COMPRESSIVE/MODULUS AT 232°C, KSI/MSI	63/4.5	
TENSILE/MODULUS AT R.T., KSI/MSI	84/5.0	
TENSILE/MODULUS AT 177°C, KSI/MSI	72/4.7	
TENSILE/MODULUS AT 232°C, KSI/MSI	78/4.2	
LAMINATE RESIN SOLIDS, % WT	27.8	
LAMINATE DENSITY, GR/CC	1.9	
LAMINATE FIBER VOLUME, %	56.5	

*WET = 30 DAYS AT 71°C, 95-100% RH

Table VII. Properties of V-378A/Hi-Tex-12K Polyimide/High Modulus Graphite Unidirectional Composite.

I. PREPREG PROPERTIES

WET RESIN CONTENT, % WT	30%
NONREACTIVE VOLATILES, % WT	0
TACK	LIGHT — MEDIUM
RECOMMENDED STORAGE TEMPERATURE	−18°C OR BELOW

II. LAYUP AND CURE (NET RESIN, NO BLEEDER)

FIFTEEN PLIES V-378A/HI-TEX12K (LOT 3W2320) 0° UNITAPE WERE LAID UP ON A FREKOTE 33 RELEASED CAUL PLATE. ONE PLY TX1040 ADDED ON TOP OF LAYUP PLUS RELEASED, TOP CAUL PLATE. COROPRENE SIDE DAMS APPLIED WITH 3 MIL TEDLAR WITH "L" SLITS AT EACH CORNER TAPED TO TOP OF COROPRENE DAMS. TWO PLIES 1581 BREATHER OVER ENTIRE LAYUP TO VENT HOLES

APPLY VACUUM PLUS 85 PSI AND HEAT FROM ROOM TEMPERATURE TO 82°C AT 3 ± 2°C/MIN. HOLD 30' AT 82°C, THEN VENT BAG TO ATMOSPHERE AND INCREASE AUTOCLAVE PRESSURE TO 100 PSI AND CONTINUE HEATING FROM 82°C TO 177°C. CURE FOUR HOURS AT 177°C. REMOVE, AND POSTCURE UNRESTRAINED FOUR HOURS AT 246°C

III. MECHANICAL AND PHYSICAL PROPERTIES, V-378A/T300·6K COMPOSITE

TEST	RESULTS
0° FLEXURE/MODULUS, KSI/MSI, R.T.	324/18.5
0° FLEXURE/MODULUS, KSI/MSI, 177°C	243/18.2
0° SHORT BEAM SHEAR, KSI, R.T.	15.3
0° SHORT BEAM SHEAR, KSI, 350°F	10.8
LAMINATE DENSITY, GR/CC	1.60
LAMINATE RESIN SOLIDS, % WT	28.7
LAMINATE FIBER VOLUME, %	64.6
LAMINATE VOIDS, %	0.2

NOTE: 5 MIN SOAK AT TEST TEMPERATURE

Table VIII. Typical Properties of V-378A/T-300 UC309, 5 mil Unitape Composites Aged at Six Months at 177°C and 232°C.

	AS-IS	COMPOSITES AGED 6 MONTHS AT 177°C	COMPOSITES AGED 6 MONTHS AT 232°C	COMPOSITES AGED 9 MONTHS AT 232°C
0° FLEXURE/MODULUS, KSI/MSI AT R.T.	298/18.2	320/20.0	275/19.0	267/21.1
0° FLEXURE/MODULUS, KSI/MSI AT 177°C	245/17.4	271/21.1	—	—
0° FLEXURE/MODULUS, KSI/MSI AT 232°C	233/17.4	—	180/19.3	205/21.4
SHORT BEAM SHEAR, KSI AT R.T.	15.8	—	—	15.5
SHORT BEAM SHEAR, KSI AT 177°C	10.1	—	—	—
SHORT BEAM SHEAR, KSI AT 232°C	7.7	—	—	8.6
	0°	0°	0°	0°
DENSITY	1.60	1.62	1.57	—
RESIN SOLIDS, %	27.3	28.5	26.6	—
FIBER VOLUME, %	66.1	63.8	64.9	—
VOIDS, %	0.2	0.2	2.6	—
WEIGHT LOSS, % WT	—	0.57	2.3	3.4

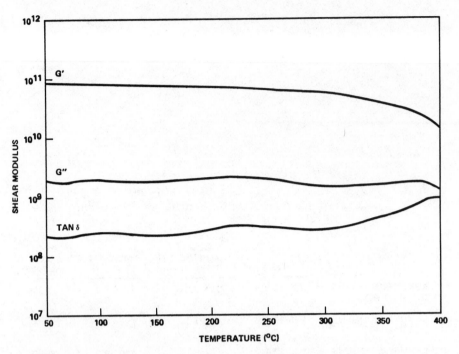

Figure 5. Tg shear modulus of cured V-378A as determined by
rheometric dynamic spectroscopy.

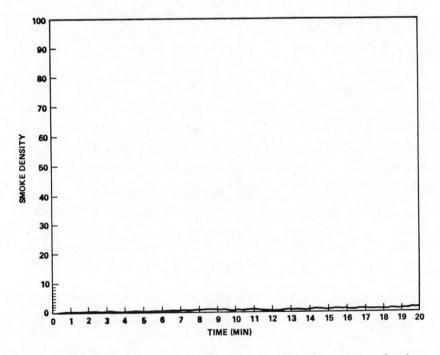

Figure 6. V-378A/graphite composite smoke density (NBS chamber).

Figure 7. View of a V-378A/T-300.6K composite wing during layup.

Figure 8. Several skins in various stages of layup.

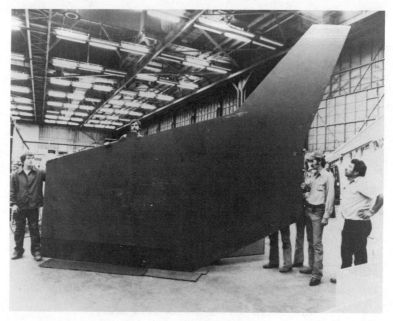

Figure 9. Completed skin removed from the tool and after postcure.

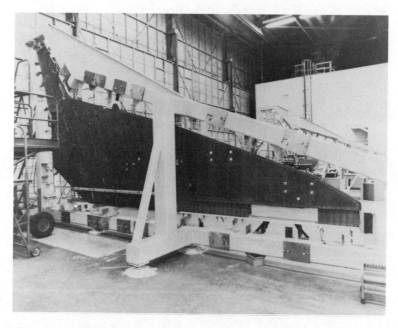

Figure 10. Completed wing structure with internal ribs mechanically
fastened to the upper and lower wing skins.

Figure 11. Completed wing attached to fuselage.

Figure 12. Both wings attached to fuselage.

Figure 13. Finished cranked arrow, delta wing F-16XL.

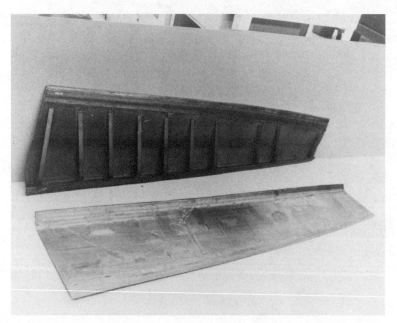

Figure 14. Wing flap from another prototype aircraft with integrally cocured V-378A/T-300 ribs bonded to a wing skin.

Table IX. F-16XL Specifications.

	F-16XL SCAMP	F-16A
WINGSPAN	32.4 FT	32.8 FT*
WING ROOT CHORD	499 IN.	195 IN.
WING AREA	646.4 SQ FT	300 SQ FT
OVERALL LENGTH	52.4 FT	47.6 FT
EMPTY WEIGHT	17,402 LB	15,137 LB
MAXIMUM TAKEOFF GROSS WEIGHT	37,500 LB	35,400 LB
FUEL CAPACITY	12,750 LB	6,972 LB
TAKEOFF ROLL (AIR-AIR COMBAT)	1,640 FT	2,425 FT
TAKEOFF ROLL (AIR-GROUND SUPPORT)	1,980 FT	3,030 FT
LANDING DISTANCE (AIR-AIR COMBAT)	1,990 FT†	2,480 FT
LANDING DISTANCE (AIR-GROUND SUPPORT)	2,230 FT**	2,830 FT
MAXIMUM SPEED	MACH 2.5	MACH 2.0
MAXIMUM CRUISE SPEED	MACH 2.2	MACH 0.93

*WITH WINGTIP MISSILES. WITHOUT MISSILES, SPAN IS 31 FT

†USING BRAKES ONLY. USING DRAG PARACHUTE, LANDING DISTANCE ESTIMATED AT 1,180 FT

**USING BRAKES ONLY. USING DRAG PARACHUTE, LANDING DISTANCE ESTIMATED AT 1,360 FT

shown in Figure 11. Figure 12 exhibits both wings attached to the fuselage and finally the finished cranked arrow, delta wing F-16XL is shown in Figure 13.

Proposed specifications for the F-16XL versus the current F-16A manufactured by General Dynamics, Ft. Worth, are listed in Table IX. Initial results from flight tests indicate performance is close to analytical predictions.

Figure 14 shows a wing flap from another prototype aircraft with integrally cocured V-378A/T-300 ribs bonded to a wing skin.

V-378A/T-300·6K is also being used to produce a firewall for an advanced concept helicopter and in numerous other applications.

SUMMARY

V-378A, a modified bimaleimide resin, has been developed for composite applications requiring greater hot-wet strength retention than currently available with 175°C curing epoxy resins.

The new resin also exhibits good prepreg parameters, facile, epoxy-like, curing characteristics and appears useful for applications at temperatures of 232°C for extended periods of time and in areas where low smoke density is required.

ACKNOWLEDGEMENT

Assistance from the U. S. Polymeric analytical, physical testing laboratories and from Lee McKague and Clarence Hart of General Dynamics for their help in providing photographs and technical assistance is gratefully acknowledged.

A THERMOPLASTIC POLYIMIDESULFONE

Terry L. St. Clair and David A. Yamaki*

NASA-Langley Research Center

Hampton, VA   23665

A polymer system has been prepared which has the excellent thermoplastic properties generally associated
with polysulfones, and the solvent resistance and thermal
stability of aromatic polyimides. This material, with
improved processability over the base polyimide, can be
processed in the 260–325°C range in such a manner as to
yield high quality, tough unfilled moldings; strong,
high-temperature-resistant adhesive bonds; and well consolidated, graphite-fiber-reinforced moldings (composites). The unfilled moldings have physical properties
that are similar to aromatic polysulfones which demonstrates the potential as an engineering thermoplastic.
The adhesive bonds exhibit excellent retention of initial
strength levels even after thermal aging for 5000 hours
at 232°C. The graphite-fiber-reinforced moldings have
mechanical properties which makes this polymer attractive
for the fabrication of structural composites.

*Present affiliation-Mobil Research and Development Corp.
Paulsboro, NJ 08066

INTRODUCTION

Aromatic polysulfones, a class of polymers processable by thermoplastic means, have a major problem in their tendency to swell and dissolve in many common solvents. This dissolution causes structural components which are fabricated from these polymers to be susceptible to damage by these solvents and precludes use of polysulfones for many applications[1,2].

Aromatic polyimides, conversely, are a class of polymers which are known to be resistant to solvents, but they are generally not processable by thermoplastic means[3]. The polyimides are exceptional in their thermal stability and, like polysulfones, their use temperature is generally governed by the softening temperature of each system.

A polymer system which possesses the processability of the polysulfones and the solvent resistance of the polyimides offers a considerable advance to the state-of-the-art. The synthesis and characterization of such a system is the subject of this paper.

Although the subject polyimide has some unusual physical properties that make it interesting from an engineering standpoint, this is not the first polyimidesulfone to be prepared. Sroog, et al., have reported the preparation of polyimides from pyromellitic dianhydride and two isomeric sulfone-containing diamines[4]. Brode, et al., prepared several polyimides from sulfone arylether diamines[5], and Acle patented a copolyimide which contained sulfone units[6].

EXPERIMENTAL

Preparation of the Polymer. The monomers used in the preparation of the thermoplastic/solvent-resistant polyimidesulfone (PISO2) were 3,3',4,4'-benzophenone tetracarboxylic dianhydride (BTDA) and 3,3'-diaminodiphenylsulfone (3,3' DDS). The BTDA was a polymer grade material used as received from Gulf Chemicals*, m.p. 215°. The 3,3'DDS was obtained from FIC Corporation and was used as received, m.p. 165-167°C.

---

*Use of trade names or manufacturers does not constitute an official endorsement, either expressed or implied, by the National Aeronautics and Space Administration.

The polymerization of the two monomers in stoichiometric quantities was carried out in reagent grade bis(2-methoxyethyl-ether). This triether which is commonly called diglyme was obtained from at least four different commercial sources and used as the medium for polymerization. The reaction was conducted at 20-25°C at a concentration of 15-25% solids by weight in the reagent grade diglyme. A typical preparation was as follows: The BTDA (25.8 g) was added to a solution of 3,3' DDS (19.9 g) in diglyme (258.6 g). This mixture was allowed to stir at room temperature until all of the BTDA had dissolved. The solution was allowed to stir for an additional two hours to allow for molecular weight build up. At this stage the polymer in solution was the polyamide acid.

Characterization. The inherent viscosity of the polyamide acid solution was determined at a concentration of 0.5 percent in N,N-dimethylacetamide at 35°C. Thermomechanical properties of the polymer were obtained by torsional braid analysis (TBA). Glass braids were coated with a 5 percent polymer solution and heated to 300°C in air before obtaining TBA spectra. Glass transition temperatures (T_g) of various films, composites, and moldings were measured by thermomechanical analysis (TMA) on a DuPont 943 Analyzer in static air at a temperature program of 5°C/min. Thermograms of the polymer were obtained by thermogravimetric analysis (TGA) by heating at a rate of 2.5°C/min in static air (dynamic TGA) or by holding the polymer at 316°C in static air (isothermal TGA). Melt flow properties were observed by use of a parallel plate plastometer accessory for the DuPont 943 Thermomechanical Analyzer. Mechanical properties of moldings, composites, and adhesive bonds were determined on a Model TT Instron Testing Machine. Solvent resistance of films was deduced from their apparent glass transition depression as measured by TMA after solvent exposure as well as by their physical behavior in the solvents.

Preparation of Adhesive Scrim. The polyamide acid in diglyme solution was brush-coated onto 112 E-glass which had an aminosilane surface treatment. This glass cloth had a nominal thickness of 0.01 cm and was used as a carrier for the adhesive as well as for bondline thickness control. Coatings of the polyamide acid were applied to build up a scrim thickness of 0.020-0.025 cm. After each coating the scrim was air dried at room temperature until tack was lost and then placed in a forced air oven and subjected to the following cure schedules:

    (1)   RT -> 100°C   hold 1/2 hour
    (2)   100 -> 150°C   hold 1/2 hour
    (3)   150 -> 200°C   hold 1/2 hour

This cure eliminated the diglyme as evidenced by TGA and also effected a conversion of the amide acid to the imide as evidenced by infrared spectroscopy. The water of imidization was lost primarily between 140-200°C and was carried out of the scrim along with the solvent.

Preparation of Molding Powder. The polyamide acid solution was poured very slowly into a mechanical blender containing distilled water. The contact with water caused the polyamide acid to precipitate and the blender blades chopped this material to a fluffy consistency. This solid polymer was washed with copious amounts of distilled water and was collected by suction filtration. The polymer was air-dried overnight. This solid was spread in a baking dish and placed in a forced air oven and heated to 100°C. The polymer was held at this temperature for one hour to drive off residual water and solvent. The temperature of the oven was then increased to 220°C and held for one hour to effect conversion of the amide acid to the imide.

Adhesive Bonding. The adhesive scrim cloth was used to bond titanium 6-4 adherends. The titanium adherends (Ti 6Al-4V) to be bonded were grit-blasted with 120 mesh aluminum oxide and treated with Pasa Jell 107* in order to form a stable oxide on the surface. A primer coating of the polyamide acid solution was applied to the adherends and they were thermally treated for one hour at 100°C and one hour at 200°C.

Single lap-shear specimens were prepared by sandwiching the scrim cloth between primed adherends using a 1.27 cm overlap. The specimens were bonded as follows:

(1) RT to 325°C at 7°C/min, apply 1.38 MPa (200 psi) at 280°C

(2) Hold 15 min. at 325°C

(3) Cool under pressure

Adhesively bonded specimens (4 per condition) were aged at various temperatures in forced air ovens and were tested according to ASTM D1002.

Preparation of Unfilled Moldings. Approximately 25 grams of molding powder was placed in a 5.72 cm diameter steel mold or 15 grams of the powder in a 19.0 cm x 2.5 cm rectangular steel mold and each cured according to the following cycle:

---

*Trade name for a titanium surface treatment available from Semco, Glendale, CA.

(1)   Heat the charged mold to 200°C without the top

(2)   Insert the top and apply 6.89 MPa (1000 psi)

(3)   Heat to 280°C

(4)   Cool under pressure

Fracture toughness testing of the 0.16 cm thick discs was performed at the Naval Research Laboratory, Washington, D.C.   Round compact tension test specimens were prepared from these discs and tested for G_{1c} (the opening-mode strain energy release rate) according to ASTM E399-78A. Flexural strength and modulus were measured according to ASTM D-790. Tensile strength and modulus were measured according to ASTM D-638 on the dogbone shaped specimens that had been polished with #85 Barnesite. The tensile bars had longitudinal and latitudinal strain gages mounted on both sides.

Preparation of Graphite-Fiber-Reinforced Moldings.
Graphite-Fiber-Reinforced moldings were prepared from the polyimidesulfone by initially applying the polyamide acid in diglyme

at 5% solids onto drum-wound Celion^R 6000 graphite fiber and subsequently coating with the same material at 15% solids. The initial coating at low solids content was employed to insure good wetting of the fibers. This prepreg was air dried on the rotary drum, cut into 7.6 cm by 17.8 cm pieces and stacked into 21-ply unidirectional preforms. The preform was next B-staged in a vacuum bag with release plies and bleeder plies on both sides. The vacuum bag assembly was heated under vacuum to 200°C and held at that temperature for four hours.

This B-staged panel with new release plies on both sides was then placed in a 7.6 cm by 17.8 cm matched metal mold with open ends. This unit was then vacuum-bagged and cured according to the following schedule:

(1)   RT to 288°C at 5°C/min

(2)   At 100°C, 1.38 MPa (200 psi) was applied

(3)   AT 200°C the pressure was increased to 2.76 MPa (400 psi)

(4)   These conditions were held for two hours

(5)   The mold was allowed to cool to below 150°C prior to removal of the molding

(6)   Postcure overnight at 250°C.

The resin content on each laminate was determined by sulfuric acid digestion of the polyimide resin. This technique leaves only the graphite fiber. The calculation of the percent resin or fiber is a simple gravimetric type as follows:

$$\% \ fiber = \frac{weight \ of \ dried \ fiber \ after \ digestion}{weight \ of \ sample \ prior \ to \ digestion} \times 100$$

% resin = 100 - % fiber

Densities were determined from the weights of the laminate in air and in water. Flexural strengths and moduli of the laminates were determined using ASTM D-790. Short beam shear strengths were determined using a span-to-thickness ratio of four and crosshead speed of 0.127 cm/min on the Instron testing machine.

RESULTS AND DISCUSSION

Resin Chemistry and Properties. The subject polyimidesulfone was synthesized according to the reaction scheme in Figure 1. The synthesis was performed in diglyme because this solvent has been shown to yield polymers with high adhesive strengths[7]. Diglyme is also easily eliminated so that moldings or laminates can be prepared with essentially no voids[8].

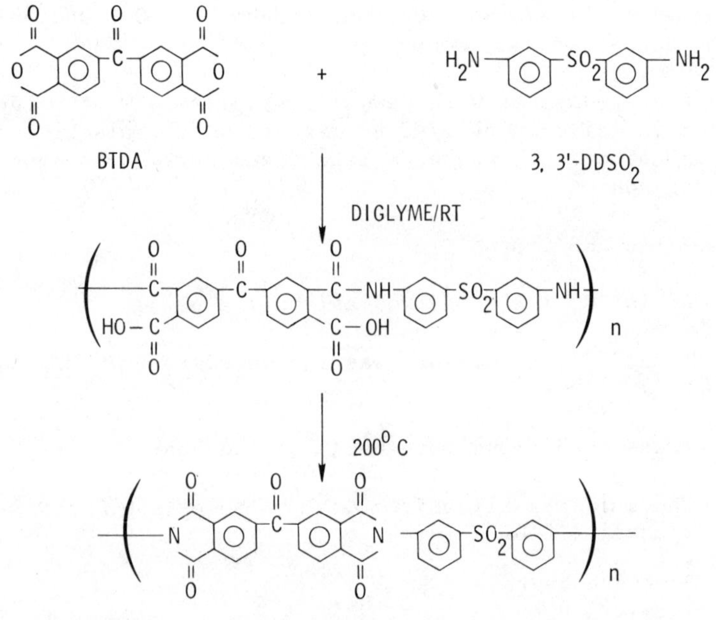

Figure 1. Polyimidesulfone preparation.

The polyamide acid from the reaction scheme had inherent viscosities from several preparations that ranged from 0.4 to 0.8
dl/g. The thermal imidization of the polyamide acids resulted in
linear, high molecular weight polyimides which had adequate flow
to allow for thermoplastic processing. This ability to be
processed as a thermoplastic relates to the flexibility of the
polymer chain due to the meta linkages in the diphenylsulfone
portion. This enhancement in thermoplastic processing due to the
incorporation of meta linkages in linear, aromatic polyimides has
been previously explained[9,10].

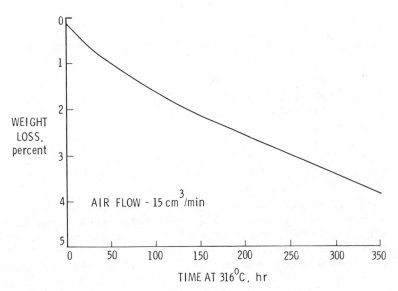

Figure 2. Polyimidesulfone weight loss at 316°C in air.

This polyimidesulfone exhibited good thermooxidative stability as evidenced by the isothermal and dynamic thermograms in Figures 2 and 3, respectively. After 350 hours at 316°C in air this
polymer had lost only 3.5% of its initial weight. The dynamic
thermogravimetric run at a heating rate of 2.5°C/min in air showed
a temperature of 590°C for 50% weight loss.

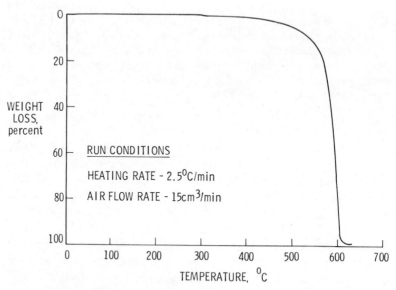

Figure 3. Polyimidesulfone weight loss in air with increasing
temperature.

The glass transition temperature (T_g) of the polyimidesulfone was determined by TMA on a film that had been cured in air
for one hour at 100°C, one hour at 200°C, and one hour at 300°C.
The T_g was found to be 273°C. The T_g as determined by TBA
after the same pretreatment was 275°C as illustrated in Figure 4.

The effects of solvent exposure on the polymer are shown in
Table 1. A 2 cm x 0.5 cm piece of 25μ thick film was immersed in
each of the solvents listed in the table. After exposure for 24
hours they were removed and their T_g was determined using TMA by
putting the solvent-laden samples under tension and subjecting
them to a heat-up rate of 10°C/min[11]. The only solvent that
caused a visible change in the polymer was N,N-dimethylformamide
(DMF). This observation was verified by the TMA tests. The only
film that showed a decrease in T_g was the sample that had been
immersed in DMF. This T_g change was a depression from 273°C to
245°C.

Polymer softening characterization using the parallel plate
plastometer showed the softening to begin at approximately 250°C

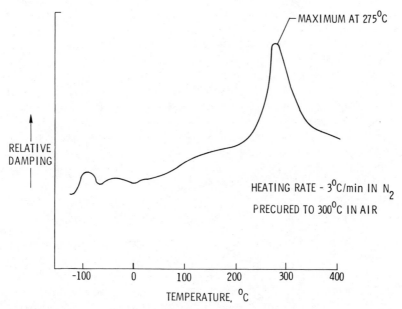

Figure 4.   Torsional braid analysis of polyimidesulfone.

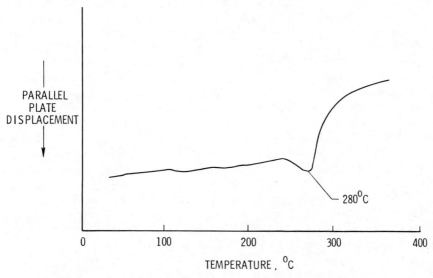

Figure 5.   Softening profile of polyimidesulfone.

Table I. Polyimidesulfone Solvent Resistance.

SOLVENT	IMMERSION EXPOSURE TIME, hours	VISUAL EFFECT	T_g AFTER EXPOSURE, $^\circ$C
NONE	—	—	273
CHLOROFORM	24	NONE	275
METHYLENE CHLORIDE	24	NONE	273
sym-TETRACHLOROETHANE	24	NONE	275
m-CRESOL	24	NONE	271
N,N-DIMETHYLFORMAMIDE	24	SHRINKAGE/SOFTENING	245
CYCLOHEXANONE	24	NONE	273
SKY JET IV	24	NONE	273

and reached an apparent minimum at 280°C (Figure 5). Beyond 280°C
there was an increase in parallel plate displacement which was due
to a swelling or bulking of the sample. From processing
experience it is apparent that the viscosity continues to decrease
with increasing temperature beyond 280°C.

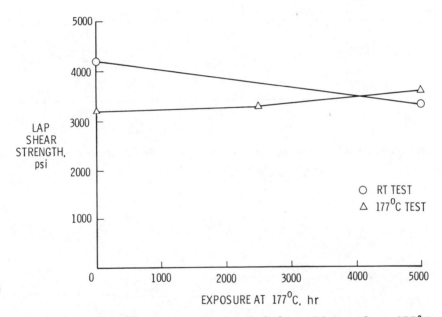

Figure 6.   Adhesive strength of polyimidesulfone after 177°C
             exposure - titanium adherends.

    Adhesives. The titanium/titanium bonds were tested before
and after aging. The data are shown in Table II and in Figures
6-8. The polyimidesulfone on the woven glass carrier was fully
imidized prior to the bonding operation; therefore, the bonds were
fabricated in a thermoplastic manner. The bondlines were examined
after failure in the lap shear test and there was no evidence of
voids in any bond. There was a decrease in lap shear strength
with increasing test temperatue (4150 psi at ambient to 2620 psi
at 232°C).

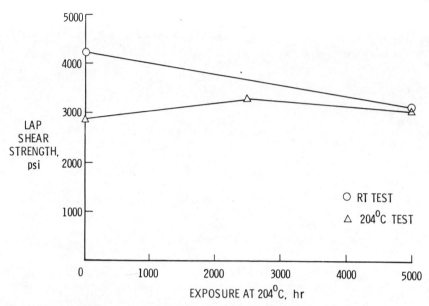

Figure 7. Adhesive strength of polyimidesulfone after 204°C
exposure - titanium adherends.

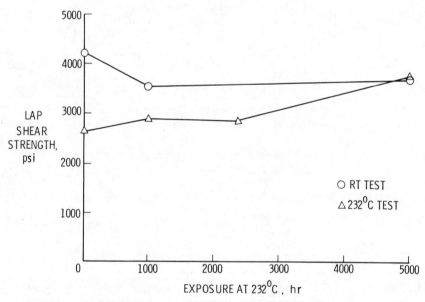

Figure 8. Adhesive strength of polyimidesulfone after 232°C
exposure - titanium adherends.

Table II. Polyimidesulfone Adhesive Properties.

TEST TEMPERATURE, ^{o}C (^{o}F)	AGING TEMPERATURE, ^{o}C	AGING TIME, hours	LAP SHEAR STRENGTH, psi *
AMBIENT	—	0	4650
	177	5000	3250
	204	5000	2980
	232	1000	3500
	232	5000	3640
177 (350)	—	0	3210
	232	1000	3230
	177	2500	3320
	177	5000	3670
204 (400)	—	0	2920
	204	2500	3180
	204	5000	2980
232 (450)	—	0	2620
	232	1000	2920
	232	2500	2790
	232	5000	3560

* PER ASTM D1002, TITANIUM ADHERENDS

The data in Table II show that this adhesive has exceptional thermal aging characteristics at temperatures to 232°C. After 5000 hours aging at 177°C, 204°C, and 232°C the ambient-temperature-tested lap shear samples exhibited pronounced decreases in strength; however, the elevated temperature samples (tested at the aging temperature) all exhibited strength increases. This behavior is indicative of a completion of cure or an annealing effect.

The ambient temperature strength of lap shear specimens (4 per condition) that had been aged for 5000 hours at 232°C was 3640 psi and the 232°C strength of specimens aged under these same conditions was 3560 psi. These data show this adhesive to have considerable potential for aerospace structural applications.

Unfilled Moldings. Four rectangular moldings of the polyimidesulfone were machined into tensile specimens according to ASTM D638. These moldings were amber in color and transparent. Latitudinal and longitudinal strain gages were mounted on both sides of these specimens and they were tested, according to the same ASTM standard, with strain gage readouts recorded.

The tensile data are summarized in Table III. Of particular interest is the tangent modulus of this polymer. The average initial modulus was 719 kpsi; 661 kpsi average at 0.005 strain level; and 603 kpsi average at 0.01 strain level. The tensile modulus reported for an unfilled polysulfone is 360 kpsi[12]. The high modulus exhibited by this polymer makes it attractive as a matrix resin for the fabrication of graphite-reinforced structures for aerospace applications. The failures on all four samples were of a flaw-initiated, brittle type. The average failure strain was 0.0133. The average Poisson's ratio was 0.38. The average tensile strength for the four samples was 9.1 kpsi with a range of 8.6 to 9.7 kpsi.

The G_{1c} value determined on this polymer system was 1400 J/m^2 (average of two tests). This result indicates that the polyimidesulfone is quite tough when compared to crosslinked systems such as addition-curing polyimides or epoxies[13].

Graphite-Fiber-Reinforced Moldings. The graphite-reinforced laminates that were prepared from the polyimidesulfone were screened by monitoring the short-beam-shear strengths of the unidirectional fabricated panels. Approximately 40 test panels with nominal dimensions of 7.5 cm x 15 cm x 0.5 cm were prepared from the solvent-impregnated prepreg. The variables studied included prepreg resin content, B-stage conditions, molding conditions and postcures. The physical and mechanical properties from the

Table III. Polyimidesulfone Tensile Properties.

SPECIMEN	TENSILE STRENGTH, ksi	INITIAL TANGENT MODULUS, ksi	TANGENT MODULUS, ksi		FAILURE STRAIN, cm/cm	POISSON'S RATIO
			AT 0.005 STRAIN	AT 0.01 STRAIN		
1	8.6	722	644	600	0.0123	0.38
2	9.3	733	689	622	0.0133	0.37
3	9.7	702	667	611	0.0141	0.38
4	8.9	720	644	578	0.0135	0.37
AVERAGE	9.1	719	661	603	0.0133	0.38

resulting laminates that were monitored were thickness, resin content, glass transition temperature, density, short-beam-shear strength at room temperature and elevated temperature, and overall weight loss during processing. These panels had considerable variability in properties initially, but this variability lessened as better cure cycles were developed. Representative data on laminates prepared according to the procedure described in the experimental section are in Table IV.

Table IV.- Properties of Graphite-Fiber-Reinforced
Polyimidesulfone Composite Panels.

Short-beam shear strength, psi

    ambient - 11,000 to 12,000

    121°C - 7,500 to 8,500

    172°C - 6,500 to 7,500

Flexural strength, kpsi	190
Flexural modulus, kpsi	21,000
Resin content, percent by weight	35 to 38
Density, g/cm^3	1.50 to 1.56
Laminate thickness, cm	0.25 (nominal)
Weight loss during cure, percent by weight	12.8 to 18.3

CONCLUSIONS

A novel polyimidesulfone which shows considerable potential as an engineering thermoplastic has been synthesized and characterized. It is a high molecular weight linear aromatic system which is flexible, tough, and thermooxidatively stable. Imidized molding powder of this polymer was fabricated into void-free neat moldings which exhibited a tensile strength in excess of 9 ksi and a modulus of 719 ksi. This tensile strength is average for engineering thermoplastics. However, the modulus is approximately double the value reported for such thermoplastics. The G_{1c} value for this polyimidesulfone is 1400 J/m^2.

Adhesive bonds prepared with the subject polymer and titanium alloy adherends had high initial lap shear strength values at room temperature (>4500 psi) and good adhesive strength was retained at test temperatures up to 232°C (>2600 psi). After aging at temperatures up to 232°C for 5000 hours the lap shear strengths were still high at room temperature (>3600 psi) and had increased when tested at 232°C (>3500 psi).

Graphite-reinforced composites were successfully prepared from the polymer system using solvent-impregnated prepreg. The short-beam-shear properties of unidirectional laminates were 11 to 12 kpsi when tested at room temperature and 6.5 to 7.5 kpsi for 177°C tests.

The solvent resistance of this polymer system also sets it apart from the more commonly used thermoplastics. There was no change in the polymer when immersed in chlorinated hydrocarbons, cresol, cyclohexanone, and aircraft hydraulic fluid (tricresyl-phosphate-base). There was slight swelling in dimethylformamide and it is expected that other amide solvents will affect the polymer in a similar fashion; however, these solvents are not likely to be encountered in most service applications.

The combination of properties that this polyimidesulfone exhibits makes it a very attractive candidate for aircraft structural applications such as adhesives and composites. Also, because of the ready availability of raw materials necessary for its preparation, the polymer has considerable commerical potential.

ACKNOWLEDGEMENT

The authors wish to express their appreciation to Dr. Robert Y. Ting, Naval Research Laboratories, Orlando, Florida for performing G_{1c} test and to James Tyeryar, Philip Robinson and Spencer Inge of NASA Langley Research Center for their technical assistance.

REFERENCES

1. J. T. Hoggatt, 20th National SAMPE Symposium Proc., 20, 606 (1975).

2. G. E. Husman, and J. T. Hartness, 24th National SAMPE Symposium Proc., 24, 21 (1979).

3. P. E. Cassidy, "Thermally Stable Polymers," Ch. 5, p. 98, Marcel Dekker, Inc., New York (1980).

4. C. E. Sroog, A. L. Endrey, S. V. Abramo, C. E. Berr, W. M. Edwards, and K. L. Olivier, J. Polym. Sci.: Part A, 3, 1373 (1965).

5. G. L. Brode, G. T. Kwiatkowski, and J. H. Kawakami, Polymer Preprints, 15 (1) 761 (1974).

6. L. Acle, U.S. Patent 3,793,281, Feb. 19, 1974.

7. T. L. St. Clair, and D. J. Progar, Polymer Preprints, 16 (1), 538 (1975).

8. A. K. St. Clair, T. L. St. Clair, SAMPE Quarterly 13 (1), 20 (1981).

9. V. L. Bell, U.S. Patent 4,094,862, June 13, 1978.

10. A. K. St. Clair, and T. L. St. Clair, NASA TM-83141, p. 7 (1981).

11. H. D. Burks, J. Appl. Polym. Sci., 18, 627 (1974).

12. J. Agranoff, "Modern Plastics Encyclopedia", Vol. 54, No. 10A, p. 482 (1977).

13. W. D. Bascom, J. L. Bitner, and R. L. Cottington, Organic Coatings and Plastics Chemistry Preprints, 38, 477 (1978).

SYNTHESIS AND CHARACTERIZATION OF A MELT PROCESSABLE POLYIMIDE

H. D. Burks and T. L. St. Clair

NASA Langley Research Center

Hampton, Virginia 23665

A melt processable polyimide, BDSDA/APB, which contains sulfur and oxygen bridges between the aromatic
rings was synthesized and characterized. Its physical,
mechanical, thermal and flow properties were determined
as was its resistance to some of the more commonly used
solvents. The melt flow properties were measured for
the temperature range 250°C-350°C and under the stress/
strain conditions encountered in commercial processes.

INTRODUCTION

Linear aromatic polyimides are a class of polymers which are generally not processable via conventional thermoplastic or hot-melt techniques. This class of polymers is, however, exceptionally thermally stable and has high glass transition temperatures.[1,2] They are also resistant to attack by common organic solvents.[3]

Linear aromatic polyphenylene oxides and sulfides, on the other hand, are more easily processed than the polyimides,[1] generally exhibit lower glass transition temperatures, and still have relatively good thermal stability (although not equal to the polyimides).[4] These PPO and PPS systems do not possess solvent resistance equal to the polyimides.[5]

A novel linear aromatic polyphenylene ethersulfideimide (BDSDA/APB) has been synthesized which has some of the favorable characteristics of each parent system. The polymer has been molded, used as a resin, and cast into thin films. A limited characterization indicates this system can be processed via conventional thermoplastic techniques and may have a wide variety of applications.

EXPERIMENTAL

Preparation

Polymer. To form the polymer, 11.0000g of 4,4'-bis(3,4-dicarboxyphenoxy)diphenylsulfide dianhydride (BDSDA) and 6.2999g of 1,3-bis(aminophenoxy)benzene (APB) were dissolved in 98.0g of bis(2-methoxyethyl)ether in a flask equipped with magnetic stirring. This solution was allowed to stir for one hour in order to build up molecular weight.

Molding Powder. The viscous polymer solution was poured into a mechanical blender containing distilled water. The contact with water caused the polyamide-acid to precipitate and the rotating blender blades chopped this material into a fluffy consistency. The solid polymer was isolated by suction filtration and allowed to air dry overnight.

The dried polymer was spread in a baking dish, placed in a forced-air oven and heated to 100°C. The polymer was held at this temperature for one hour to drive off residual water and solvent. The temperature was increased to 200°C to effect conversion of the amide-acid to the imide. The reaction is shown in Figure 1.

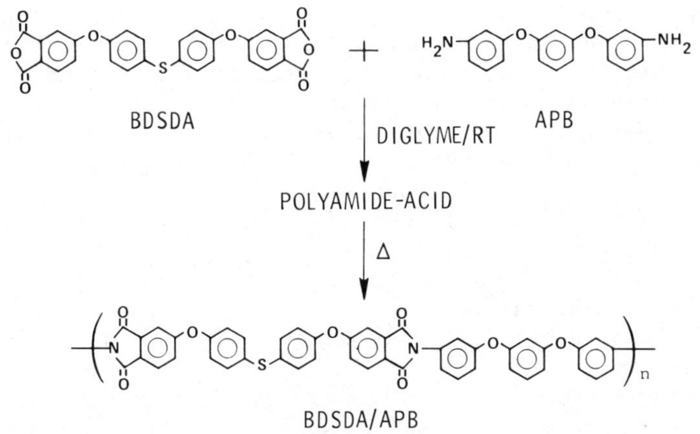

Figure 1. Polymer synthesis scheme for BDSDA/APB.

Unfilled Moldings. The polyimide powder was molded according
to the following procedure. The imidized powder was placed in a
matched-metal molding die which was preheated to 160°C. A pressure
of 1.38-2.07 MPa (200-300 psi) was applied to effect consolidation. This temperature and pressure were held for one-half hour.
The mold and molding were allowed to cool to approximately 100°C
and the molding was removed. This molding was light brown and
transparent when prepared in discs up to 0.635 cm (0.250 inches) in
thickness.

Adhesive Bonds. Two sets of adhesive bonds were prepared at
different heat-up rates. Duplicate 2.54 cm-wide strips of 0.127 cm
(0.050 inch) thick titanium alloy were grit blasted with 120 mesh
aluminum oxide, washed with methanol, and treated with Pasa Jell
107*. These strips were washed with water and dried in a forced-
air oven at 100°C for 15 minutes. Each strip was coated with the

---

*Pasa Jell 107 is a commercial product of American Cyanamid Co.
(Use of trade names or manufacturers does not constitute an offi-
cial endorsement, either expressed or implied, by the National
Aeronautics and Space Administration.)

polyamide-acid solution. A piece of woven glass cloth (0.01 cm thickness) was laid into the wet polymer on one of the titanium panels. The coated panels were allowed to air dry for approximately one hour and then placed in a forced-air oven and heated to 160-275°C in order to drive off solvent and to convert the polyamide-acid to the polyimide. Imidization occurs at temperatures above 160°C and the degree of conversion is a function of time and temperature. An example of a cure is as follows:

> 1 hour at 100°C
> 1 hour at 200°C
> 1 hour at 275°C

This treatment was repeated several times on the panel with the glass cloth in order to build up sufficient adhesive thickness for bonding.

The panels were then overlapped according to ASTM D 1002-72 and bonded in a hydraulic press under 2.07 MPa (300 psi) pressure. This sample was heated to 316°C at 5°C/min and held at this temperature for fifteen minutes. The bonded specimen was allowed to cool to 100°C before removal from the press.

In order to achieve a higher heat-up rate another pair of adherends were placed into the press which had been previously heated to 343°C. The temperature rise on these panels was monitored, as in the previous case, using a thermocouple which was spot-welded to the titanium adherend at the edge of the bondline. This bonding procedure afforded an average 22°C/min heat rise to 316°C. A bonding pressure of 2.07 MPa was also used for this sample.

Film. A 15% solution of BDSDA/APB in diglyme was used to cast a 381μm thick wet film on plate glass using a doctor blade. The film was then cured in an air oven as follows:

> 1 hour at 100°C
> 1 hour at 200°C
> 1 hour at 300°C

The cured film, approximately 40μm thick, was removed from the glass for future testing.

Characterization Methodology

Flow Properties. Melt flow properties for BDSDA/APB in the range 250°C-350°C were determined using a capillary rheometer (Instron Model 3211).[6] The capillaries used had length-to-diameter ratios of 33 and 66; therefore no end corrections were required. Melt flow properties were not measured below 250°C because the pressure needed to force the polymer through the capillary was greater than the rheometer load cell was capable of measuring. Above 350°C, the flow measurements became very erratic.

Physical Properties. Water absorption at ambient temperature was determined using four 0.635 cm x 0.635 cm x 3.493 cm bar samples cut from molded stock. The samples were weighed and oven dried in air for 24 hours at 100°C. They were then cooled to ambient in a desiccator and reweighed. They were immersed in distilled water for 24 hours, removed and blotted dry and reweighed. Samples were reimmersed for an additional 48 hours, removed and blotted dry and again weighed.

Density determinations (g/cm^3) were made on two molded discs 6.350 cm diameter x 0.318 cm thick and the average value reported.

Inherent viscosity measurements were made using a Cannon-Ubbelohde viscometer in a 35°C water bath. A 10 ml solution of 0.5% solids in DMAc was made and filtered. The average of five runs of this solution was reported.

The number-average molecular weight ($\overline{M}_n$) and weight-average molecular weight ($\overline{M}_w$) were determined at room temperature for the polyamide-acid dissolved in tetrahydrofuran.* A Knauer Membrane Osmometer was used for the $\overline{M}_n$ measurements and a Brice-Phoenix Light Scattering Photometer for the $\overline{M}_w$ measurements.

Mechanical Properties. All mechanical properties were determined at room temperature using an Instron Testing Machine Model TT-C.

The flexural strength and modulus of three samples 0.635 cm x 0.254 cm x 3.175 cm, cut from a molded disc, were run in three-point bending using a span of 2.54 cm and a crosshead speed of 0.127 cm/min. Due to a lack of material, these samples had a span-to-depth ratio of 10 and not 16 as specified by ASTM Standard D 790-71. The average of three samples was reported.

The compressive strength was determined for four samples nominally 0.645 cm x 0.709 cm x 1.224 cm. They were run at 0.127 cm/min crosshead speed and the average value reported. ASTM

---

*Molecular weights determined by ARRO Laboratory, Joliet, Illinois.

Standard D 695-69 was used as a guide, but due to lack of
material, sample size was decreased from the recommended ASTM
standard size (1.27 cm x 1.27 cm x 2.54 cm).

The fracture energy value (G_{Ic}), the opening mode strain
energy release rate, was determined from two compact tension samples. The samples were machined from discs 5.715 cm diameter and
nominally 0.127 cm thick, precracked, and run at a crosshead
speed of 0.127 cm/min according to ASTM Standard E 399-78A. The
average of two samples was reported.

The lap shear tests were performed in accordance with ASTM
Standard D 1002-72. Four samples bonded at a low heating rate
(5°C/min) and four bonded at a high heating rate (22°C/min) were
measured for lap shear strength and the average value reported.

Thermal Properties. The coefficient of thermal expansion was
determined for the range 30°C-125°C using a DuPont Model 941 Thermomechanical Analyzer (TMA) operating in static air at a programmed heating rate of 10°C/minute. The glass transition temperature (T_g) was determined calorimetrically with a DuPont Model
990 Thermal Analyzer/Differential Scanning Calorimeter in static
air at a programmed heating rate of 20°C/minute. The apparent
glass transition temperature was determined using a DuPont Model
943 Thermomechanical Analyzer in static air at 5°C/minute, and
also on a DuPont Model 1090 Dynamic Mechanical Analyzer under the
same run conditions. Thermooxidative stability (weight loss vs.
temperature) was determined using a Perkin-Elmer Model TGS-2 thermogravimetric system using a heating rate of 2.5°C/minute and an
air flow of 15 cc/minute.

A thermomechanical spectrum of the polymer was obtained by
torsional braid analysis (TBA). A glass braid was coated with a
10% diglyme solution of the polyamide-acid and precured to 300°C
in air. This spectrum was obtained at a heating rate of 3°C/min
in a nitrogen atmosphere.

Chemical Resistance. Six film samples approximately 40μm
thick were used for T_g measurement (apparent) using a DuPont
Model 941 Thermomechanical Analyzer.[7] Each of the six samples
was immersed in one of six commonly used solvents at room temperature for a period of 72 hours. Their physical condition was noted
and they were removed and blotted dry. When completely dry, they
were again measured for T_g(apparent) and any change was noted.

RESULTS AND DISCUSSION

Synthesis

BDSDA/APB was synthesized according to the reaction scheme shown in Figure 1. Reaction of the monomers in the ether solvent diglyme produced a highly viscous polyamide-acid with an inherent viscosity of 0.66 dl/g. The thermal imidization of the poly-amide-acid was carried out on the powder and resulted in a linear high molecular weight polymer which could be processed as a hot-melt thermoplastic due to its novel molecular structure. This ease of processing may be due to the incorporation of oxygen and sulfur linkages in the polymer backbone as well as the <u>meta</u> orientation of some of these linkages.

After thermal imidization at 200°C in air, the polymer exhibited a relatively low T_g of 161°C as shown by the TBA spectrum (Figure 2). Further imidization was accomplished by curing at successively higher temperatures. This further imidization was indicated by the increase in melt flow viscosity with increased cure temperature (Figure 3).

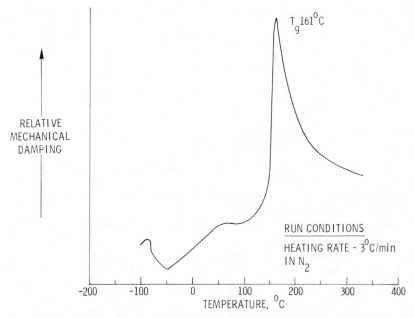

Figure 2.   Torsional Braid Analysis (TBA) spectrum of imidized BDSDA/APB (braid precured to 300°C in air).

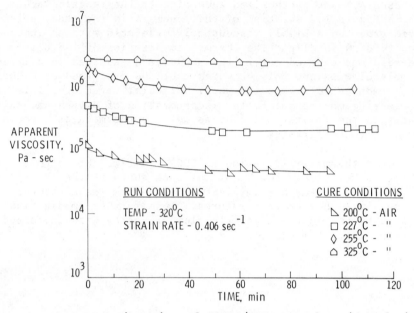

Figure 3. Apparent viscosity of BDSDA/APB as a function of time
and cure temperature.

Characterization

Flow Properties. The melt flow properties of a polymer are
important in determining how it should be processed. Since
compression molding, milling, calendering, extrusion, and
injection molding are some of the more commonly used processing
methods,[8] the melt flow properties of BDSDA/APB were determined
for the shear strain rate region (10^{-1}–10^3 sec^{-1})
encountered in these processing methods.

The stress vs. strain rate data (Figure 4) obtained using the
capillary rheometer indicates the polymer to be pseudoplastic, a
non-Newtonian shear thinning flow property, typical of molten
polymers. The pressures associated with commercial molding
presses correlate, generally, with stresses in the 10^4–10^5 Pa
range. Most of the measured stresses for this polymer were within
or exceeded this range. Consequently, high temperatures and low
strain rates would be required for molding of BDSDA/APB (i.e.,
long times).

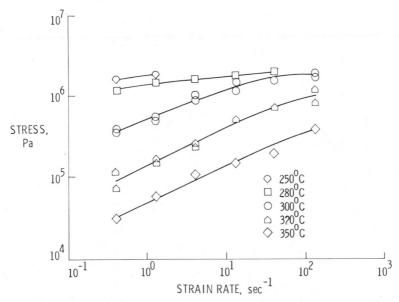

Figure 4.   Capillary rheometer stress vs. strain rate data for
            BDSDA/APB in the 250°C-350°C temperature range.

The apparent viscosity as a function of strain rate at various temperatures (Figure 5) is shown for the strain rates encountered in different industrial processes. The apparent viscosity
was calculated by dividing the flow stress by the strain rate. As
the strain rate was calculated from the volumetric flow data and
was not corrected to obtain the wall rate, the viscosity is an
apparent rather than a true viscosity.[9] The BDSDA/APB polymer
should be processable via compression molding and calendering
techniques. However, no conclusions can be drawn concerning the
extrudability of the polymer above a strain rate of 135 sec^{-1}
due to the stress and strain rate limitations of the rheometer in
its present configuration.

Figure 6 compares the change in apparent viscosity with
strain rate at the midrange processing temperature for this polymer (BDSDA/APB), commercially available Torlon*, and a typical
widely used ABS resin.[10] This comparison is made because no
data have been generated on a linear aromatic polyimide system
prior to this BDSDA/APB study. At low strain rates the BDSDA/APB
exhibits a considerably lower melt viscosity (i.e., lower processing pressure) than Torlon or ABS resin and maintains this

---

*Torlon is the registered trademark for an Amoco poly(amide-
imide).

125

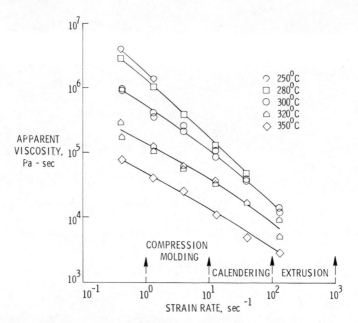

Figure 5. Apparent viscosity as a function of strain rate for BDSDA/APB in the 250°C-350°C temperature range.

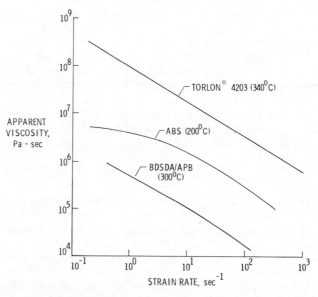

Figure 6. Comparison of apparent viscosity vs. strain rate curves for BDSDA/APB, ABS and Torlon at their midrange processing temperatures.

relationship even at the higher strain rates. Thus, BDSDA/APB appears to be the most processable of the materials compared.

Curves of apparent viscosity as a function of temperature for selected strain rates corresponding to specific processing methods (Figure 7) were constructed by cross-plotting the apparent viscosity-strain rate data. The curves show fairly uniform flow behavior with no apparent processing "windows". However, they do indicate a processing advantage, a well defined decrease in viscosity with increased temperature for the different processing strain rates.

A sample of BDSDA/APB was cured to a temperature of 200°C in air and allowed to cool to ambient. This procedure was repeated for three additional samples at 227°C, 255°C and 325°C. Figure 3 shows the apparent viscosity for these four samples at 320°C as a function of elapsed time at that temperature. The viscosity increase with increasing cure temperature may be due either to a continuing buildup of molecular weight of the polymer in the melt form or to crosslinking taking place. The viscosity at the three lower cure temperatures does not assume a constant value until the polymer has been at its run temperature of 320°C for 40-55 minutes. The viscosity of the sample cured at 325°C did not

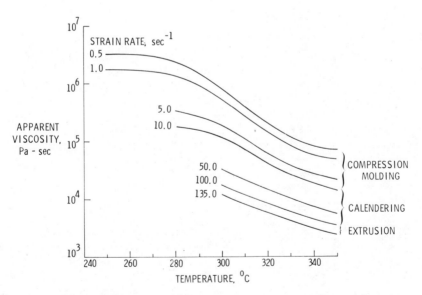

Figure 7. Apparent viscosity vs. temperature at various strain rates for BDSDA/APB for specific processing methods.

change with time. Samples were then cured at 227°C and 255°C in
nitrogen and the viscosities compared with the air-cured values
(Figure 8). The viscosity for the 227°C nitrogen-cured sample was
1.50 MPa-sec, which is 1.31 MPa-sec higher than that for the air
cured. The viscosity stabilized after approximately 30 minutes at
the 320°C run temperature. The viscosity for the 255°C nitrogen-
cured sample was 2.40 MPa-sec, which is 1.6 MPa-sec higher than
that for the air-cured. The viscosity of that sample was constant
from the beginning of the 320°C run. Further studies would be required to determine the specific mechanisms that cause this
effect.

Rubbery type flow of molten polymers during extrusion, especially in the 2×10^5 Pascal shear stress region, can cause distortion or "melt fracture" in the extrudate.[11,12] The BDSDA/APB
polymer exhibited this characteristic during melt flow determinations. Three samples of the polymer extruded at 300°C and generated at increasingly higher stresses were compared (Figure 9).
Extrudate (a) formed at the lowest stress (3.57×10^5 Pa) exhibited a relatively smooth uniform surface, indicative of a low
degree of melt fracture. Extrudates (b) and (c) generated at

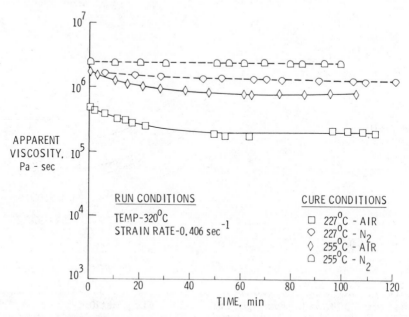

Figure 8. Comparison of apparent viscosity vs time for air and
nitrogen-cured BDSDA/APB.

128

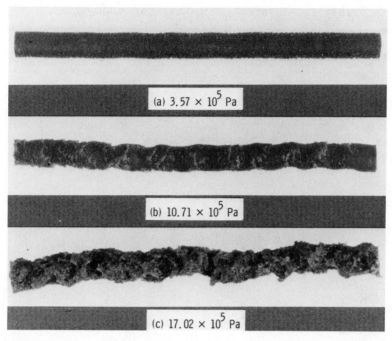

Figure 9. Effect of increasing stresses on the melt fracture of BDSDA/APB extrudate at 300°C.

successively higher rates of 10.71 x 10^5 Pa and 17.01 x 10^5 Pa exhibited successively more pronounced melt fracture characteristics. Although higher extrusion temperatures did not materially alter the melt fracture, lower molecular weight polymer might exhibit less melt fracture.[11]

Physical Properties. Some physical properties for BDSDA/APB are listed in Table I. The ambient moisture content of 0.15% by weight was determined by oven drying the sample in air for 24 hours at 100°C. A 72 hour water soak restored the sample to its original weight. An interim check at 24 hours showed the sample had regained one-half of its ambient moisture content. These values are an average of four samples taken at room temperature. The density was determined to be 1.34 g/cm^3, typical for an amorphous, linear aromatic polymer. The inherent viscosity of 0.66 dl/g for the polyamide-acid reflects the molecular weight of the polyamide-acid. Because the polymer melts during subsequent processing to the imide form, there is a possibility for either an increase or decrease in its molecular weight. The imide form of the polymer is insoluble, thus precluding the determination of its inherent viscosity. The number-average molecular weight ($\overline{M}_n$) for the polyamide-acid was 13,900. This was

Table I. Physical Properties of BDSDA/APB.

WATER ABSORPTION[a]
    24 hr SOAK   0.07%
    72 hr SOAK   0.15%

DENSITY[b]   1.34 g/cm^3

INHERENT VISCOSITY[c]   0.66 dl/g

MOLECULAR WEIGHT[d]
    NUMBER AVERAGE ($\overline{M}_n$)   13 900
    WEIGHT AVERAGE ($\overline{M}_w$)   27 500

(a) BAR SAMPLES NOMINALLY 0.635 cm × 0.635 cm × 3.493 cm
    CUT FROM MOLDED STOCK
(b) MOLDED DISKS 6.350 cm DIAMETER × 0.318 cm THICK
(c) SOLUTION OF 0.50% SOLIDS IN DMAc AT 35°C
(d) VALUES DETERMINED ON THE AMIDE ACID BY ARRO LABS,
    JOLIET, ILLINOIS

the average of nine determinations (three measurements for each of three solution concentrations). The weight-average molecular weight ($\overline{M}_w$) was 27,500 (four measurements for each of two solution concentrations).

Mechanical Properties. The mechanical properties determined for BDSDA/APB at room temperature are listed in Table II. The flexural strength and flexural modulus of 75.1 MPa and 3.48 GPa, respectively, represent the average of three samples. A compressive strength of 153 MPa was the average for four samples. All three values compare favorably with engineering thermoplastics.

The critical strain energy release rate (G_{Ic}) for this polymer was determined to be 4,100 J/m^2, a very high value compared to addition-curing polyimides which have G_{Ic} values of approximately 100 J/m^2.[13] Even an elastomer toughened addition polyimide has been reported to have a value of only 387 J/m^2.[14]

Adhesive bonds were fabricated at heating rates of 5°C/min and 22°C/min using 6Al-4V titanium adherends. In both cases the room temperature lap shear values of 40.3 MPa and 43.4 MPa, respectively, were quite high compared to conventional polyimide adhesives.[15]

130

Table II.   Mechanical Properties of Unfilled BDSDA/APB.

FLEXURAL STRENGTH	75.1 MPa (10.9 ksi)
FLEXURAL MODULUS	3.48 GPa (505 ksi)
COMPRESSIVE STRENGTH	153 MPa (22.2 ksi)
CRITICAL RATE OF RELEASE OF STRAIN ENERGY, G_{Ic}	4100 J/ m^2
LAP SHEAR STRENGTH (Ti/Ti)*	
LOW HEATING RATE (5^0C/min)	40.3 MPa (5.85 ksi)
HIGH HEATING RATE (22^0C/min)	43.4 MPa (6.30 ksi)

* BONDED AND TESTED ACCORDING TO ASTM STANDARD D 1002-72

Thermal Properties. Some of the more common thermal properties were determined for BDSDA/APB and are listed in Table III.
The coefficient of thermal expansion, $5.14 \times 10^{-5}°C^{-1}$, is
typical for state-of-the-art polyimides. The glass transition
temperature (T_g), measured calorimetrically, was determined to
be 160°C. The T_g (apparent) values for a dry molded sample and
for one water soaked for 72 hours at room temperature were
measured thermomechanically. The T_g of the water soaked sample,
run wet, was 145°C, 7°C below that for the dry sample (152°C).

BDSDA/APB film exhibits thermooxidative stability similar to
Kapton® film as shown in the thermogram Figure 10. The dynamic
TGA curves obtained at a heating rate of 2.5°C/min and an air flow
rate of 15 cc/min indicate that both films undergo essentially no
weight loss below 400°C. The BDSDA/APB exhibits a 50% weight loss
temperature of 561°C, only 19°C below that for the Kapton, while
both films undergo complete degradation at approximately 610°C.

Chemical Resistance. The chemical resistance of BDSDA/APB
thin film (40μm thick) to six common solvents was determined and
the results listed in Table IV. Methylethyl ketone, cyclohexanone, xylene, and tricresylphosphate had no visible effect on the
film and there was no change in T_g (apparent). Methylene chloride and cresol caused severe swelling and T_g measurements were
not possible, although the methylene-chloride-soaked film did
maintain sufficient integrity as a film to allow mounting in the
T_g fixture.

131

Table III. Thermal Properties of Unfilled BDSDA/APB.

COEFFICIENT OF THERMAL EXPANSION (30°C - 125°C)	$5.14 \times 10^{-5}\,^\circ C^{-1}$
GLASS TRANSITION TEMP (T_g), CALORIMETRIC	$160^\circ C$
THERMOMECHANICAL T_g (APPARENT)	
TMA - DRY	$152^\circ C$
WET	$145^\circ C *$
DMA	$159^\circ C$
DECOMPOSITION TEMP	$561^\circ C **$

* 72 hr WATER SOAK, TESTED WET (0.15% BY WT AMBIENT WATER CONTENT)
** 2.5°C/min HEATING RATE, 15 cc/min AIR FLOW, 50% WT LOSS

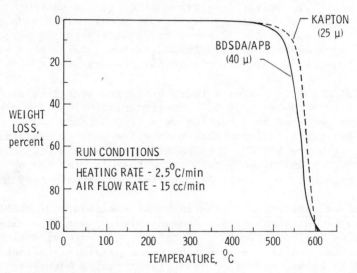

Figure 10. Comparison of thermooxidative stability of BDSDA/APB and Kapton® polyimide film.

Table IV. Chemical Resistance of BDSDA/APB Film.*

SOLVENT	EFFECT	CHANGE IN T_g, (APPARENT), ^oC**
METHYLETHYL KETONE	NONE	NONE
CYCLOHEXANONE	NONE	NONE
XYLENE	NONE	NONE
TRICRESYLPHOSPHATE	NONE	NONE
METHYLENE CHLORIDE	SWELLED	NOT DETERMINED
CRESOL	SWELLED	NOT DETERMINED

*40 μ THICK FILM
**THERMOMECHANICAL ANALYSIS OF SOLVENT-LADEN FILMS
SOAKED 72 hrs AT ROOM TEMPERATURE AND BLOTTED DRY

CONCLUSIONS

A novel linear aromatic polyphenylene ethersulfideimide (BDSDA/APB) that has characteristics of linear aromatic polyimides and both linear aromatic polyphenylene oxides and sulfides was synthesized and characterized. Thermal imidization of the polyamide-acid at 200°C resulted in a linear, high molecular weight polymer with a T_g of 161°C that could be processed as a hot-melt thermoplastic.

BDSDA/APB exhibited a considerably lower melt viscosity at its midrange processing temperature (300°C) than did either commercially available Torlon® or a widely used ABS resin, both measured at their midrange temperature, 340°C and 200°C respectively. It also exhibited a major well defined decrease in viscosity with increased temperature, a definite processing advantage.

The density of 1.34 g/cm^3 is typical for an amorphous, linear aromatic polymer. The inherent viscosity of 0.66 dl/g and molecular weights of 13,900 ($\overline{M}_n$) and 27,500 ($\overline{M}_w$) were determined for the polyamide-acid because the imide form of the polymer was insoluble.

The G_{Ic} value of 4,100 J/m^2 was very high compared to addition-curing polyimides which have values of approximately 100 J/m^2 or an elastomer toughened addition polyimide with a

133

reported value of 387 J/m^2. Room temperature lap shear adhesive strength values of 40.3 MPa and 43.4 MPa were quite high compared to conventional polyimide adhesives.

The coefficient of thermal expansion value 5.14 x 10^{-5}°C^{-1} is typical of state-of-the-art polyimides. The T_g (apparent), 152°C, decreased by only 7°C when the polymer was subjected to a 72 hour water soak and tested wet. BDSDA/APB is a very thermooxidatively stable polymer showing essentially no weight loss below 400°C and 50% weight loss at 561°C in a dynamic TGA. Complete degradation did not occur until 610°C.

Because of its melt-flow properties, high G_{Ic}, chemical and moisture resistance, and thermooxidative stability, this novel polymer exhibits considerable attractiveness as an engineering thermoplastic. In particular, due to the degree of melt-flow, this polyethersulfideimide shows potential as a composite matrix resin and as a hot-melt adhesive.

ACKNOWLEDGEMENTS

The authors wish to thank Dr. R. Y. Ting, Naval Research Laboratory, Orlando, Florida for his invaluable contribution of BDSDA/APB G_{Ic} data. They also wish to thank Mr. Robert Ely, Mr. James Tyeryar, and Mr. Spencer Inge, NASA-Langley Research Center, Hampton, Virginia for their excellent technical assistance.

REFERENCES

1. A. H. Frazer, "High Temperature Resistant Polymers," in Polymer Reviews, Vol 17, p. 315, John Wiley and Sons, New York, 1968.
2. C. E. Sroog, Macromolecular Reviews, 11, 161 (1976).
3. R. D. Deanin, "Polymer Structure Properties and Applications," Ch. 8, p. 457, Cahners Publishing, Boston, 1972.
4. J. M. Cox, B. A. Wright and W. W. Wright, J. Appl. Polym. Sci., 9, 513 (1965).
5. V. V. Korshak, "Heat Resistant Polymers," (English Translation), Ch. 3, p. 128, Keter Press, Jerusalem, 1971.
6. L. E. Nielsen, "Polymer Rheology," Ch. 2, p. 12, Marcel Dekker, Inc., New York, 1977.
7. H. D. Burks, J. Appl. Polym. Sci., 18, 627 (1974).
8. R. M. Ogorkrkiewicz, "Thermoplastics," Ch. 11, p. 171, John Wiley and Sons, New York, 1974 .

9.   J. R. Van Wazer, J. W. Lyons, K. Y. Kim and R. E. Colwell, "Viscosity and Flow Measurements," Ch. 4, p. 193, Interscience Publishers, New York, 1963.

10. Torlon Applications Guide, p. 11, Amoco Chemical Corp., Chicago, (Feb. 1979).

11. R. D. Deanin, Op. Cit., p. 166.

12. F. W. Billmeyer, Jr., "Textbook of Polymer Science," Ch. 6, p. 190, John Wiley and Sons, New York, 1962.

13. W. D. Bascom, J. L. Bitner and R. L. Cottington, Organic Coatings and Plastics Chemistry Preprints, 38, 477 (1978).

14. R. Y. Ting and R. L. Cottington, 12th National SAMPE Technical Conference Preprints, 12, 725 (1980).

15. A. K. St. Clair and T. L. St. Clair, "A Review of High-Temperature Adhesives," NASA TM 83141 (1981). (Available from the National Technical Information Service, Springfield, Virginia 22161).

NEW PROCESSABLE POLYIMIDE BASED ADHESIVES

D. Landman

Hysol Division, The Dexter Corporation
P. O. Box 312
Pittsburg, California 94565

The development of a broad line of 232°-280°C
service adhesives which are easily processed is described. The first generation film adhesives, EA9655
and EA9655.1 typically perform best up to about 232°-
260°C. Adhesive properties using various substrates
including aluminum, titanium and graphite composites
are described. EA9351 and EA9367 are one-component
pastes with excellent adhesive strength to 260°C.
LR 100-581 is a sprayable 232°-260°C service primer
which complements these products. Preliminary data
for a second generation film adhesive called LR 100-573
indicates that its service capabilities are in the
range 260°-280°C. In all cases, the processing of
each adhesive is very simple. Typical cure cycles are
one hour at 177°C with minimum autoclave pressure (0.14
to 0.3MPa) followed by an unrestrained post-cure for
two hours at 246°C.

INTRODUCTION

Many applications, particularly in warmer sections of jet engine nacelles and high performance military aircraft, have created a need for 232°-280°C service adhesives. High performance systems such as LARC's and others[1], have been introduced. Typically, these materials have very good retention of physical properties with heat aging. However, they also require processing conditions which are expensive or limited to small parts because specialized equipment is needed for the temperatures and pressures that are required. We have taken an alternative approach to adhesives for these temperature requirements, always keeping in mind that processability which is similar to that for conventional epoxy resins is highly desirable. The resin systems described in this work are bismaleimide modified epoxy systems.

DISCUSSION

Film Adhesives with 232°-260°C Service

Tensile Shear Properties: The first generation of film adhesives we have developed for 232°-260°C service are called EA9655 and EA9655.1. EA9655.1 uses the same resin as EA9655, but also contains an aluminum filler. Table I shows typical single-overlap tensile shear strengths for EA9655 (293 gm/m² weight) using FPL-etched bare aluminum (2024 T3) substrates. The tensile shear specimens were not restrained in any fashion to prevent failures due to peeling effects.

Results for two different post-cure cycles are shown in Table I. The post-cure with higher temperature (260°C) gives improvement in tensile shear strength at higher test temperature, but lower tensile shear strength at lower temperatures. The tensile shear strength of EA9655 tested at 232°C after aging at 232°C (in air) drops by 31% after 500 hours, 41% after 1000 hours and 59% after 3000 hours. However, these data are reported on aluminum substrates where it is known[2] that excessive thermal history can alter the nature of this substrate. Further thermal aging using more suitable substrates is in progress.*

Table I also shows the effect of moisture on specimens bonded with EA9655. The tensile shear strength at 25°C is improved (probably due to the plasticizing effect of the water), while the 232°C and 260°C strengths are decreased (probably because the wet Tg of EA9655 is exceeded).

---

*Thermal aging of EA9655 as a resin on a glass prepreg substantially confirms the aluminum data. These data will be reported in further publications.

Table I. Tensile Shear Strength for Aluminum Adherends* using 293 gm/m^2 EA9655 Film.

Test Temperature (°C)	Tensile Shear Strength (MPa)	
	Cure A	Cure B
25°	18.6	13.1
25°/after 30 days @49°C/100% RH	25.5	-
149°	-	17.9
232°	20.0	17.9
232°/after 30 days @49°C/100% RH	12.4	-
232°/aged 500 hours @232°C	13.8	-
232°/aged 1000 hours @232°C	11.7	-
232°/aged 3000 hours @232°C	8.3	-
260°	12.4	13.1
260°/after 30 days @49°C/100% RH	9.6	-
288°	6.2	9.6

*Adherends: 2024 T3 bare (FPL etched)

Cure A: 1 hour @177°C/0.3 MPa + 2 hours @232°C (unrestrained)

Cure B: 1 hour @177°C/0.3 MPa + 2 hours @260°C (unrestrained)

Note: Test results in this and remaining Tables are the average of at least 3 test coupons.

Table II shows single-overlap tensile shear strengths for other substrates including 6A1-4V titanium, 301 stainless steel, PMR-15 (NASA)/Celion (Celanese) woven graphite and V378 (U.S. Polymeric)/T300/3K (Union Carbide Corp.) unidirectional graphite, using EA9655 as the film adhesive. The titanium data are rather scattered, probably because of the difficulty in attaining good surface preparation for this substrate. The V378 data at 177° and 232°C show a large range too. The lower end of this range represents tensile shear data for specimens that were not restrained from failing by peel modes. The high end values represent specimens that were restrained to fail in shear only. Since we are still evaluating EA9655 on these and other substrates, the data shown in Table II should be taken as preliminary in nature.

Table III shows tensile shear strength data for EA9655.1 using the same aluminum substrates as for EA9655. The areal weight of EA9655.1 for these tests was 391 gm/m². Generally, EA9655.1 shows similar properties to those for EA9655, except that it is aluminum filled.

Honeycomb Peel and Flatwise Tensile Properties: To date, limited data for honeycomb peel of EA9655 have been obtained. When tested at ambient temperature, using 2024 T3 aluminum face-sheets (bare, FPL etched) and 5052 aluminum honeycomb core (4.8mm cell size, 128 Kg/m³ density), the metal-to-metal honeycomb climbing drum (HCCD) torque was 19N for 488 gm/m² film weight and roughly half this value for 293 gm/m² film weight. We have developed a film adhesive called LR 100-580 which has HCCD peel torque values of 31 N. However, the thermal performance in tensile shear strength is less than for EA9655. Table IV summarizes the adhesive data we have obtained for LR 100-580.

For most applications using high performance adhesives, flatwise tensile strength is a more important consideration. Table V summarizes our data for EA9655 (488 gm/m² weight) on aluminum, graphite and quartz fabric laminate substrates. In both the graphite and quartz fabric cases, the facesheets were precured prepregs of F178 resin system from Hexcel.

Paste Adhesives with 232°-260°C Service

The first product to be described is called EA9351. This is a one-component paste adhesive which is stored at 4° to -20°C, and has a shelf-life of at least six months. EA9351 can be troweled with some difficulty at 25°C, and very easily at 45°C. The applications envisioned for EA9351 include edge-filling and potting where slump during cure needs to be avoided. The cure cycle for EA9351 is 1 hour @177°C/0.2 MPa + 2 hours @246°C (unrestrained) and no slump is observed.

Table II. Tensile Shear Strength for Various Adherends using 293 gm/m^2 EA9655 Film.[a]

Substrate	Test Temperature (°C)	Tensile Shear Strength (MPa)
6Al4V Titanium[b]	260°	9.6 – 17.9
301 Stainless Steel	25°	12.1
	204°	12.1
PMR-15[c]	25°	12.1
	177°	13.8
	260°	10.3
V378[d]	25°	14.5
	177°	12.1 – 22.7
	232°	13.0 – 16.5

a) Cure is 1 hour @177°C/0.1 to 0.3 MPa + 2 hours @246°C (unrestrained)

b) Surface preparation of the titanium: Degrease, alkali clean, HF/HNO$_3$ etch, phosphate treat, wash and dry

c) Prepreg of Celion/Woven graphite fiber (Celanese)

d) Prepreg of T300/3K/unidirectional graphite (Union Carbide)

Table III. Tensile Shear Strength for Aluminum Adherends* using 391 gm/m^2 EA9655.1 Film.

Test Temperature (°C)	Tensile Shear Strength (MPa)
25°	17.9
25°/after 30 days @49°C/100% RH	24.8
232°	20.3
232°/after 30 days @49°C/100% RH	11.7
232°/aged 500 hours @232°C	12.4
232°/aged 1000 hours @232°C	12.1
260°	12.4
260°/after 30 days @49°C/100% RH	9.0

*Adherends: 2024T3 bare (FPL etched). Cure cycle is 1 hour @177°C/0.3 MPa + 2 hours @232°C (unrestrained)

Table IV. Adhesive Properties for Aluminum Adherends* using 488 gm/m^2 LR 100-580 Film.

Property Measured	Test Temperature (°C)	Value
Tensile Shear Strength	25°	20.7 MPa
	232°	8.3 MPa
	260°	4.8 MPa
Honeycomb climbing drum peel	25°	31 N

*Adherends: 2024T3 bare (FPL etched) and 5052 honeycomb core (4.8mm cell size, 128 Kg/m^3 density)

Cure cycle is 1 hour @177°C/0.3 MPa + 2 hours @232°C (unrestrained)

Table V. Flatwise Tensile Strength for Various Substrates using 488 gm/m^2 EA9655 Film.

Substrates	Test Temperature (°C)	Flatwise Tensile Strength (MPa)
Aluminum[a]	25°	6.9
	177°	5.5
	216°	5.5
F178 quartz[b]	25°	4.7
	177°	4.8
F178 graphite[c]	25°	4.8
	149°	4.8

a) Substrate: 2024T3 aluminum facesheet (bare, FPL etched) and 5052 aluminum honeycomb core (4.8mm cell size, 128 Kg/m^3 density).

Cure cycle is 1 hour @177°C/0.2 MPa + 2 hours @246°C.

b) Substrate: F178 resin (Hexcel Corp.) woven quartz fabric facesheets with HRP honeycomb core (4.8mm cell size, 12.5mm thick, 64 Kg/m^3 density).

Cure cycle is 1-1/2 hours @177°C/0.3 MPa + 12 hours @196°C.

c) Substrate: F178 resin (Hexcel Corp.) woven graphite fabric (T300/3000 Union Carbide Corp.) facesheets with the same core as in b), except that the core is 6.3mm thick.

Cure cycle is 1 hour @177°C/0.3 MPa + 2 hours @246°C (unrestrained).

143

Table VI shows tensile shear strength data for EA9351. As can be seen, it performs well throughout the range 25°-260°C. Two different cure cycles are also compared in Table VI. It is interesting that with even as short a cure as 2 hours @177°C very acceptable adhesive properties are obtained with EA9351 throughout the range 25°-260°C.

The second paste to be described is called EA9367. This is a syntactic core paste which was also developed for potting and edge-filling applications where lighter weight is desirable. Typically, the density of EA9367 is 0.92 gm/cm^3 compared to 1.3 gm/cm^3 for shear strength on aluminum substrates (2024 T3, bare, FPL etched). Typical tensile shear strengths are 14.5 MPa at 25°C and 13.1 MPa at 260°C using a 1 hour @177°C + 2 hours @246°C cure cycle.

Primer with 232°-260°C Service

In the course of our work to develop high temperature service adhesives, we decided that it would be desirable to include a primer system. We thought such a system would be particularly suitable for spraying over titanium surfaces as protection against environmental damage prior to bonding. This primer is designated LR 100-581 and should also be useful as a protective coating over graphite composites. Conventional spray equipment is used to apply LR 100-581 and cure is achieved by heating the sprayed substrate between 177°C and 246°C for one hour. Table VII compares the effect of using LR 100-581 primer combined with EA9655 on aluminum and titanium substrates, as well as the effect of primer cure cycle on final adhesive properties. Although a cure at 177°C

Table VI. Tensile Shear Strength for Aluminum Adherends* using EA9351 Paste.

Test Temperature (°C)	Tensile Shear Strength (MPa)	
	Cure A	Cure B
25°	20.0	13.8
177°	----	19.3
232°	18.6	----
260°	12.7	6.89

*Adherends: 2024 T3 bare (FPL etched)

Cure A: 1 hour @177°C/0.2 MPa + 2 hours @246°C (unrestrained)

Cure B: 2 hours @177°C/0.2 MPa

Table VII. Comparison of Tensile Shear Strengths of Adherends Primed with LR 100-581 using 488 gm/m² EA9655 Film.

Adherend	Test Temperature (°C)	Tensile Shear Strength (MPa)		
		Primed		Unprimed
		Cure A	Cure B	Cure A
Aluminum[a]	25°	16.3	9.4	13.4
	260°	13.0	14.7	14.4
Titanium[b]	260°	-	10.9	9.6 – 17.9

a) Substrate is 2024 T3, bare, FPL etched.

b) Substrate is 6Al4V; coupons were cleaned by wire brushing and then sandblasted, followed by the etching procedure described in b) of Table II.

Cure A: Primer is cured 1 hour @177°C. Adhesive is cured 1 hour @177°C/0.3 MPa + 2 hours @246°C (unrestrained).

Cure B: Primer is cured 1 hour @177°C + 1 hour @246°C. Adhesive is cured as for Cure A.

145

for one hour gives adequate strength, one hour post cure at 246°C gives improved properties at elevated temperatures. We feel that the values for titanium substrates are lower because of poor surface preparation prior to priming. Further work on LR 100-581 primer is in progress.

Film Adhesive with 260° to 280°C Service

LR 100-573 is a second generation film adhesive which appears to have good 260° to 280°C adhesive properties. The chemistry for LR 100-573 is also bismaleimide modified epoxy resin with an unique curing mechanism. This has higher Tg and thermal performance than for EA9655. The cure conditions are the same as those for EA9655. Results for tensile shear strength using aluminum substrates are shown in Table VIII. The tensile shear coupons were not restrained to avoid peel failure. Examination of Table VIII shows that LR 100-573 has a fairly flat profile for tensile shear strength from 25° to 280°C. In addition, the outlife of LR 100-573 appears to be excellent as evidenced by the tensile shear strength of material left open to the atmosphere at approximately 25°C for 48 days. The tensile shear results for LR 100-573 are also very repeatable with a standard deviation of 0.3 to 0.7 MPa over many tests and after long time periods between tests (3 to 6 months). Table IX shows tensile shear strength results for LR 100-573 using V378 (U.S. Polymeric) and PMR-15 (NASA) resins on graphite fibers. The strengths are comparable to those for aluminum. In addition, the effect of an elevated cure cycle (at 288°C) is shown in Table IX. It appears that at least for short periods of time LR 100-573 may have useable properties even at 316° to 360°C.

Table VIII. Tensile Shear Strength for Aluminum* Substrates using 488 gm/m^2 LR 100-573 Film.

Test Temperature (°C)	Tensile Shear Strength (MPa)
25°	13.8
260°	15.5
288°	13.2
316°	3.4
25°/after 27 days outlife	11.6
260°/after 27 days outlife	16.1
25°/after 48 days outlife	15.0
260°/after 48 days outlife	16.1

*Adherends: 2024 T3 bare (FPL etched). Cure is 1 hour @177°C/ 0.3 MPa + 2 hours @246°C (unrestrained).

Table IX. Tensile Shear Strength for Various Substrates using 488 gm/m^2 LR 100-573 Film.

Adherend	Test Temperature (°C)	Tensile Shear Strength (MPa)	
		Cure A	Cure B
V-378[a]	177°	12.4	
	232°	15.4	
	260°	12.7	
	288°	12.5	
PMR-15[b]	22°		10.7
	204°		10.7
	260°		10.1
	316°		2.1
	360°		1.1

a) Substrate is V-378 (U.S. Polymeric) on T300/3K unidirectional graphite (Union Carbide Corp.)

b) Substrate is PMR-15 (NASA) on Celion (Celanese) Woven graphite.

Cure A: 1 hour @177°C/0.3 MPa + 2 hours @246°C (unrestrained)

Cure B:. 1 hour @177°C/0.3 MPa + 2 hours @288°C (unrestrained)

Finally, some limited honeycomb peel strength data have been obtained using aluminum (2024 T3, bare, FPL etched) facesheets and aluminum honeycomb core (5052, 4.8 mm cell size, 128 Kg/m^3 density). The peel torque value obtained at 25°C was 19.1 N with a range extending to 20.8 N. We are still evaluating the properties of LR 100-573 including the long-term thermal aging charcteristics.

CONCLUSIONS

We have developed a broad line of polyimide based adhesives with processability similar to conventional epoxy systems. These new adhesives span the temperature service range of 232° to 280°C. Additional characterization work is still in progress for these adhesives, and they hold the promise that we will be able to extend their serviceability to 316°C. Work is in progress in our labs to reach the 316°C service adhesive strength plateau and above.**

ACKNOWLEDGEMENTS

Support of this work by Hysol Division, The Dexter Corporation and numerous colleagues at our Pittsburg facility is gratefully acknowledged.

REFERENCES

1. A. K. St. Clair and T. L. St. Clair, "A Review of High-Temperature Adhesives", NASA Technical Memorandum 83141 (1981).
2. See for example, "Alcoa Aluminum Handbook", Alcoa Company of America (Copyright 1959, 1962), p. 29 in which tensile strength, modulus, and elongation as a function of temperature are listed for aluminum alloys.

**Early indications for one formulation are that improvement in the 316°C tensile shear strength values have occurred. This will be reported in further publications.

POLYETHERIMIDE: A VERSATILE, PROCESSABLE THERMOPLASTIC

I. William Serfaty
General Electric Company
Specialty Plastics Division
Pittsfield, MA 01201

Polyetherimide is a new resin introduced in 1982 by General Electric Company under the Tradename ULTEM®. An amorphous, high performance engineering thermoplastic polyetherimide is based on the regular repeating units of ether and imide linkages. The aromatic imide units provide the high performance properties while the flexible ether linkages allow for good melt flow characteristics and easy processability.

The polymer exhibits: • High heat resistance, with a DTUL of 392°F (200°C) at 264 psi (1.82 N/mm^2) and a Tg of 419°F (215°C); • Excellent mechanical properties, with a tensile strength at yield of 15,200 psi (100 N/mm^2). These properties remain high at elevated temperatures [6,000 psi (41 N/mm^2) and 350,000 psi (2,400 N/mm^2), respectively, at 350°F (175°C)]; • Inherent flame resistance [the polymer has been listed* V-0 at 0.025 inch (0.64 mm) under UL Bulletin 94], extremely low smoke evolution in the NBS smoke chamber test, low flame spread index in the radiant panel test, and very high oxygen index (47%); • Outstanding electrical properties which remain stable over a wide range of temperatures and frequencies.

®A Registered Trademark of General Electric Company

*This listing is not intended to reflect hazards
 presented by this or any material under actual
 fire conditions.

INTRODUCTION

Polyetherimide (PEI) resin recently introduced by the Plastics Group of General Electric Company under the Tradename ULTEM® resin is a new thermoplastic based on completely new chemistry.

The introduction of PEI was based on more than ten years of effort. In 1970, Wirth and his team, working on exploratory chemistry, discovered a new synthetic pathway, nitro displacement polymerization.[1] Ensuing efforts refined the chemistry,[2] and other articles have described studies of the displacement reactions.[3-5]

In February of 1982, ULTEM® resin was commercially introduced, and is now being used in electrical, electronics, transportation, and appliance applications.

The resin has the following chemical structure:

The aromatic imide units provide stiffness, and high heat resistance, while the ether linkages allow for good melt flow characteristics and processability.

PEI exhibits:

- High physical strength
- Inherent flame resistance with extremely low smoke evolution
- Outstanding electrical properties over a wide frequency and temperature range
- Chemical resistance to aliphatic hydrocarbons, acids, and dilute bases
- High temperature stability
- UV stability
- Ready processability on conventional equipment

®Registered Trademark of General Electric Company

MECHANICAL PROPERTIES

The mechanical properties of unmodified and three glass-reinforced PEI compounds are shown in Table I.

Table I. Mechanical Properties of Unmodified and Three Glass-reinforced PEI Resins.

Property	ASTM Test	Units	Unmodified	10% GR	20% GR	30% GR
Tensile strength yield	D638	psi	15,200	16,500	20,100	24,500
Tensile modulus, 1% secant	D638	psi	430,000	650,000	1,000,000	1,300,000
Tensile elong., yield	D638	%	7-8	5	-	-
Tensile elong., ultimate	D638	%	60	6	3	3
Flexural strength	D790	psi	21,000	28,000	30,000	33,000
Flexural modulus, tangent	D790	psi	480,000	650,000	900,000	1,200,000
Compressive strength	D695	psi	20,300	22,500	24,500	23,500
Compressive modulus	D695	psi	420,000	450,000	515,000	550,000
Izod impact	D256	ft-lb/in				
Notched 1/8"			1.0	1.1	1.6	2.0
Unnotched 1/8"			25	9.0	9.0	8.0

Polyetherimide exhibits a room temperature flexural modulus of 480,000 psi (3,300 N/mm^2), outstanding for an unmodified thermoplastic resin. PEI is exceptionally strong with a tensile strength at yield in excess of 15,000 psi (100 N/mm^2) and a flexural strength of 21,000 psi (145 N/mm^2). In load-bearing applications, non-reinforced PEI provides the rigidity frequently associated with other glass-reinforced polymers. Even more impressive is the good retention of these mechanical properties at elevated temperatures. Figures 1 and 2 depict the tensile and flexural behavior of PEI as a function of temperature. Even at 180°C the tensile strength and flexural modulus of PIE remain in excess of 6,000 psi (41 N/mm^2) and 300,000 psi (2,100 N/mm^2), respectively,

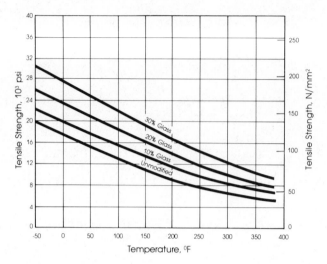

Figure 1. Tensile strength of PEI resins as a function of temperature.

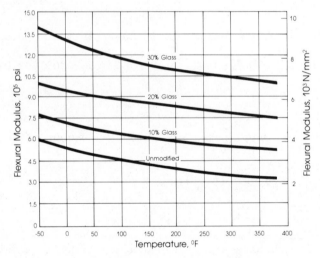

Figure 2. Flexural modulus of PEI resins as a function of temperature.

Polyetherimide exhibits classical mechanical behavior in accordance with Hookes' law. Under load, it displays a linear stress-strain relationship below the proportional limit. The proportional limit for PEI ranges from approximately 8,000 psi (69 N/mm^2) for unreinforced resin to 18,000 psi (138 N/mm^2) for 30 percent glass-reinforced compositions at 73°F (23°C). The stress-strain curves for natural and glass-reinforced PEI are shown in Figure 3.

152

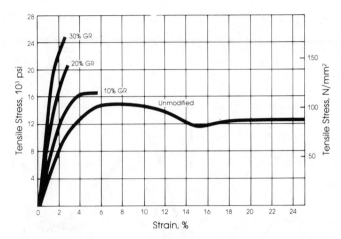

Figure 3. PEI Stress-strain curves in tension at 73°F (23°C)

The reinforcing effects of glass fibers in PEI are demonstrated by the substantial increase in tensile yield strength
which increases smoothly from an initial value of 15,000 psi
to as much as 24,000 psi as glass fiber content is increased
to 40 percent. This results from a combination of the polymer's
adhesion to the glass fibers, and the high tensile strength of
the fibers themselves (micrographs taken of sample fracture
surfaces show excellent adhesion of PEI to glass).[6] Beyond 40
weight percent of glass, the tensile strength drops as the
maximum packing volume for the fibers is approached and as the
extensibility of the composite lowers.

ELECTRICAL PROPERTIES

The electrical characteristics of unmodified and 30 percent
glass-reinforced PEI are given in Table II.

Table II. Electrical Properties of PEI.

Property	Test	Units	Unmodified	30% GR
Dielectric strength 1/16"	D149	Volts/mil	710	630
Dielectric constant @1 kHz	D150	–	3.15	3.7
Dissipation factor @1 kHz, 50% RH	D150	–	0.0013	0.0015
Volume resistivity	D257	ohm-cm	6.7×10^{17}	3.0×10^{16}

The resin exhibits an excellent balance of electrical properties which remain stable over a wide range of environmental conditions. Its stable dielectric constant and low dissipation factor make it suitable for use at frequencies in excess of 10^9 Hz even at elevated temperatures, as shown in Figure 4.

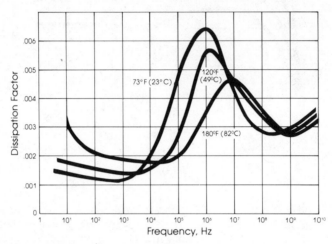

Figure 4.   Dissipation factor vs frequency of PEI at 50% RH.

THERMAL PROPERTIES

One of the most outstanding properties of polyetherimide is its ability to withstand exposure to elevated temperatures. PEI exhibits a glass transition temperature of 423°F (217°C). The high Tg of PEI, which provides a heat deflection temperature under load (DTUL) of 392°F (200°C), accounts for its excellent retention of physical properties at elevated temperatures. The DTUL profile for PEI is given in Figure 5.

This value provides a useful comparison of material deformation under constant stress conditions at a given temperature. In recognition of its inherent thermal stability, a continuous use temperature listing of 338°F (170°C) with impact has been granted PEI by Underwriters Laboratories Inc.

FLAMMABILITY

Polyetherimide is flame resistant without the incorporation of additives.[7] PEI is listed V-0 at 0.025 inch (0.64 mm) according to UL Bulletin 94, and exhibits a limiting oxygen index of 47, among the highest of any thermoplastic. Smoke generation as monitored in the NBS chamber (ASTM E-662) is extremely low in comparison with other thermoplastics, as seen in Table III. The combination of low smoke evolution and high oxygen index is unique and rarely seen among engineering thermoplastics.

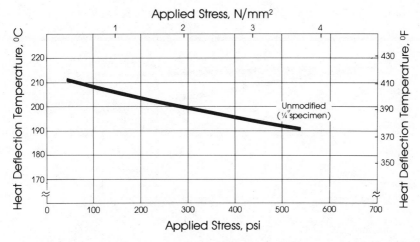

Figure 5. Heat deflection temperature of PEI vs applied stress.

Table III. Combustion Characteristics. NBS Smoke Chamber (ASTM E-662). Nominal 0.060" (1.52 mm) Thickness.

Condition	Polyetherimide		Polyethersulfone		Polycarbonate	
D (1.5 min)	0[b]	0[c]	0[b]	0[c]	13[b]	0.1[c]
D (4.0 min)	0.7	0	1.7	0	127	1.2
D (20 min)	31	0.4	37	0.6	130	4

[a] D is the specific optical density at time shown
[b] Flaming mode
[c] Smoldering mode

ENVIRONMENTAL RESISTANCE

Unlike the majority of amorphous resins, polyetherimide exhibits resistance to a wide range of chemical media. PEI resin is unharmed by most hydrocarbons (including gasolines and oils), alcohols and fully halogenated solvents. Only partially halogenated hydrocarbons such as methylene chloride and chloroform will dissolve PEI. Polyetherimide has shown exceptional resistance to mineral acids and is tolerant to short-term contact with dilute bases.

Hydrolytically stable, PEI retains high tensile strength after 10,000 hours immersion in water at 212°F (100°C), as shown in Figure 6.

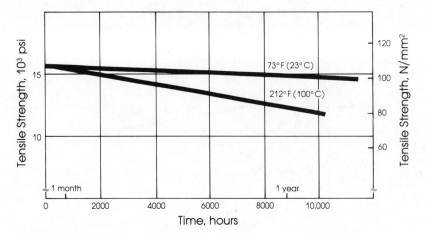

Figure 6. Effect of water exposure on the tensile strength of PEI.

Polyetherimide's physical properties remain stable following autoclave exposure and drying in vacuum at room temperature.

Polyetherimide is resistant to UV radiation without the addition
of stabilizers. Exposure to 1000 hours of Xenon arc weatherometer
irradiation (0.35 watt/m^2 irradiance at 340 nm, 63 °C) has produced a
negligible change in the tensile strength of PEI. QUV exposure
(accelerated UV weathering at 280-320 nm) has confirmed the outstanding retention of physical properties exhibited by PEI versus conventional engineering thermoplastics.

PEI also demonstrates excellent resistance to physical property
degradation following ionizing radiation exposure. A decrease of less
than 5 percent in tensile strength was observed following 500 megarads
of Cobalt 60 exposure at the rate of 1 megarad per hour, as seen in
Figure 7.

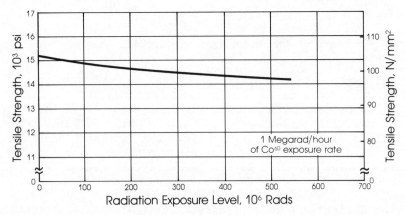

Figure 7. Effect of gamma radiation exposure on tensile strength of
PEI.

PROCESSING CHARACTERISTICS

Polyetherimide can be easily processed on conventional injection molding, extrusion, or blow molding equipment. The generally recommended injection molding conditions for PEI are given in Table IV.

Table IV. Molding Conditions for PEI.

RESIN DRYING	300°F, 4 hours, trays, (hopper dryer preferred while molding)
MELT TEMPERATURE	640°F
MELT PROFILE	Nozzle 620 to 775°F Front Zone 610 to 760°F Center Zone 600 to 740°F Rear Zone 590 to 700°F
MOLD TEMPERATURE	150 to 350°F water heating is sufficient for most applications; increased mold temperature will increase flow and improve surface
MOLDING PRESSURE	Booster Pressure 10000 to 18000 psi (1st stage) Holding Pressure 8000 to 15000 psi (2nd stage) Back Pressure 500 to 2000 psi
SCREW DESIGN	1.5 to 4.0:1 Compression Ratio, 16 to 24:1 L/D
RAM SPEED	Medium to Fast
CLAMP PRESSURE	3 to 6 tons per square inch
SCREW SPEED	50/400 rpm
PURGE	HDPE, glass reinforced polycarbonate or ground acrylic*. Begin purging at processing temperature and reduce barrel temperature to about 500°F while continuing to purge. *Use only at lower barrel temperatures

(continued)

TABLE IV. (continued)

SHRINK RATE   .005-.007 in/in

MOLD RELEASE   Normally not required, if part design causes difficulty in removal, use any standard mold release agent

PEI exhibits a resin viscosity vs. shear rate profile comparable to polycarbonate or polysulfone (rather than conventional polyimides) although the processing temperature is higher. In addition, due to its thermal stability, PEI exhibits a much wider processing window than most engineering plastics.

Figure 8 depicts the effect of wall thickness and injection pressure upon the flow length of PEI injection molding resin. The excellent flow characteristics of polyetherimide permit its use in complex, thin wall parts.

Because of its high melt strength, PEI is readily converted into film, sheet, and profiles via melt extrusion. This melt integrity also allows PEI to be easily thermoformed or blow molded using injection or extrusion techniques.

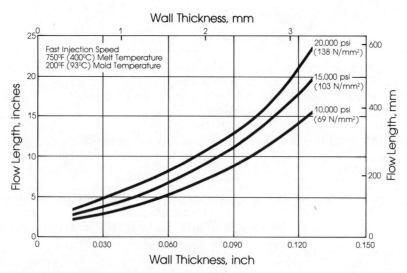

Figure 8. Effect of wall thickness on melt flow of PEI at various injection pressures.

APPLICATION AREAS

Economic and environmental changes, such as stricter regulations for smoke evolution and flame resistance, energy-efficiency requirements, and electronics miniaturization offer expanded opportunities for polyetherimide resin.

Automotive--Polyetherimide resin's advantages in this segment are high heat resistance; resistance to lubricants, coolants, and fuels; creep resistance and mechanical strength; dimensional stability under load; and ductility. Application areas include under-the-hood, electrical, and heat-exchange components.

Appliances--Key material features are chemical resistance to foods such as oils, greases, and fats; microwave transparency; heat resistance up to 400°F; good practical impact strength; flame resistance, and high gloss/colorability.

Electrical and electronics--Attributes include heat resistance to wave and vapor-phase soldering, intricate design potential, dimensional stability (allowing close point-to-point tolerances without warpage), flame resistance, and ductility. Application areas include circuit boards, connectors, switches and controls, and integrated circuit test devices.

Printed wiring boards--PEI's combination of electrical, thermal, physical, and chemical properties makes the material particularly suitable for high frequency printed wiring boards. The key feature for microwave circuit designers is the resin's low dissipation factor at high frequencies.

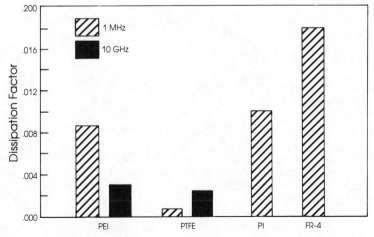

Figure 9. Dissipation factor of PEI vs other circuit board substrates.

Figure 9 compares the dissipation factor of PEI with
conventional substrates at 1 MHz and 10 GHz. In the microwave
range, polyetherimide has a slightly higher dissipation factor
than PTFE/glass (.0030 vs .0022) but is far lower than FR-4
or polyimide.

PEI's dielectric constant ranges from 3.05 to 3.15, as shown
in Figure 10. This property is unusually stable, varying only 3
percent from 10 Hz to 10 GHz at room temperature. Even less
variation over this frequency occurs at 180°F.

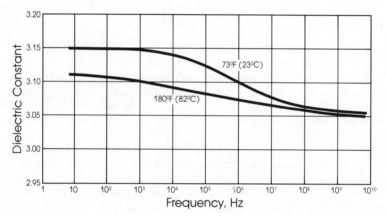

Figure 10. Dielectric constant vs frequency of PEI at 50% RH.

PEI's high continuous use temperature rating of 338°F
(170°C) is important to circuit board designers requiring
high temperature capabilities above the 130°C rating of FR-4;
PEI-molded boards can indeed withstand wave soldering
operations.

Aerospace--The main strengths of polyetherimide for this
segment are flame resistance, low smoke evolution, potential
weight savings vs. metals and other heavier materials, chemical
resistance, and stain resistance. Applications include lighting,
seating, wiring, electrical hardware. Even engine components
comprised of graphite fiber laminates are presently under investigation by various end-users.

160

Other uses--Fibers represent an emerging area for
polyetherimide resin. Applications include protective clothing,
(e.g. for fire fighters, race drivers), dry filtration (fume
bags), aircraft fabrics (upholstery, carpets), and drying
screens (used in papermaking). Additional market areas include
film (dielectric films, insulating tapes), and wire insulation.

CONCLUSION

Polyetherimide is an amorphous thermoplastic characterized
by a high heat deflection temperature, excellent mechanical
strength, good chemical resistance, and stable electrical
properties over a wide range of temperature and frequency. The
flame resistance and UV stability of PEI are excellent without
the addition of flame retardants or stabilizers. Polyetherimide
exhibits this exceptional property balance without sacrificing
processability, providing a material capable of meeting the difficult
design requirements of many applications in the electrical/electronic,
transportation and industrial sectors.

REFERENCES

1. J.G. Wirth and D.R. Heath, U.S. Patent 3,730,946;
    May 1, 1973.
2. T. Takekoshi, J.G. Wirth, D.R. Heath, J.E. Kochanowski,
    J.S. Manello, and M.J. Webber, paper presented at the
    177th ACS Meeting, Honolulu, Hawaii, April 1-6, 1979.
3. F.J. Williams and P.E. Donahue, J. Org. Chem., 42, 3414
    (1977).
4. F.J. Williams, H.M. Relles, P.E. Donahue, and J.S. Manello,
    J. Org. Chem., 42, 3425 (1977).
5. F.J. Williams, H.M. Relles, J.S. Manello, and P.E. Donahue,
    J. Org. Chem., 42, 3419 (1977).
6. R.O. Johnson and E.O. Teutsch, Polymer Composites, 4, 162
    (1983).
7. D.E. Floryan and G.L. Nelson, J. Fire Flammability, 11, 284
    (1980).

THE SYNTHESIS OF POLYAMIC-ACIDS WITH CONTROLLED MOLECULAR WEIGHTS

W. Volksen and P. M. Cotts

IBM Research Laboratory
San Jose, California 95193

A series of polyamic-acid polymers with varying molecular weight have been prepared by condensation of pyromellitic dianhydride (PMDA) and oxydianiline (ODA) in N-methylpyrrolidone (NMP). Polyamic-acid solutions between 10 - 20 wt% total solids content were prepared under highly controlled conditions.

Molecular weights, which were determined by light scattering, were in good agreement with those calculated from the monomer stoichiometric imbalance according to classical step growth behavior. This behavior held for polymerizations carried out at solids contents of less than 10 wt% and was unaffected by the monomer addition sequence, provided the functionality used in excess was initially present as a homogeneous solution. At higher solids contents, noticeable deviation from the theoretically calculated molecular weights were observed. Possible mechanisms for this behavior are discussed along with ways to remedy this problem.

INTRODUCTION

Polyimides are classically prepared by a two-stage synthesis, which involves solution condensation of an aromatic dianhydride with an aromatic diamine to yield an intermediate polyamic-acid and subsequent transformation to the polyimide via chemical or thermal dehydration[1,2]. Synthetically, the initial stage involving formation of the polyamic-acid is the most critical. Thus, the molecular weight of the polyamic-acid can vary significantly depending on the monomer addition sequence and the solution viscosity can decrease markedly with time. This latter behavior has been attributed to hydrolytic effects as a result of trace amounts of water present in the highly polar solvents employed and/or produced by partial imidization of the polyamic-acid[3-5].

Since the reproducible preparation of polyamic-acids with controlled molecular weights is of extreme importance with respect to processing considerations, we have investigated the effect of carefully controlled reaction conditions, i.e., exclusion of atmospheric moisture and oxygen as well as rigorous purification of the starting materials, on the ability to synthesize polyamic-acids with controlled molecular weights.

EXPERIMENTAL

N-methylpyrrolidone (NMP) purchased from Burdick & Jackson was additionally distilled at reduced pressure and Argon atmosphere from P_2O_5. Pyromellitic dianhydride (PMDA) was obtained by dehydration and sublimation of twice recrystallized pyromellitic acid, followed by zone refining. Technical grade oxydianiline (ODA) was sublimed and zone refined.

All polymerizations were performed in a glove box under a Helium atmosphere, using a jacketed resin kettle equipped with mechanical stirrer, thermocouple, and in some cases a temperature controlled addition funnel. Polymerizations were conducted at 20 ± 5 °C using liquid nitrogen as a coolant.

Number average molecular weights were calculated according to classical step growth behavior. Thus, the degree of polymerization, $\overline{X}_n$, was calculated according to Equation (1):

$$\overline{X}_n = (1 + r)/(1 - r) \qquad (1)$$

where r is the ratio of the total # of A functionalities over the total # of excess B functionalities. The number average molecular weight, $\overline{M}_n$, is then defined as in Equation (2):

$$\overline{M}_n = \overline{X}_n \times M_0 \qquad (2)$$

where M_0 is the mean of the molecular weights of the two structural units and in the case of PMDA/ODA is 209.

Weight average molecular weights were obtained by low angle light scattering determinations using a Chromatix KMX-6.

RESULTS AND DISCUSSION

Molecular weight control for polyamic-acid preparations, particularly those based on PMDA/ODA, has been an area of great confusion. Generally, this has been attributed to the sensitivity of polyamic-acids toward hydrolytic degradation [3-5] and the unusual solubility characteristics of PMDA in the polymerization medium[6], Therefore, exclusion of moisture from the polymerization reaction and use of relatively low monomer concentrations should culminate in a polyamic-acid synthesis that follows classical step growth behavior. According to step growth kinetics the breadth of the molecular weight distribution, as reflected by the ratio $\overline{M}_w/\overline{M}_n$, should be approximated by a most probable distribution.[7] Thus, the polydispersity of a polymer sample derived from stoichiometrically imbalanced monomer ratios is given by Equation (3):

$$\overline{M}_w/\overline{M}_n = 1 + pr^{1/2} \qquad (3)$$

Assuming that the anhydride/amine reaction proceeds to completion, i.e., $p \rightarrow 1$, the ratio $\overline{M}_w/\overline{M}_n \rightarrow 2$ for small stoichiometric imbalances. Comparison of experimentally determined weight average molecular weights with number average molecular weights calculated from the monomer stoichiometry should readily verify the above behavior.

Table I, which allows comparison of the calculated weight average molecular weights according to the above treatment with those determined experimentally by light scattering, clearly shows this to be the case. It is important to keep in mind that the experimental molecular weights have an inherent uncertainty of ± 10%. In addition, any error in the stoichiometric imbalance as introduced by weighing errors, would be most noticable at higher molecular weights. This may account for the slightly larger deviations at high degrees of polymerization. The fact, that a relatively high molecular weight was obtained for the equivalent monomer stoichiometry, confirms a very high extent of reaction. Thus, a weight average molecular weight of 250,000 corresponds to a reaction extent of 99.8 %.

Table I. Molecular Weight Data for PMDA/ODA Polymerizations at < 10 wt% Total Solids Content.

PMDA/ODA	% Solids	$\bar{M}_w$(calc.)	$\bar{M}_w$(exp.)
0.8182	9.27	4,200	4,500
0.9259	9.55	11,000	10,000
0.9804	9.31	42,000	37,000
0.9921	7.55	115,000	80,000
1.000	3.28	-	250,000

Traditionally, polyamic-acid preparations are performed by addition of powdered dianhydride to a solution of diamine. The primary reasons for this addition mode are the higher solubility of the diamine in the reaction medium and the sensitivity of dianhydride solutions toward hydrolytic degradation. However, in the case of stoichiometrically imbalanced formulations, this addition mode leads to amino-terminated polymer chains, resulting in oxidatively sensitive polyamic-acid solutions. This oxidative sensitivity can be greatly minimized by reversing the addition mode and adding the diamine to a solution of excess dianhydride, yielding anhydride end groups.

Contrary to literature data[3-5], this reversed addition mode has no effect on the ultimate molecular weight of the polyamic-acid solution as illustrated in Table II. It is quite possible that this may be a result of the rigorous exclusion of moisture from the polymerization system. From a practical standpoint, this type of addition mode is only suitable for polyamic-acid formulations with a total solids content of approximately 10 wt% due to the limited solubility of PMDA at ambient temperatures.

In the case of polyamic-acid solutions with molecular weights below 50,000, solids contents greater than 10 wt% may be preferable. As indicated in the preceding discussion, this

Table II. Molecular Weight Data for PMDA/ODA Polymerizations as a Function of Monomer Addition Sequence.

r	% Solids	$\bar{M}_w$(calc.)	$\bar{M}_w$(exp.)
Addition of Solid PMDA to ODA Solution			
0.9259	9.55	10,800	9,000
0.9804	9.31	42,200	37,000
Addition of Solid ODA to PMDA Solution			
0.9259	9.58	10,000	10,500
0.9804	9.32	42,000	35,000

necessitates the use of an addition mode consisting of adding powdered dianhydride to a solution of excess diamine. Upon preparing polyamic-acid solutions of up to 25 wt% total solids, we observed very unusual behavior. Polymerization time to achieve a completely homogeneous system increased from 2 hours (typical at % Solids ≤ 10 wt%) to an average of 24 hours. Solution viscosities initially became very high and then equilibrated to a lower value. Experimental weight average molecular weights are illustrated in Table III and are consistently higher than predicted from the monomer stoichiometry. The effect of monomer concentration on the polymer molecular weight is consistent with literature reports[8], but no further explanation of this phenomenon is reported. Since the solubility behavior of PMDA in the polymerization medium (NMP) appears to be highly influenced by the presence of polyamic-acid [6], we felt it to be advantageous to go to a completely homogeneous reaction system. However, due to the low solubility of PMDA in NMP it was necessary to go to higher temperatures (~50 °C) to achieve dissolution of sufficient PMDA to allow for an overall solids content of approximately 20 wt%. The higher reactivity of the PMDA solution was controlled by going to a lower polymerization temperature of approximately 10°C. Experimental weight average molecular weights obtained for polyamic-acid preparations made in this manner were again in good agreement with those calculated from the stoichiometric imbalance as shown in Table III.

167

Table III. Molecular Weight Data for PMDA/ODA Polymerizations at
> 10 wt% Total Solids Content.

PMDA/ODA	% Solids	$\overline{M}_w$(calc.)	$\overline{M}_w$(exp.)
Addition of Solid PMDA to ODA Solution			
0.8182	25.6	2,000	6,700
0.9231	17.5	10,000	22,000
			(10,000)[a]
0.9725	18.6	30,000	43,000
			(29,000)[b]
Addition of PMDA Solution to ODA Solution			
0.8182	20.0	2,000	2,400
0.9259	20.4	10,000	11,000
0.9717	16.7	28,000	26,000

[a] After aging a 0 °C for 60 days.
[b] After aging at 25 °C for 10 days.

The main differences between the heterogeneous and homogeneous
high solids content polymerizations are the much more efficient
mixing of the monomers and the short residence time of the added
PMDA in the latter case. Thus, possible explanations for the
formation of higher than predicted molecular weights for the
heterogeneous case may be directly linked to high local
dianhydride concentrations. The known tendency of PMDA to act as
a strong charge-transfer complexing agent with the diamine in a
1:1 type complex could account for high molecular weight polymer
formation inspite of a non-equivalent stoichiometry.
Furthermore, the heterogeneous nature of the polymerization could
be controlled by diffusional effects, yielding a
"pseudo-interfacial" polymerization with similar results on the
overall weight average molecular weight. At this point it is
impossible to distinguish between these two mechanisms and their
influence on the polymerization to yield molecular weight
distributions skewed toward the high molecular weight end.

Rather surprising results were obtained when we reexamined
samples that had been prepared heterogeneously at high solids
content after they had aged for a period of time. It was found
that the weight average molecular weight had reequilibrated to a
value consistent with the monomer stoichiometry and was stable at
this molecular weight, see Table III. The existence of an
equilibrium between the amic-acid group and anhydride and diamine
has been proposed in the literature[9] and would provide a
mechanism for a broad molecular weight distribution to
equilibrate to assume a most probable distribution in accordance
with classical step growth kinetics. Furthermore, this
observation questions the previously reported decay in the
concentrated solution viscosity as being a sole result of
hydrolytic molecular weight degradation.

CONCLUSIONS

In summary, the results of the above work indicate classical
step growth behavior for the hydrogen transfer polymerization of
pyromellitic dianhydride and oxydianiline to yield the
corresponding polyamic-acid. This applies both to heterogeneous
polymerizations at low solids content ($\leq$ 10 wt%) irrespective of
the monomer addition sequence and to homogeneous polymerizations
at high solids content (> 10 wt%). Data are presented for the
first time indicating the decay of a skewed molecular weight
distribution to a most probable one, as a result of a
polyamic-acid/anhydride-amine equilibrium. Further work
concerning this particular issue as well as the cause of the
apparently skewed molecular weight distribution is in progress
and should ultimately provide a complete and consistent picture
of the parameters effecting a reproducible synthesis of PMDA/ODA
based polyamic-acids.

ACKNOWLEDGEMENT

The authors wish to express their gratitude to G. Castro for
supplying the zone refined pyromellitic dianhydride and
oxydianiline.

REFERENCES

1. C. E. Sroog, A. L. Endrey, S. V. Abramo, C. E. Berr,
    W. M. Edwards and K. L. Olivier, J. Polym. Sci.,
    A-3, 1373 (1965).

2. C. E. Sroog, J. Polym. Sci., Macromolecular Reviews, 11, 161 (1976).
3. G. M. Bower and L. W. Frost, J. Polym. Sci., A-1, 3135 (1963).
4. R. A. Dine-Hart and W. W. Wright, J. Appl. Polym. Sci., 11, 609 (1967).
5. R. A. Dine-Hart and W. W. Wright, Die Makrom. Chem., 143, 189 (1971).
6. S. A. Zakoshchikov, I. N. Ignat'eva, N. V. Nikolayeva, and K. P. Pomerantseva, Vysokomol. Soyed., A11, No. 11, 2487 (1969).
7. P. J. Flory, "Principles of Polymer Chemistry," p. 323, Cornell University Press, Ithaca and London, 1953.
8. T. M. Birshtein, V. A. Zubkov, I. S. Milevskaya, V. E. Eskin, I. A. Baranovskaya, M. M. Koton, V. V. Kudryavtzev, and V. P. Sklizkova, Eur. Polym. J., 13, 375 (1977).
9. J. B. Tingle and H. F. Rolker, J. Amer. Chem. Soc., 30, 1882 (1908).

INVESTIGATION OF THE REACTIVITY OF AROMATIC DIAMINES AND DIANHYDRIDES OF TETRACARBOXYLIC ACIDS IN THE SYNTHESIS OF POLYAMIC ACIDS

M. M. Koton, V. V. Kudriavtsev and V. M. Svetlichny

Institute of High Molecular Compounds of the
Academy of Sciences of the USSR
Leningrad, USSR

The kinetics of acylation of aromatic diamines by dianhydrides of aromatic tetracarboxylic acids in amide solvents have been investigated.

The reaction rate constants for various dianhydrides are compared with their electron acceptor properties. It is shown that the reaction rate increases regularly with increasing affinity for the electron of the dianhydride.

The reaction rate constant for various diamines is compared with their electron donor and basic properties. It is shown that a linear relationship exists between the pK_a values for diamines and the logarithms of rate constants. The correlation between these logarithms and the ionization potentials for diamines is less marked.

For polyamic acids based on different dianhydrides the isomerism of the position of amide and carboxyl groups at the dianhydride radical was investigated. The ratio of units with para- and meta-positions of amide bonds formed during acylation is determined experimentally. It is shown that for binuclear bridging aromatic dianhydrides this ratio increases with increasing electron – acceptor character of the bridge.

The dependences presented here are explained by using the parameters of the electronic structure of dianhydrides and diamines calculated by the methods of quantum chemistry.

INTRODUCTION

At present several tens of dianhydrides of tetracarboxylic acids and an even greater number of aliphatic, aromatic and heterocyclic diamines are being used in the synthesis of polyimides. In this connection it is of interest to study the relative reactivity of monomers in polyacylation reactions on which various methods of synthesis of polyimides are based.

The present paper is concerned with some results of our investigations on the kinetics of acylation of aromatic diamines by acid dianhydrides and the determination of isomeric composition of polyamic acids formed during acylation.

The kinetic data are correlated with the characteristics of electron acceptor properties of dianhydrides and electron donor and basic properties of diamines. The totality of data presented here suggests that in the synthesis of polyamic acids, relationships exist between the chemical structure of the monomers and their relative reactivities.

It should be noted that the investigation of the kinetics of synthesis of polyamic acids started in 1964[1] in which the interaction between pyromellitic dianhydride and m-phenylene diamine was considered. In many papers published on this subject[2-15] the chemical structure of both components and the reaction media have been varied widely.

EXPERIMENTAL

In most studies kinetic investigations have been carried out by IR spectroscopy by following a decrease in the concentration of anhydride functional groups. It should be borne in mind that IR spectroscopy may be used[16] if the extinction coefficients of anhydride rings in the initial monomers and intermediate polyacylation products (such as anhydride functional groups of oligomers) do not differ greatly. In this respect a convenient monomer for kinetic investigations is a dianhydride of 3,3¦4,4' -tetracarboxybenzophenone. Calorimetric methods have also found extensive usage in the investigation of polyacylation kinetics.

Poly(o-carboxamides) are known to undergo reversible degradation. However, taking into account a high value of the equilibrium constant of acylation[17-20] it is permissible to calculate the rate constant (k) according to the equation for an irreversible second-order reaction.

We have investigated the acylation kinetics by IR spectroscopy. The methodological problems have been considered earlier[5,6]. Dried amide solvents were used. In the case of acylation of monoamines by dianhydrides of 3,3',4,4'-tetracarboxydiphenyl and 3,3',4,4' -tetracarboxydiphenyloxide a slight retardation of reaction is observed after 50% of functional anhydride groups have reacted. A slight increase in reaction rate is observed for the dianhydride of 3,3',4,4'-tetracarboxybenzophenone. In pyromellitic dianhydride, both anhydride groups could not be distinguished kinetically. In the case of acylation of diamines with two benzene rings the reaction rate decreases slightly after 50% conversion has been attained. These observations were taken into account in plotting the dependences by using the rate constants calculated up to the conversion of functional groups of about 50-60%. The problem of the change in the reactivity of second functional groups in diamines and dianhydrides occurring after the first groups have reacted has been considered in greater detail elsewhere[21].

RESULTS AND DISCUSSION

The relative reactivities of various dianhydrides may be determined on the basis of the data presented in Table I in which the rate constants are listed for the acylation of two amines: 4-aminodiphenyl ether (ADPhE) and 4,4'-diaminodiphenyl ether (DADPhE) in N,N-dimethylformamide (DMF) at 25°C. In the series of aromatic dianhydrides investigated, the differences in rate constants are within two orders of magnitude. The values of constants (Table I) are close to those found for similar systems[7,9].

It is noteworthy that the reactivity of dianhydrides of aliphatic and alicyclic tetracarboxylic acids is much lower than that of aromatic tetracarboxylic acids. The rate constants for the acylation by dianhydrides of 1,2,3,4-butane tetracarboxylic[2], 1,2,3,4-cyclopentane tetracarboxylic[5], tricyclodecene tetracarboxylic[10] and 1,2,3,4-cyclohexane tetracarboxylic[13] acids are lower by two or three orders of magnitude than those for the acylation by pyromellitic dianhydride.

The chemical structure of dianhydrides affects their acceptor properties. The study of charge transfer complexes formed by aromatic dianhydrides with N,N,N',N'-tetramethyl-p-phenylene diamine in tetrahydrofuran at -90° has shown[22] that the position of charge-transfer bands (Table I) in the electronic spectra of complexes distinctly corresponds to the effect of bridge bonds on the acceptor properties of dianhydrides. The spectra of charge transfer complexes were used to evaluate the values, E_a, of the affinity of dianhydrides for an electron. The value of E_a increases with the increasing acceptor character of the bridge.

Table I. Reactivity of Dianhydrides of Aromatic Tetracarboxylic Acids.

No.	Dianhydride	$k \cdot dm^3$ mol^{-1} s^{-1}		ν ct * cm^{-1}	E_a, eV
		ADPhE	DADPhE		
1		2.32	9.0	15000,21000	0.85
2	SO$_2$ bridged dianhydride	6.72	11.1	18000,24000	0.52
3	CO bridged dianhydride	1.48	4.6	18500,24000	0.48
4	C=O–O–...–O–C=O bridged dianhydride	1.33	2.12	19000,25000	0.43
5	biphenyl dianhydride	0.47	1.34	21000–21500	0.21
6	O bridged dianhydride	0.35	0.88	21500	0.18
7	O...O bridged dianhydride	0.21	0.48	22000	0.14
8	SO$_2$, O linked dianhydride	0.43	0.91	22500–23000	0.07
9	O–(CH$_2$)$_4$–O dianhydride	0.05	0.16	23000–23500	0.02

* νct is the maximum of the charge transfer band.

174

A comparison of the E_a values and the acylation rates shows (Figure 1) that these characteristics exhibit a pronounced correlation: the higher the E_a value, the faster the acylation. Hence, the effect of the acceptor properties of dianhydrides on the progress of acylation was proved experimentally.

Some characteristics of the electronic structure of molecules reflect their acceptor properties. In this case they include, first, the energy of the lower vacant orbital, E_{1v}, of the dianhydride. This is due to the fact that in the formation of chemical bonds during acylation the interaction between the upper occupied molecular orbital (UOMO) of amine on which the lone pair of electrons is located and the lower vacant molecular orbital (LVMO) of the anhydride is of great importance because their energies are closest to each other (Figure 2).

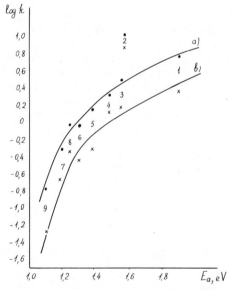

Figure 1. Relationship between the logarithm of the acylation rate constant of a) 4,4' diamino diphenyl ether and b) 4-amino-diphenyl ether on the one hand and the affinity E_a of dianhydrides for an electron on the other. Numeration of dianhydrides is the same as in Table I.

175

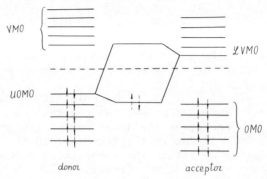

VMO

UOMO

donor

ℒVMO

OMO

acceptor

Figure 2. Interaction between two overlapping orbitals: UOMO and LVMO.

In fact, the quantum chemical calculation of E_{1v} for a series of dianhydrides has made it possible to establish a close correlation between E_{1v} and the experimental values of E_a and the log k[23]. Figure 3 shows that E_{1v} is proportional to the affinity E_a of dianhydride for an electron. Consequently, the value of E_a should be regarded as the measure of the acylating ability of anhydrides.

The values of E_a calculated from the results of polarographic reduction are reported[21-24] for a large series of bis-phthalic anhydrides of the same type. These values distinctly correspond to the structural changes in dianhydride molecules and may be used to compare their reactivities in acylation. Moreover, it should be borne in mind that the values of E_a for the same dianhydride (as those of pK_a for diamine) obtained by different methods may differ. This difference is due to calibration procedures and is not of major importance.

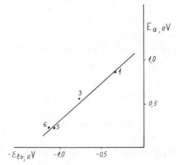

Figure 3. Relationship between the calculated energy of the lower vacant orbital of dianhydries ($-E_{1v}$) and the affinity, E_a, of dianhydrides for an electron. Numeration of dianhydrides is the same as in Table I.

According to Svetlichny et al.[4] the well-known Hammet's equation may be used for quantitative evaluation of the effect of structural changes in the molecules of aromatic diamines on the acylation rate

$$\log k = \log k_o + \rho\sigma$$

where k and k_o are the rate constants for the acylation of substituted and unsubstituted aniline, respectively, ρ is the constant for the reaction series and σ is the constant for the substituent. The parameters of correlation equations for the acylation of aniline and its p-Br, p-CH$_3$, p-OC$_6$H$_5$ and m-NO$_2$ derivates by dianhydrides of pyromellitic acid (PMAD) and 3,3',4,4'-tetracarboxybenzophenone (BPhAD) are given in Table II.

Table II. Correlation of Reactivities of Substituted Anilines in the Acylation of Amines by PMAD and BPhAD-Dianhydrides.

Dianhydride	Solvent	$\log k_o$	$-\rho$	r correlation coefficient
PMAD	DMF	0,24 ± 0,04	3,4 ± 0.1	0,998
BPhAD	DMA	0,19 ± 0,04	3,4 ± 0,1	0,998
BPhAD	DMF	0,032 ± 0,02	3,32 ± 0,06	0,999
BPhAD	MP	-0,22 ± 0,08	3,0 ± 0,2	0,992

It is noteworthy that acylation rates (Table II) increase with the solvents in the following order: N-methyl-pyrrolidone (MP) < N,N-dimethylformamide (DMF) < N,N-dimethylacetimide (DMA).

The carboxylic groups of polyamic acid are potential catalysts for acylation. However, in DMF the autocatalysis is negligible and in DMA and MP it is not observed at all.

The negative sign of the constant ρ (Table II) shows that the reaction is accelerated by electron donor substituents, whereas the high absolute value of ρ indicates that the reaction is sensitive to the structural changes in the amine molecule. In fact, the investigation of the kinetics of acylation of a great[4] number of aromatic diamines by PMAD and BPhAD in DMF has shown that the reaction rate constants vary by five orders of magnitude (Table III).

Aromatic diamines are represented as aniline derivatives, $X-C_6H_4-NH_2$, with complex substituent X. Since $\log k_o$ and ρ are known (Table II) and $\log k$ is determined, it is possible to calculate σ, the constants of complex substituents[4]. The constants σ listed in Table III have the meaning of nucleophilic constants.

The data given in Table III show that the values of constants σ for complex substituents calculated from two reaction series are relatively close to each other. The values of constants σ for complex substituents corresponding to two-ring bridge aromatic diamines obtained separately from kinetic experiments according to pK_a values for diamines in nitromethane are reported in Ref. 25. The constants σ determined in Refs. 4 and 25 differ by the same value (equal to about 0.09), which corresponds to the $\log 2/\rho$ ratio. Hence, it may be concluded that the constants σ determined by different methods are in fairly good agreement. The applicability of Hammet's equation to the acylation of aromatic diamines by acid dianhydrides allows the approximate calculation of reaction rates.

It is of interest to correlate the acylation rates of aromatic diamines with their electron donor and basic properties.

The chemical structure of diamines affects their donor properties, in particular, the value of ionization potentials I[26]. Figure 4 shows the dependence of the constant k for the acylation of a series of aromatic diamines by pyromellitic anhydrides on the value of I. With increasing I, i.e. with weakening of electron donor properties of diamine, acylation rate decreases monotonically. A considerable scattering of points in Figure 4 shows, however, that the correlation of the acylation rate constant with the ionization potential of diamines is less pronounced than that with the affinity of dianhydrides for an electron (Figure 1).

The values of pK_a for diamines characterize their basic properties. The value of $\log k$ decreases in the same manner as I (Fig. 4), whereas a direct proportionality is observed between $\log k$ and pK_a (Fig. 5). A comparison between the values of pK_a for diamines and the σ-constants for substituents calculated from kinetic data leads to the following linear relationship

$$pK_a - \log 2 = (4.69 \pm 0,03) - 3.04\sigma$$

at the correlation coefficient r = 0.989.

Table III. Rate Constants k for Acylation of Diamines by PMAD and BPhAD Dianhydrides in DMF at 25°C and σ Constants for Complex Substituents X in Diamine Molecules of the $XC_6H_4NH_2$ Type.

No.	Complex substituent	PMAD		BPhAD	
		$dm^3 \cdot mol \cdot s^{-1}$	σ	$dm^3 \cdot mol \cdot s^{-1}$	σ
1	$p-H_2N-C_6H_4-$	2.33	0.051	1.34	0.062
2	$p-H_2N-C_6H_4O-$	5.97	-0.069	3.09	-0.047
3	$p-H_2N-(C_6H_4O)_2-$	2.87	0.025	1.83	0.021
4	$p-H_2N-C_6H_4NHOCC_6H_4CONH-$	4.46	-0.032	2.57	-0.023
5	$p-H_2N-C_6H_4CH_2-$	3.59	-0.004	2.27	-0.007
6	$p-H_2N-C_6H_4S-$	0.560	0.234	0.308	0.254
7	$p-H_2N-C_6H_4SO_2-$	0.0022	0.946	0.0023	0.898
8	$p-H_2N-C_6H_4P(O)C_6H_5-$	$3.4 \cdot 10^{-4}$	1.18	$1.78 \cdot 10^{-4}$	1.13

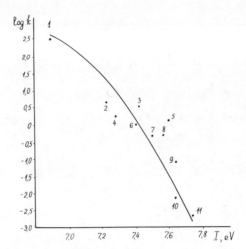

Figure 4. Relationship between the rate constant for acylation of diamines by pyromellitic dianhydride and ionization potential I of diamines (1-11).

Numeration of diamines (for Fig. 4):

1. $H_2N - \langle \rangle - NH_2$

2. $H_2N - \langle \rangle - O - \langle \rangle - NH_2$

3. $H_2N - \langle \rangle - O - \langle \rangle - O - \langle \rangle - O - \langle \rangle - NH_2$

4. $H_2N - \langle \rangle - \langle \rangle - NH_2$

5. $H_2N - \langle \rangle - O - \langle \rangle - \langle \rangle - O - \langle \rangle - NH_2$

6. $H_2N - \langle \rangle - NH_2$

7. $H_2N - \langle \rangle - S - \langle \rangle - NH_2$

8. $H_2N - \langle \rangle - O - \langle \rangle - SO_2 - \langle \rangle - O - \langle \rangle - NH_2$

9. $H_2N - \langle \rangle - O - \langle \rangle - SO_2 - \langle \rangle - O - \langle \rangle - NH_2$

180

Numeration of diamines (for Fig. 4) (cont.):

10. H_2N—⟨_⟩—CO—⟨_⟩—NH_2

11. H_2N—⟨_⟩—SO_2—⟨_⟩—NH_2

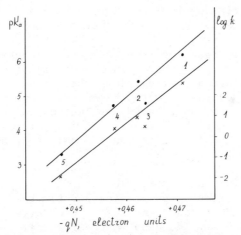

Figure 5. Relationship between the experimental values of log k, the rate of acylation of diamines by pyromellitic anhydrides and pK_a and the calculated values, q_N, of charges on the nitrogen atoms of diamines. Points refer to pK_a and the crosses correspond to log k. Numeration of diamines:

1. H_2N—⟨_⟩—NH_2

2. H_2N—⟨_⟩—O—⟨_⟩—NH_2

3. H_2N—⟨_⟩—NH_2

4. H_2N—⟨_⟩—⟨_⟩—NH_2

5. H_2N—⟨_⟩—CO—⟨_⟩—NH_2

When diamines are varied (in contrast to the variation of dianhydrides), the reaction rate is also affected by changes in the type of electrostatic interaction between the reagents which are also caused by the chemical structure of diamines. In fact, the quantum chemical calculations for a series of aromatic diamines[27] have shown that a close correlation exists between the charges on nitrogen atoms of diamines, q_N, and log k for acylation (Fig. 5). The negatively charged nitrogen atom having a lone-pair of electrons is the natural location for the protonation of amine. This may account for the existing linear relationship between q_N and pK_a and, hence, for the aforementioned linear relationship between pK_a and log k. The values of E_{uo} – the energy of the upper occupied molecular orbital of diamines – characterizing their donor properties correlate with log k but this correlation is not as good as that of q_N with log k.

This acylation reaction is a case of nucleophilic substitution at the carbonyl carbon atom in which amine plays the part of an electron donor and anhydride plays that of an electron acceptor. However, in accordance with the correlation dependences shown in Fig. 5, when quantitative evaluations of the reactivity of various diamines in the formation of polyamic acids are carried out, it is advisable to use the values of pK_a reflecting their basic properties.

The reasons for the correlation of reactivity of dianhydrides with their acceptor properties and for that of reactivity of diamines with their donor and basic properties have been considered in detail[27-30]. The quantum chemical calculations of components of the energy of intermolecular interaction between model amines and anhydrides in the initial stage of acylation (at the inter-plane distance of molecules of 0.24 nm) have shown[27-30] that for anhydrides differing in structure (at a fixed amine) the reaction is controlled by orbital and, possibly, induction interactions. The changes in the charge distribution do not correlate with the acylation rate constant. In the case of different diamines (at a fixed anhydride) the reaction is controlled by charge and orbital interactions. Moreover, the changes in the energy of Coulombic interaction are responsible for the variations in reaction rates to a greater extent than those in the charge-transfer energy. These factors explain the reasons for the correlations observed.

As a result of nucleophilic attack of the amino function at various carbonyl atoms of the dianhydride ring (C^2 or C^5, $C^{2'}$ or

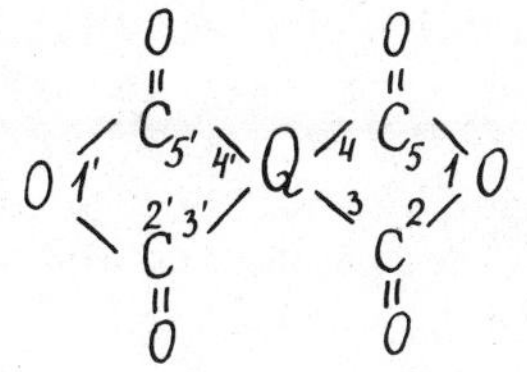

$$O \overset{1'}{\underset{\overset{\parallel}{O}}{\overset{C}{\diagdown}}} \overset{O}{\overset{\parallel}{\underset{5'}{C}}} \overset{4'}{\underset{3'}{\diagup}} Q \overset{4}{\underset{3}{\diagdown}} \overset{C}{\underset{\underset{\overset{\parallel}{O}}{C}}{\overset{5}{\diagup}}} \overset{O}{\overset{\parallel}{}} O$$

$c^{5'}$) during polyacylation isomeric units may be formed with the para- and meta- positions of amide bonds.

If the radical Q contains a bridge group X, units with the following three isomeric structures may be formed:

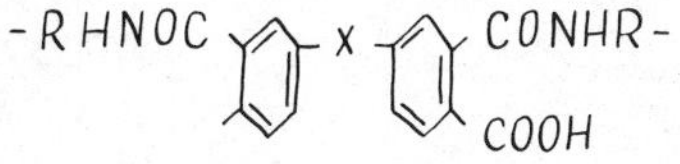

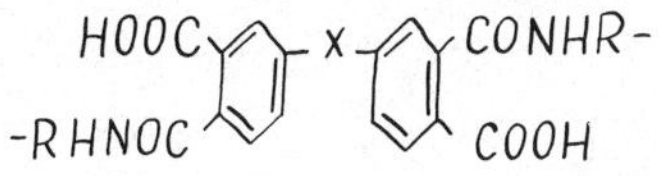

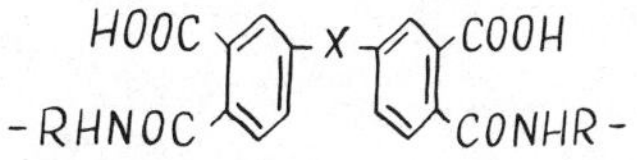

In the case of pyromellitic anhydride, units with the following two isomeric structures are formed:

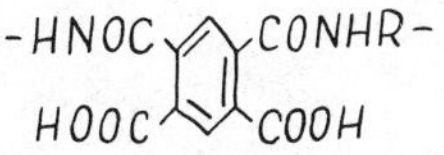

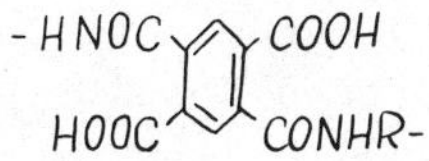

The ratio of para- and meta-isomeric units may be determined by ^{13}C NMR[31,32]. ^{13}C NMR spectra of polyamic acids obtained by polycondensation of dianhydrides of pyromellitic, 3,4,3',4'-benzophenonetetracarboxylic and 3,4,3',4'-diphenyloxide tetracarboxylic

acids with p-phenylene diamine and benzidine have been completely
interpreted[32]. The results of the determination of isomeric units
are given in Table IV. It was found that the isomeric composition
of polyamic acids is independent of the nature of diamine and is
determined by the dianhydride component.

Table IV. Isomeric Composition of Polyamic Acids as Determined
by ^{13}C NMR Spectroscopy.

Dianhydride	Fraction of units		f_{para}	$-f_{meta}$
	para-	meta-	di-anhydride	- MAD
(structure)	0.40	0.60	0.000	-0.018
(structure)	0.37	0.63	-0.043	-0.053
(structure)	0.50	0.50	-0.008	0.067
(structure)	0.55	0.45	0.037	0.067

The quantum chemical interpretation of isomerism of polyamic
acids has been considered[23,30] in Refs. 21, 23, 30 and 33. The quantum
chemical calculation has shown that the isomeric composition
is in qualitative agreement with the limiting density f_r of the
electron in the carbonyl carbon atom in the lowest vacant
orbital. The values of the f_2/f_5 (or f_{para}/f_{meta}) ratios and the
differences $f_2 - f_5$ (or $f_{para} - f_{meta}$) indicate the relative
preference of nucleophilic attack on the carbonyl atom C^2 ($C^{2'}$)
or C^5($C^{5'}$). The values of $f_{para} - f_{meta}$ have also been calculated for the products of the opening of one functional group of
dianhydrides by ammonia (MAD)[21].

$$
\begin{array}{c}
\quad\quad O \quad\quad\quad O \\
\quad\quad \| \quad\quad\quad \| \\
O\!\!\diagup\!\!\overset{C}{\diagdown}\!\diagdown_O\!\diagup\!\overset{C}{\diagup}\!-OH \\
\quad\diagdown\!\overset{C}{\diagup}\!\diagup\overset{Q}{\diagdown}\!\diagdown C - NH_2 \\
\quad\quad \| \quad\quad\quad \| \\
\quad\quad O \quad\quad\quad O
\end{array}
$$

MAD

These compounds represent the electronic structure of functional anhydride groups of the growing chains of polyamic acids.

As can be seen in Table IV, the ratio of para- and meta-isomers in polyamic acid chain changes in the same direction as the value of $f_{para} - f_{meta}$ on passing from one dianhydride to the other. When the electron acceptor character of the bridge in dianhydride increases, the content of amine groups in the meta-position in the bridge group of the polymer decreases. The data obtained for pyromellitic dianhydride (Table IV, line 1) suggest that the differences ($f_{para} - f_{meta}$) referring to MAD are in better agreement with the isomeric composition of polyamic acids than those for dianhydrides. It is quite probable that the isomeric composition of polyamic acids is determined by the specific features of interaction between functional end groups.

SUMMARY

In this paper we have summarized the main lines of approach used in the last decade for the investigation of the relationships between the chemical structure of dianhydrides and diamines and the kinetics of acylation on the one hand and the physical characteristics of monomers reflecting their donor-acceptor and basic properties and the characteristics of the electronic structure of their molecules on the other. In our opinion these relationships are very useful for understanding the mechanism for the reactions involved in polyimide formation.

REFERENCES

1. W. Wrasidlo, P. M. Hergenrother and H. H. Levine, Amer. Chem. Soc., Polymer Preprints, 5, No. 1, 141 (1964).
2. K. K. Kalninsh, E. F. Fedorova, I. V. Novozhilova, B. C. Belenkii and M. M. Koton, Dokl. Akad. Nauk SSSR, 195, No. 2, 364 (1970).
3. V. D. Moiseev, N. G. Avetisian, A. G. Chernova and A. A. Atrushkevich, Plast. massy, No. 3, 12 (1971).
4. V. M. Svetlichny, V. V. Kudriavtsev, N. A. Adrova and M. M. Koton, Zhur. Organ. Khim., 10, No. 9, 1896 (1979).

5. M. M. Koton, V. V. Kudriavtsev, N. A. Adrova, K. K. Kalninsh, A. M. Dubnova and V. M. Svetlichny, Vysokomol. Soedin., A16, No. 9, 2081 (1974).

6. B. A. Zhubanov, P. E. Messerle, V. A. Solomin and S. R. Rafikov, Dokl. Akad. Nauk SSSR, 220, No. 2, 362 (1975).

7. Ya. S. Vygodsky, M. N. Spirina, P. P. Nechaev, L. I. Chudinova, C. E. Zaikov, V. V. Korshak and S. V. Vinogradova, Vysokomol. Soedin., A19, No. 7, 1516 (1977).

8. V. A. Solomin, P. E. Messerle and B. A. Zhubanov, Dokl. Akad. Nauk SSSR, 234, No. 2, 397 (1977).

9. S. V. Vinogradova and Ya. S. Vygodsky, Faserforschung und Textiltechnik, Zeitschrift fur Polymerforschung, 28, No. 11/12, 581 (1977).

10. B. A. Zhubanov, P. E. Messerle and V. A. Solomin, Vysokomol. Soedin., B17, No. 7, 560 (1975).

11. B. A. Zhubanov, V. A. Solomin, P. E. Messerle, N. G. Avetisian and V. D. Moiseev, Vysokomol. Soedin., A19, No. 11, 2500 (1977).

12. V. A. Solomin, I. E. Kardash, Yu. S. Snagovsky, P. E. Messerle, B. A. Zhubanov and N. N. Pravednikov, Dokl. Akad. Nauk SSSR, 236, No. 1, 139 (1977).

13. A. I. Volozhin, E. T. Krutko, A. M. Shishko and Ya. M. Paushkin, Izv. Akad. Nauk SSSR, ser. khim., No. 2, 40 (1978).

14. M. M. Koton, V. M. Svetlichny, V. V. Kudriavtsev, V. E. Smirnova, T. A. Maricheva, E. P. Aleksandrova, G. S. Mironov, V. A. Ustinov and Yu. A. Moskvichev, Vysokomol. Soedin., A22, No. 5, 1058 (1980).

15. B. F. Malichenko and A. E. Borodin, Vysokomol. Soedin., B21, No. 4, 356, (1979).

16. A. Ya. Ardashnikov, M. E. Kardash and A. N. Pravednikov, Vysokomol. Soedin., B19, No. 12, 195 (1977).

17. A. Ya. Ardashnikov, I. E. Kardash and A. N. Pravednikov, Vysokomol. Seodin., A13, No. 8, 1863 (1971).

18. N. V. Kariakin, N. G. Bazhan, V. E. Sapozhnikov, K. G. Shvetsova, G. L. Berestneva, A. N. Lomteva, Ya. B. Zimin and V. V. Korshak, Vysokomol. Soedin., A19, No. 7, 1541 (1977).

19. N. V. Kariakin, I. B. Rabinovich and N. G. Paltseva, Vysokomol. Seodin., A20, No. 9, 2025 (1978).

20. A. I. Volozhin, I. I. Globa, A. M. Shishko and Ya. M. Paushkin, Dokl. Akad. Nauk SSSR, 237, No. 6, 1365 (1977).

21. V. V. Kudriavtsev, M. M. Koton, V. M. Svetlichny and W. H. Subkov, Plaste und Kautschuk, 28, No. 11, 601 (1981).

22. V. M. Svetlichny, K. K. Kalninsh, V. V. Kudriavtsev and M. M. Koton, Dokl. Akad. Nauk SSSR, 237, No. 3, 612 (1977).

23. V. A. Zubkov, V. M. Svetlichny, V. V. Kudriavtsev, K. K. Kalninsh and M. M. Koton, Dokl. Akad. Nauk SSSR, 240, No. 9, 862 (1978).

24. D. V. Pebalk, B. V. Kotov, O. Ya. Neiland, I. V. Mazere, B. Zh. Tilika and A. N. Pravednikov, Dokl. Akad. Nauk SSSR, 236, No 6, 1379 (1977).

25. B. A. Korolev, Z. B. Gerashchenko and Ya. S. Vygodsky, in coll. "Reaktsionnaya sposobnost organicheskikh soedinenii", Tartu, Vol. 8, No. 3, p. 681 (1971).

26. K. K. Kalninsh, G. I. Solovieva, B. G. Belenkii, V. V. Kudriavtsev and M. M. Koton, Dokl. Akad. Nauk SSSR, 204, No. 9, 876 (1972).

27. V. A. Zubkov, M. M. Koton, V. V. Kudriavtsev and V. M. Svetlichny, Zhur. Organ. Khim., 17, No. 8, 1682 (1981).

28. V. A. Zubkov, V. V. Kudriavtsev and M. M. Koton, Zhur. Organ. Khim., 15, No. 5, 1009 (1969).

29. V. V. Korshak, V. A. Kosobutsky, A. I. Bolduzev, A. Ya. Rusanov, V. K. Beliakov, I. B. Dorofeeva, A. M. Berlin and F. I. Adyrkhaeva, Izv. Akad. Nauk SSSR, ser. khim., No. 7, 1553 (1980).

30. V. A. Zubkov, M. M. Koton, V. V. Kudriavtsev and V. M. Svetlichny, Zhur. Organ. Khim., 16, No. 12, 2486 (1980).

31. S. G. Alekseeva, S. V. Vinogradova, V. D. Vorobieva, Ya. S. Vygodsky, V. V. Korshak, I. Ya. Slonim, T. I. Spirina, Ya. G. Urman and L. I. Chudina, Vysokomol. Seodin., B18, No. 11, 803 (1976).

32. V. M. Denisov, V. M. Svetlichny, V. A. Gindin, V. A. Zubkov, A. I. Koltsov, M. M. Koton and V. V. Kudriavtsev, Vysokomol. Soedin., A21, No. 7, 1498 (1978).

33. V. V. Korshak, V. A. Kosobutsky, A. I. Rusanov, V. K. Beliakov, A. N. Gusarov, A. I. Bolduzev and I. Batirov, Vysokomol. Soedin., A22, No. 9, 1931 (1980).

SYNTHESIS OF TIN-CONTAINING POLYIMIDE FILMS

Stephen A. Ezzell and Larry T. Taylor*

Department of Chemistry
Virginia Polytechnic Institute and State University
Blacksburg, Virginia 24061

A series of tin-containing polyimide films derived from either 3,3',4,4'-benzophenone tetracarboxylic acid dianhydride or pyromellitic dianhydride and 4,4'-oxydianiline have been synthesized and their electrical properties examined. Highest quality materials (i.e. homogeneous, smooth surface, flexible) with the best electrical properties were doped with either $SnCl_2 \cdot 2H_2O$ or $(n-Bu)_2SnCl_2$. In all cases, extensive reactivity of the tin dopant with water, air or polyamic acid during imidization is observed. Lowered electrical surface resistivities appear to be correlatable with the presence of surface tin oxide on the film surface.

INTRODUCTION

For several years our laboratory has searched to find metal complex/polyimide combinations which will offer a high temperature resistant material with semi-conductive or conductive properties. Traditionally the approach has been to study the effect of a variety of compounds of a single element on a few select polyimide systems. Lithium,[1] copper[2] and palladium[3] have previously been investigated in this manner. A wide variety of materials were synthesized in these studies. The nature of the polyimide films that were produced depended on the chemical nature of the additive. Migration of the additive to the surface of one side of the film was typical for most polymer-additive formulations during thermal imidization. Carefully chosen additives which possess relatively low volatility and high solubility in the polymerization solvent (N,N-dimethylacetamide) are required. In our studies todate, the integrity of the additive is seldom maintained during conversion from the polyamic acid to the polyimide stage. Partial or complete reduction of palladium(II) to the metallic state is observed; while, hydrolysis followed by dehydration is the rule for most lithium(I) and copper(II) additives.

Previous study by other workers in the area of polymer modification have examined the effect of chemically incorporating tin complexes into polyethyleneimine, polyvinyl alcohol and polyacrylic acid. A lowering, in certain instances, of the electrical resistivity of the polymer into the semiconductor range was observed.[4] Another study has briefly examined $SnCl_2 \cdot 2H_2O$ as a dopant for polyimides. This preliminary study found that incorporation of $SnCl_2 \cdot 2H_2O$ into the polyimide matrix resulted in a slight modification of thermal stability and electrical resistivity of the polymer.[5] This investigation has now been extended to include a number of other tin-containing dopants. A fair sampling of different tin structural types has been possible due to the solubility of numerous tin compounds in the N,N-dimethylacetamide polyamic acid solution. The following report is an account of the modification of polymer properties attained through the incorporation of various tin complexes into several polyimide systems.

EXPERIMENTAL

Starting Materials

3,3´,4,4´-benzophenone tetracarboxylic acid dianhydride (BTDA), I, pyromellitic dianhydride (PMDA), II, and 4,4´-oxy-bis[aniline] (ODA), III, were all obtained from commercial sources and purified by sublimation at 200–230°C and less than 1 torr pressure. N,N-Dimethylacetamide (DMAC) was obtained from Burdick and Jackson as reagent grade, distilled in glass, N_2 packed and was used as-received. All tin compounds were secured from commercial sources and used as-received.

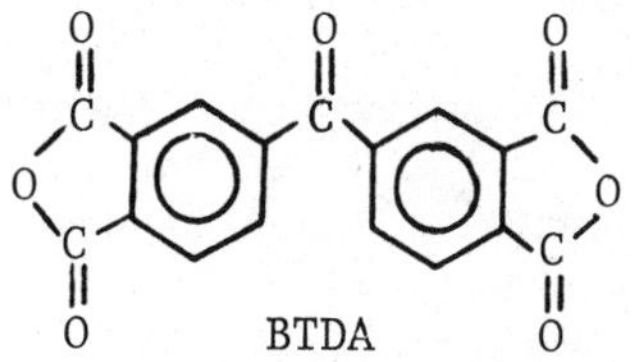

BTDA
I

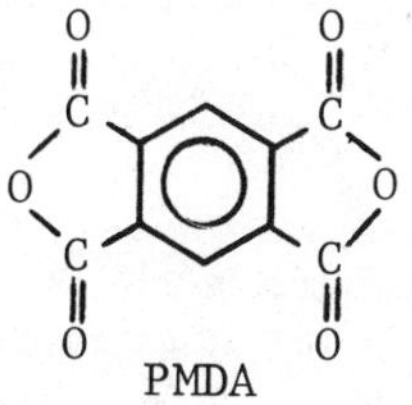

PMDA
II

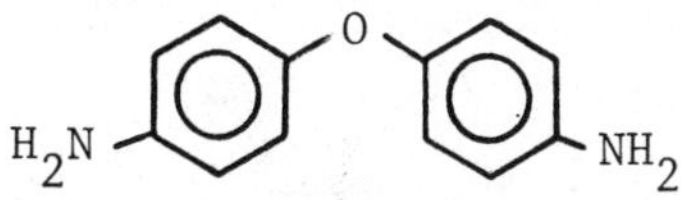

ODA

III

Polymer Synthesis/Modification

Two polyimide systems (PMDA-ODA and BTDA-ODA) were chosen for modification via doping with tin complexes. These systems were modified by synthesis of the polyamic acid from the respective dianhydride and ODA followed by addition of the DMAC-soluble tin complex to the polyamic acid.

Specific details regarding the preparation of the polyamic acid/metal complex solution are as follows. First the polyamic acid was synthesized by dissolving the diamine (ODA) in DMAC and then adding the stoichiometric quantity of dianhydride. Solutions were made at 15–20% solids, 4×10^{-3} mole of polymer. Stirring of the polyamic acid solution under a N_2 atmosphere was maintained for 4–6 hours. The tin compound was then added either neat to the polyamic acid or dissolved in DMAC. The molar ratio of polymer to metal complex was 4:1. Inert atmosphere handling techniques were employed for the weighing and transfer of air sensitive tin complexes. Polyamic acid/tin complex solutions were stirred for 5–10 hours and then cast as films or frozen for later casting.

Film Preparation

Polyamic acid/tin complex solutions were centrifuged at ~1700 rpm for 5 minutes in order to remove any particulate matter prior to casting. Solutions were then spread on acid-cleaned, dust free soda lime glass plates with a doctor blade. Blade gap ranged from 18-23 mils depending on solution concentration.

Thermal treatment of the films began immediately after casting by drying for 2 hours at 60°C in static air. Imidization of the polyamic acid was allowed to occur via curing in a forced air oven for 1 hour each at 100°, 200°, and 300°C. The thin films (~1 mil) were allowed to cool to room temperature under forced air. Removal from the glass plates was accomplished by soaking the film at room temperature in distilled water. Films were then air-dried and cleaned repeatedly with a 1:1 (v/v) anhydrous methanol/ether mixture.

Materials Characterization

Galbraith Laboratories, Inc., Knoxville, TN performed elemental analysis on representative tin-modified polyimide films. Thermomechanical analyses (TMA) were performed with an E.I. Dupont Model 990 Thermomechanical Analyzer on film samples in static air at a 5°C/min temperature program. Thermogravimetric analyses (TGA) were conducted in static air at a 2.5°C/min temperature program.

Surface and volume resistivities were measured following the ASTM standard method of test for electrical resistance of insulating materials (D257-66) utilizing a Keithley voltage supply, electrometer, and electrode assembly. Elevated temperature resistivity data were obtained for one film employing a Keithley electrode assembly modified to accommodate cartridge heaters and thermocouples. The assembly was enclosed to sustain a dynamic vacuum of one torr. Heating and temperature monitoring were manually controlled. In one instance a N_2 purged, H_2O-free glove bag was utilized as a dry environment for the electrode assembly in order to measure the resistivity of a film evacuated and heated free of moisture.

X-ray photoelectron spectra (XPS) were obtained on either a DuPont 650 B spectrometer or Physical Electronics 550 instrument, both equipped with a Mg anode (Mg K_α = 1253.6 eV) target. Binding energies of all electrons were measured relative to the instrumental background carbon ($1s_{1/2}$) photopeak assumed to have a value of 284.0 eV.

192

RESULTS AND DISCUSSION

Table I presents a list of polyimide/dopant combinations with either tin(II) or tin(IV) additives which produced films of fair to good quality. All films were somewhat flexible and tough enough to undergo electrical resistivity measurement. The overall appearance and physical properties of these materials varied greatly. In order to compare general physical properties of the films and also to describe their individual unique properties, a comparison of all data will be presented initially followed by a more specific discussion of each dopant/polymer pair individually.

Electrical resistivity measurements on the tin-modified materials are presented in Table I, along with resistivity measurements on the undoped BTDA-ODA and PMDA-ODA systems. Films 3 and 7 exhibited the lowest resistivities of the polymers modified with Sn(IV) complexes. Volume resistivity was somewhat similar for most all the Sn(IV) modified materials. Significant differences, however, were found in the surface resistivity mesurement for the air side of the film (side facing up during cure) opposed to that of the glass side (side facing down during cure). The lowest surface resistivities measured were $1x10^5$ ohm for film 3 and $8x10^7$ ohm for film 7. These values are lower by a factor of approximately 10^6 over that of the next most surface conductive material doped with a different Sn(IV) complex. Both films have been doped with di-n-butyltin dichloride. Film 11 exhibited the lowest resistivity of the Sn(II) doped films. Again, lowest resistivity was realized for the surface/air-side measurement with a value about 10^{12} times lower than the undoped polymer.

Other general trends regarding the resistivity data should be mentioned. Volume resistivity was lowered to a minimum value of about 10^{14} ohm-cm regardless of the dopant. In no case was surface resistivity for the glass side of any film lowered below 10^{12} ohm via doping. The same dopant in different polymer systems produced materials with different resistivity values. For example, comparison of films 3/7 and 10/11 show appreciable differences particularly in surface resistivity values for the air-side of the films.

Elemental analyses (Table II) revealed that nearly total vaporization of dopant must have occurred during the curing process for films 6 and 8. Tin-containing species were still present in all other films analyzed.

Thermal behavior of the tin-modified films (Table II) results in an increase in the softening temperature (AGT) of nearly all materials analyzed. Thermomechanical analysis (TMA) of some BTDA-

Table I. Electrical Resistivities of Tin-Containing Polyimide Films.

Film No.	Monomers	Tin Additives	Resistivity[c] Volume (ohm-cm)	Surface (ohm) Air Side	Surface (ohm) Glass Side
1	BTDA/ODA	1,2-bis(triphenyltin)acetylene (ϕ_3SnC)$_2$	8×10^{16}	5×10^{14}	4×10^{14}
2	BTDA/ODA	triphenyltinbenzpyrazole ϕ_3Sn($C_7H_5N_2$)	$>10^{18}$	$>10^{18}$	$>10^{18}$
3	BTDA/ODA	di-n-butyltin dichloride (n-Bu)$_2$SnCl$_2$	1×10^{15}	1×10^{5}	2×10^{13}
3a			2×10^{15}	1×10^{4}	2×10^{13}
3b			5×10^{15}	$>10^{18}$	$>10^{18}$
4	BTDA/ODA	tri-n-butyltin hydride (n-Bu)$_3$SnH	$>10^{18}$	$>10^{18}$	$>10^{18}$
5	BTDA/ODA	dimethyltindichloride (CH$_3$)$_2$SnCl$_2$	5×10^{15}	4×10^{11}	3×10^{12}
6	BTDA/ODA	phenyltintrichloride ϕSnCl$_3$	$>10^{18}$	$>10^{18}$	$>10^{18}$
7	PMDA/ODA	di-n-butyltindichloride (n-Bu)$_2$SnCl$_2$	1×10^{16}	8×10^{7}	5×10^{14}
8	PMDA/ODA	tetramethyltin (CH$_3$)$_4$Sn	2×10^{16}	5×10^{12}	4×10^{13}
9	PMDA/ODA	triphenyltin hydroxide ϕ_3SnOH	3×10^{16}	4×10^{13}	7×10^{15}
10	BTDA/ODA	SnCl$_2\cdot$2H$_2$O	1×10^{15}	1×10^{8}	6×10^{12}
11	PMDA/ODA	SnCl$_2\cdot$2H$_2$O	5×10^{14}	2×10^{6}	6×10^{12}

a Post-drying, inert atmosphere measurement
b 200° cure (maximum temperature)
c BTDA-ODA and PMDA-ODA (no additive) surface and volume resistivies equal approximately 10^{17}

Table II. Thermal Data for Some Tin-Containing Polyimides.

Film No.	AGT[a] (°C)	PDT[b] (°C)	% Tin Found	% Tin Calc'd
BTDA/ODA	286	550	–	–
PMDA/ODA	405	580	–	–
1	310	514	–	–
2	357	517	5.21	4.56
3	404	548	3.47	4.92
7	Decomp.	534	4.05	5.97
8	363	558	<0.01	6.46
9	472	516	–	–
10	280	544	4.24	5.23
5	–	–	6.20	5.05
6	–	–	<0.01	5.00
3c	–	–	5.92	4.92

[a]Apparent Glass Transition Temperature (Thermomechanical Analysis)

[b]Polymer Decomposition Temperature (Thermal Gravimetric Analysis)

[c]Cured at 200°C, 3 hours

ODA materials indicated polymer cross-linking (probably promoted by the tin additive) had occurred (Figure 1). Decomposition of the PMDA-ODA materials near their AGT made it difficult to ascertain if the same effect was taking place for these materials.

Thermal gravimetric analysis (TGA) revealed generally some decrease in polymer stability after addition of the tin dopants. Film 9 had the lowest decomposition temperature measured, 64° below that of the undoped polymer. All films tested had decomposition temperatures above 500°C indicating that the tin dopants overall had a minimal effect on the elevated temperature stability of BTDA-ODA and PMDA-ODA polyimide materials.

XPS data were obtained (Table III) on both sides of selected films. Information regarding (1) the relative binding energies of Sn found on air and glass sides of the same film, (2) the presence of chlorine on the surfaces of films doped with a chlorine-containing complex and (3) the relative amount of tin present on each film's surface was of interest. All materials that were measured

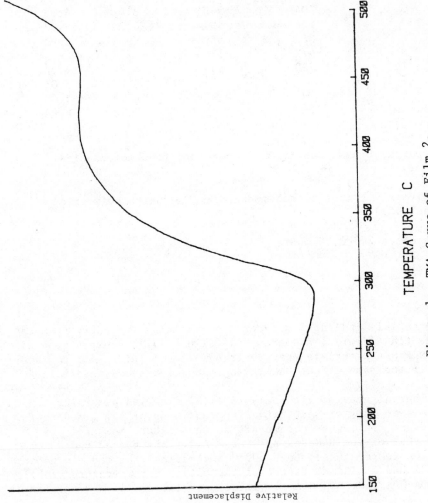

TEMPERATURE °C

Figure 1. TMA Curve of Film 2.

196

Table III. XPS Binding Energy Data for Some Tin-Containing Polyimides.

Film No.	Sn $3d_{5/2}$ Photopeak (eV)	
	Air Side	Glass Side
1	486.3	486.5
2	485.1	486.8
3	486.6	--
10	486.3	486.8
11	485.8	486.2

show a slightly higher binding energy for tin present on the glass side of the film relative to tin on the air side. Chlorine, in no instance, is present on the surface of any side of any film doped with a chlorine containing complex. Relative amounts of tin present on the air and glass sides of these materials varied greatly between the Sn(IV) additives (Films 1 and 2) and the Sn(II) additives (Films 10 and 11). Films 1 and 2 which have high surface resistivities have about the same amount of tin on either side; while, films 10 and 11 which exhibit low surface resistivities on the air-side have about a factor of 10^2 more tin on the air side than the glass side of the film. Tin binding energies are essentially identical regardless of the tin dopant and monomer combination. Unlike numerous elements, XPS binding energies for tin are not highly definitive concerning oxidation state. Tin compounds in divalent and tetravalent states have been shown to have quite similar binding energies.[6] Considering the aerobic imidization conditions, tin(IV) is anticipated on the surface of these films. The absence of chlorine from films doped with chlorine-containing complexes indicates that some mechanism for loss of chlorine must exist at some stage in the film preparation process. It is notable that low surface resistivities for the air side of both Sn(II) and Sn(IV) materials occurred only with dopants containing chlorine.

Individual Film Properties

Film 1: BTDA-ODA/(ϕ_3SnC)$_2$

This material has a yellowish-brown color and is transparent.

The film is somewhat brittle. It is homogeneous except for a number of small holes unevenly dispersed through the film which apparently were made by escaping gas during imidization. The nature of the size and quantity of the holes suggests that they may have been formed via degradation and/or volatilization of small particles of tin complex which may have precipitated from the polyamic acid solution at some stage. A likely mechanism can be envisioned. The tin-acetylene bond is readily hydrolyzed,[7] and may have reacted with water of imidization (released from the polymer near 180°C) as follows:

$$\phi_3Sn-C\equiv C-Sn\phi_3 + 2H_2O \rightarrow 2\phi_3SnOH + 2HC\equiv CH_{(g)}$$

$$x\phi_3SnOH \xrightarrow{\Delta} (\phi_3Sn-O)_x + x/2\ H_2O$$

The aryltin oxide may have precipitated from solution and volatilized thereby causing the film peculiarities mentioned earlier. Not all the tin has vaporized from the polymer because XPS shows appreciable tin on both sides of the film. This material had relatively poor polymer properties and was of interest primarily for its unique, "pocked" surface.

Film 2: BTDA-ODA/$\phi_3Sn(C_7H_5N_2)$

The dopant utilized for the preparation of this material was the only tin-nitrogen compound studied. It produced a very clear yellow, homogeneous film with no surface anomalies. The material was tough, flexible and possessed a static charge (e.g. seemingly more so than the undoped polymer film). Elemental analysis verified that more tin is present in the film than theoretically predicted. The tin-nitrogen bond is known to be hydrolytically unstable and may likely have been cleaved at some point during the film preparation. The amine residue, therefore, may have volatilized accounting for the higher % Sn. The other product of hydrolysis may be the triphenyltin oxide postulated in the formation of film 1. Based on results obtained with film 2, the imperfections in film 1 would therefore have come from acetylene evolution which should be more readily evolved than benzimidizole during thermal imidization. This however is not likely since the Sn-acetylide bond should have cleaved upon dopant addition to the polyamic acid. Alternately, a different form of tin may have been produced in this film. XPS of this material showed about equal amounts of tin on each side of the film.

Film 3:   BTDA-ODA/(n-Bu)$_2$SnCl$_2$

This material had the overall lowest resistivity properties
of all the materials in this study. It also had the most unusual
physical features. The material is dark-brown in color and
opaque. Large raised areas in the film give it a "bumpy" appearance. It is very tough and appears to be cross-linked. Small
amounts of a blackish material are present in some areas.

Electrical properties of this film were studied in some
depth. The film was dried under vacuum and resistivity measured
under H$_2$O-free conditions to determine if H$_2$O played a role in its
electrical properties. Resulting volume and surface resistivity
measurements were about the same (Table I). The surface resistivity was also measured at elevated temperature (max. 160°C)
under vacuum (0.1 torr) and found to vary little from the room
temperature value (Table IV). This result is quite different from
previous resistivity studies with lithium-doped polyimides.[1]
Resistivity in the lithium case markedly increased on cyclic
heating and pumping in vacuo because moisture was being removed
from the film. A similar film was cast and cured at 200°C rather
than 300°C. This film was found to be a rather smooth, homogeneous
material with much poorer electrical properties (Table I). No
raised surface (presumably due to gas generation) was found with
the 200°C cured film. Evidently decomposition of the (n-Bu)$_2$SnCl$_2$
complex is necessary for this material to acquire its enhanced
surface electrical properties. Butane or possibly butene may be
evolved during thermal curing at 300°C, due to the instability of
the Sn-C bond above 200°C.[7] Chlorine is lost during imidization
probably via evolution of HCl. The final metal-containing product
may be a form of tin-oxide which may be dispersed primarily on the
air side of the film. A faint surface deposit in fact appears on
the air side of the film.

Film 4:   BTDA-ODA/(n-Bu)$_3$SnH

This dopant produced a clear, homogeneous film with high
resistivity. Raised areas on the film surface were likely due to
H$_2$ evolution arising from reaction of the hydride with H$_2$O.
Percent tin found and calculated were practically identical.

Film 5:   BTDA-ODA/(CH$_3$)$_2$SnCl$_2$

Use of (CH$_3$)$_2$SnCl$_2$ as a dopant produced a film somewhat
similar in appearance to Film 3. Raised areas were present which
may be due to methane and/or HCl generation during cure. Volume

Table IV. Elevated Temperature Resistivity Measurements for Film 3.

| | Surface Resistivity |
Temperature($^\circ$ C)	Air Side (ohm)
25	7.5×10^5
44	7.1×10^5
61	6.4×10^5
80	5.5×10^5
123	3.2×10^5
143	2.7×10^5
154	2.4×10^5
160	2.3×10^5

resistivity values for both materials were the same. Glass side surface resistivities were also comparable for films 3 and 5. Air side values differed significantly with film 5 being much poorer.

Film 6: BTDA-ODA/$\phi SnCl_3$

Elemental analysis revealed this additive totally vaporized during the cure process. Boiling point of $\phi SnCl_3$ is 128°C, 15 torr.[8] Evidently there was no interaction of additive with the polymer solution. Electrical resistivity values were comparable to those of undoped BTDA-ODA.

Film 7: PMDA-ODA/$(n-Bu)_2SnCl_2$

In contrast to film 3, which employed the same dopant but in BTDA-ODA, this dopant/polymer combination produced a relatively homogeneous, dark brown film of high quality. Resistivity was not as low as was the case with the BTDA-ODA material. For PMDA-ODA doped films the air side surface resistivity was lower here than with any other Sn(IV) additives. Some black material was present in the film in small quantities, which may be a decomposition product of the additive. Overall, this is the highest quality Sn(IV) doped material prepared in this study with significant lower resistivity values.

Film 8: PMDA-ODA/$(CH_3)_4Sn$

The dopant utilized for this material also totally vaporized apparently due to its low boiling point (76.6 °C).[9] Elemental analysis indicated tin was present at less than 0.01% concentration. Properties of the PMDA-ODA polyimide, nevertheless, have been slightly modified. Softening and decomposition temperatures have been slightly reduced along with electrical resistivity.

Film 9:   PMDA-ODA/ϕ_3SnOH

This material was homogeneous and slightly darker in color than the undoped PMDA-ODA polymer. The film was more brittle than other tin-containing films. Evidently the mechanical properties of the polymer were degraded to some extent by this additive. T_g has increased but the polymer decomposes ~60°C lower than the undoped material. Degradation of the polymer could have occurred through reaction of the additive hydroxyl group with the carboxyl moiety of the polyamic acid thereby altering the extent of imidization as pictured below.

$$\phi_3\text{SnOH} + \text{HO-}\overset{\displaystyle \overset{O}{\|}}{C} \quad \rightarrow \quad \phi_3\text{SnO-}\overset{\displaystyle \overset{O}{\|}}{C} \quad + \text{H}_2\text{O}$$

〜〜〜〜–N–C   〜〜〜〜–N–C
          | ‖ | ‖
          H O   H O

This could be followed by cleavage of the tin-carboxyl linkage either thermally or hydrolytically, resulting in a loss of the structural integrity of the polymer.

Film 10:   BTDA-ODA/SnCl$_2$·2H$_2$O

This material is fairly smooth, tough and flexible. Coloration is dark yellow with darker brown splotches running through the film. This material offered a somewhat lower resistivity in comparison to most of the Sn(IV) additives. Lowest resistivity was determined for the air side of the film. XPS data showed several orders of magnitude more tin on the air side than the glass side, suggesting migration of the tin during cure. No chlorine was detectable on either side of the film. Tin is likely to be present on the surface in some oxide form, as a result of hydrolysis and air oxidation as shown below.

$$SnCl_2 + 2H_2O \rightarrow Sn(OH)_2 + 2HCl$$

$$2Sn(OH)_2 + O_2 \rightarrow 2SnO_2 + 2H_2O$$

Thermomechanical properties for this polyimide have been uniquely modified. T_g has decreased relative to the polymer-alone unlike all other tin additives examined in this study. As in the $(n-Bu)_2SnCl_2$ cases, approximately 1% tin has been lost during imidization.

Film 11: PMDA-ODA/SnCl₂·2H₂O

This film was more homogeneous in appearance than the previous film and overall was several orders of magnitude lower in resistivity. XPS data showed a factor of 10^2 more tin on the air side than the glass side of the film, which probably is the reason for its much lower air side surface resistivity. XPS also indicated no chlorine present on either side of the film.

SUMMARY

A number of thermally stable tin containing polyimides with enhanced electrical properties have been prepared. Resistivity of one surface-semiconductive film was found to be independent of atmosphere and humidity by elevated temperature measurements under dynamic vacuum and measurements taken in a desiccated, inert atmosphere glove bag following cyclic heating and pumping. XPS measurements confirmed the presence of 100 times more tin on the air as opposed to the glass side of films exhibiting air side surface resistivities in the 10^5-10^8 ohm range. Lowered surface resistivities, into the semiconductor region, therefore appears to be correlatable with large surface concentrations of tin, most likely in the form of SnO_2. The results of a further study of the SnO_2-like surfaces of these films will be soon forthcoming.[10]

ACKNOWLEDGEMENT

We thank the National Aeronautics and Space Administration (Grant NSG 1428) for sponsoring this research and Anne K. St. Clair for supplying the thermal data.

REFERENCES

1. E. Khor and L. T. Taylor, Macromolecules, 15, 379 (1982).
2. S. A. Ezzell, T. A. Furtsch, E. Khor and L. T. Taylor, J. Polym. Sci. Polym. Chem. Ed., 21, 865 (1983).
3. T. L. Wohlford, J. Schaaf, L. T. Taylor, T. A. Furtsch, E. Khor and A. K. St. Clair, in "Conductive Polymers", R. B. Seymour, Editor, p. 7, Plenum Publishing Corp., New York, 1981.
4. C. E. Carraher, in "Modification of Polymers", C. E. Carraher and M. Tsuda, Editors, ACS Symp. Ser., No. 121, p. 60, American Chemical Soc., Washington, D.C., 1980.
5. L. T. Taylor, A. K. St. Clair, V. C. Carver and T. A. Furtsch, ibid, p. 7.
6. A. W. C. Lin, N. R. Armstrong and T. Kuwana, Anal. Chem., 49, 1228 (1977).
7. W. P. Newman, "The Organic Chemistry of Tin", p. 33, John Wiley, New York, 1970.
8. H. Zimmer and H. Sparmann, Chem. Ber., 87, 645 (1954).
9. W. F. Edgell and C. H. Ward, J. Amer. Chem. Soc., 76, 1169 (1954).
10. S. A. Ezzell and L. T. Taylor, manuscript in preparation.

PART II. PROPERTIES AND CHARACTERIZATION

DIELECTRIC AND CHEMICAL CHARACTERIZATION OF THE POLYIMIDE, LARC-160

D. E. Kranbuehl, S. E. Delos, P. K. Jue, and
R. K. Schellenberg

Department of Chemistry
College of William and Mary
Williamsburg, Virginia 23185

ABSTRACT

Dynamic dielectric measurements of the imidizing LARC-160 resin have been made at a series of frequencies from 5 to 5×10^6 Hz. Capacitance (C), Dissipation (D), and Resistance (R) were recorded. These results have been correlated with the observed cure chemistry of the LARC-160 system. Changes in these measurements with both resin age and resin composition have been monitored. The cross-linking reaction has also been followed.

INTRODUCTION

The imidization reactions of the polyimide LARC-160 have been studied both chemically and dielectrically. LARC-160 is a cross-linked polyimide prepared from a solventless monomer mixture composed of diethyl-3,3',4,4'-benzophenone tetracarboxylate (BTDE), ethyl-5-norbornene-2,3-dicarboxylate (NE), and a mixture of aromatic amines (Jeffamine) of which methylene dianiline is the primary component. In the initial reaction, the resin imidizes, forming short polyimide chains terminated by the unsaturated nadic groups. Above 280°C these nadic groups begin to react via an addition reaction to form a cross-linked network. Scheme I shows the reaction sequence for the LARC-160 system. Carbon 13 nuclear magnetic resonance (^{13}C-NMR), infrared spectroscopy (IR), and high pressure gel permeation chromatography (HPLC-GPC) have been used to follow the imidization reactions of NE and BTDE. NE and BTDE have been found to imidize at significantly different rates. The effect of monomer mixture age on the imidization of LARC-160 has also been examined.

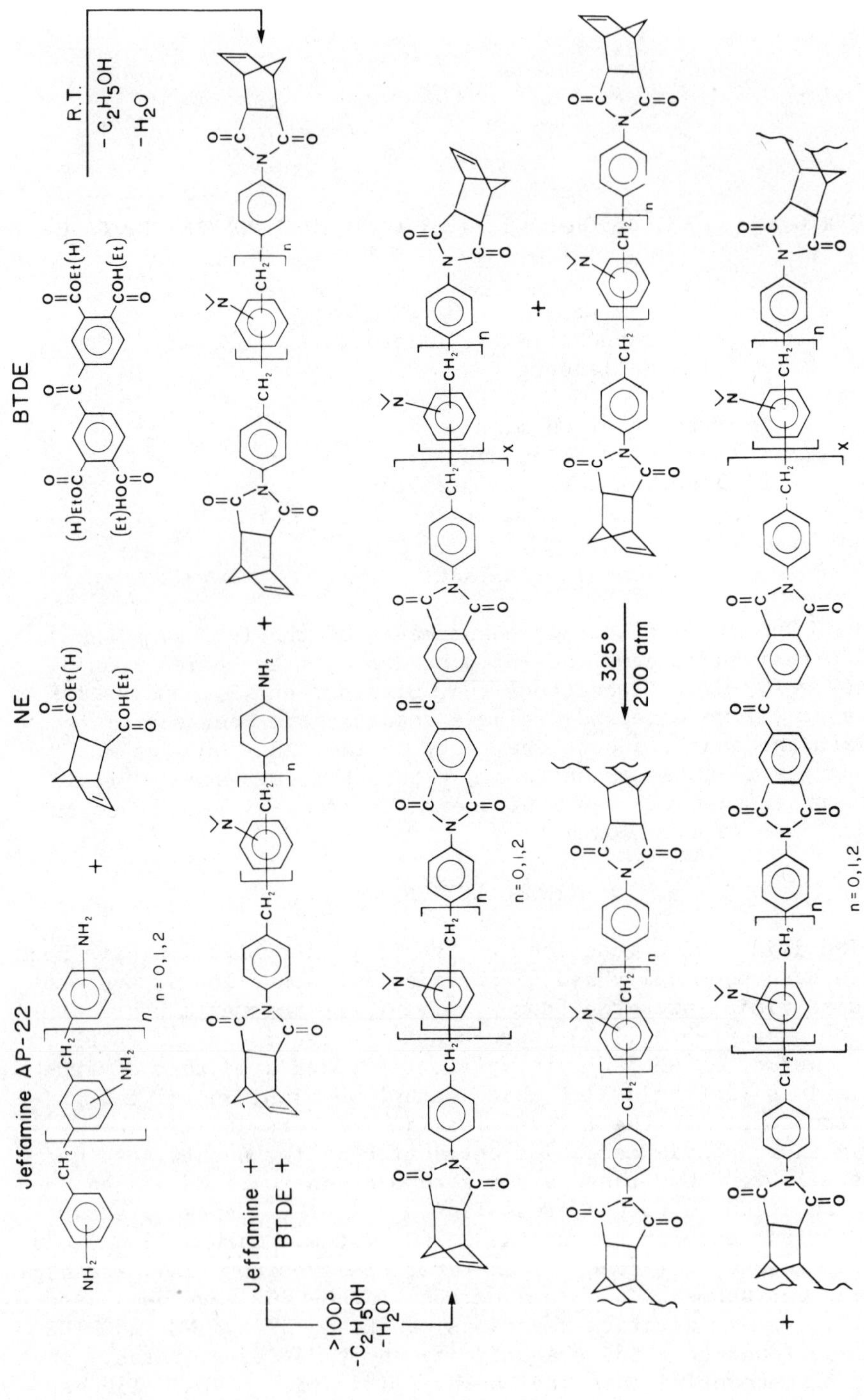

Scheme 1

Dynamic dielectric measurements of the imidizing LARC-160 resin have been made at a series of frequencies from 5 to 5×10^6 Hz. Capacitance (C), Dissipation (D), and Resistance (R) were recorded. These results have been correlated with the observed cure chemistry of the LARC-160 system. The cross-linking reaction has also been followed.

A LARC monomer mixture rich in NE has been prepared. DDA measurements have been made during the imidization of this resin and the results compared with those obtained for the LARC-160 resin.

A detailed chemical characterization of the imidization of LARC-160 has been reported elsewhere[1]. This paper briefly reviews those results, then extends the work to dynamic dielectric analysis (DDA) of this reaction in the aging LARC-160 system.

EXPERIMENTAL

The solventless LARC-160 resin was prepared as prescribed by St. Clair and Jewell using 3,3',4,4'-benzophenone tetracarboxylic anhydride (BTDA), 5-norbornene-2,3-dicarboxylic anhydride (NA) and a commercially available mixture of aromatic amines, Jeffamine AP-22, which is composed primarily of p,p-methylenedianiline (MDA).[2] BTDE, the diethylester of BTDA, was prepared by dissolving the anhydride in a 5 mol % excess of ethyl alcohol and refluxing for one hour. NE, the monoethyl ester of NA, was similarly prepared. These mixtures were then cooled and mixed with the appropriate amine molar equivalent of Jeffamine AP-22. The resulting resin mixture was stored in a refrigerator at 12°C. LARC-160 has a NE/BTDE ratio of 1.8.[2] Another LARC resin with an NE/BTDE ratio 8.5 was also prepared and stored as described above.

High Pressure Liquid-Gel Permeation Chromatography (HPLC-GPC) was performed on a Waters Associates Liquid Chromatograph equipped with a model M-6000A chromatography pump, model U6K injector, and a Model E 401 differential refractometer. Samples were dissolved in tetrahydrofuran (THF), UV grade, from Burdick and Jackson, then eluted on a four column bank consisting of two 500 A° and two 100 A° μ-styragel columns.

Carbon 13 Nuclear Magnetic Resonance (^{13}C-NMR) studies were done on a Varian FT-80A NMR Spectrometer. Samples of the resin were dissolved in DMSO-d_6, with TMS as an internal standard. A spectral width of 5000 Hz was used.

^{13}C chemical shift assignments of the carbonyl resonances of BTDE and NE were made by comparison with the monomers and the fol-

lowing model compounds: 3-[(phenylamino)carbonyl]-Bicyclo[2.2.1]-
hept-5-ene-2-carboxylic acid; 2-[(phenylamino)carbonyl]-benzoic
acid; N-phenyl-5-Norbornene-2,3-dicarboximide; N,N'-(methylenedi-
p-phenylene)di-5-norbornene-2,3-dicarboximide (Bisnadimide or
BNI); and p,p'-methylenedianiline (MDA). These compounds were
supplied by NASA-Langley Research Center.

A Perkin-Elmer 337 Grating Infrared Spectrophotometer was
used for IR studies. Films of fresh resin were prepared on KBr
plates and the same sample was taken through a given cure cycle.
IR spectra were taken of the fresh resin and at selected points
along the cure cycle. The percentage of imidization was
calculated using the equation:[3]

$$\text{Percentage Imidization} = \frac{\dfrac{A(1)}{A(2)}\ (t) - \dfrac{A(1)}{A(2)}\ (t=0)}{\dfrac{A(1)}{A(2)}\ (t=\infty) - \dfrac{A(1)}{A(2)}\ (t=0)}$$

A(1) = Absorbance of Imide Peak at 1790 cm^{-1}
A(2) = Absorbance of Standard Reference Peak at 1490 cm^{-1}
t = ∞ was taken as the time beyond which no further changes in the
imide peak were observed at 220°C.

Dynamic dielectric measurements were made using a Hewlett-
Packard 4192A LF Impedance Analyzer controlled by a 9826 Hewlett-
Packard computer. The resin was cured in a 3" mold placed on a
hot plate. The layup consisted of a number of layers of fiberglass cloth to insulate the probe from the metal mold, a piece
of Kapton to hold the melting resin and keep it near the gauge, a
probe which can be inserted directly between plies in the laminate,
one bleeder ply (also fiberglass cloth), a thin pancake of the
resin monomer mixture, a piece of Kapton and a further layer of
bleeder plies between this and the mold top (Figure 1). An iron-
Constantan thermocouple was attached directly to the mold and the
temperature measured by a Kiethley 179 TRMS Digital Multimeter.
Measurements of Capacitance (C), Dissipation (D), and Resistance
(R) at each of six frequencies from 5 to 5 x 10^6 Hz were taken
every 2.5 min. during the cure cycle and stored on a computer disk.
The temperature was also recorded for each measurement. Plots of
the results were prepared from the stored data and were printed
using a Hewlett-Packard 2671G graphics printer. Figure 2 is a
block diagram of the DDA apparatus.

RESULTS AND DISCUSSION

Uncured LARC-160 is a solventless mixture of the monomers
diethyl-3,3'4,4'-benzophenone tetracarboxylate (BTDE), ethyl-5-

norbornene-2,3-dicarboxylate (NE) and a mixture of aromatic amines primarily composed of methylenedianiline (Jeffamine). Upon being heated, this resin was found to proceed directly to the imide. No significant amount of amide, as determined by ^{13}C-NMR, was observed at any stage during the imidization process. Table 1 presents a summary of the chemical shifts in the carbonyl region of ^{13}C-NMR spectra of LARC-160 resin after various curing times and temperatures. By comparison with spectra of the monomers and the appropriate model compounds (see experimental) the following assignments were made: (a) nadic imide, (b, c) nadic acid and ester, (e-h) benzophenone tetracarboxylate acid and ester, (i) benzophenone tetracarboxylate amide (present only as a minor shoulder on the imide peak where listed), (j) benzophenone tetracarboxylate imide. The major observation to be made from this table is that NE reacts at a lower temperature and much faster than does BTDE.

IR data support the ^{13}C-NMR evidence that NE reacts faster than BTDE. LARC-160 resin was cured as a thin film between two salt plates and IR spectra were taken at various times during the curing process. Table II is a summary of these results.

No imidization was observed in the resin before heating. When the sample was heated at 70°C for up to 660 minutes, the imide peak grew to a constant value of approximately 40% of its final size. The sample was then heated at 180° for up to 360 minutes. This second cure stage brought about a significant increase in the amount of imidization, to approximately 85% of the final imidization. The sample was then cured at 220°C for up to 330 minutes. The amount of imidization increased slightly indicating that the imidization reaction was not quite complete after curing at 180°C. As shown in Table II, slightly less than 1/2 of the imide is formed below 100°C, consistent with the hypothesis that only the nadic group, which represents 48% of the carbonyl carbon in the monomer mixture, is reacting at this lower (70°) temperature. The remainder of the imidization, which involves BTDE, takes place at the higher temperature (180°).

HPLC analyses were performed on samples of resin after monomer mixture storage at 12°C for various times. Table III summarizes the results of this study. Even at this low temperature the nadic group imidizes with time. After about 1 year, all of the NE has imidized. This complete imidization of NE is corroborated by ^{13}C-NMR. Figure 3 is a ^{13}C-NMR spectrum of LARC-160 monomer mixture after storage for 14 months at 12°C. In this spectrum all of the nadic carbonyl absorption occurs at 176.7 ppm (a). Peaks at 173.15 and 171.95 (b and c), the acid and ester, have completely disappeared. BTDE, on the other hand, remains stable during this storage period (peaks e-h). There is no evidence of either the amide (166.3 ppm) (i) or the imide (166.2 ppm) (j) in this spectrum.

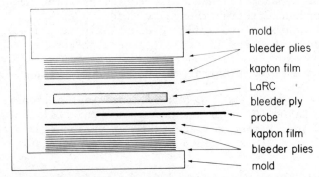

Figure 1. Lay-up of dielectric probe, bleeder plies and LARC in mold.

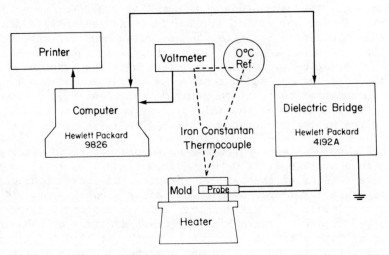

Figure 2. Block diagram of dynamic dielectric apparatus.

Table I. ^{13}C-NMR Shifts of Observed Carbonyl Carbons of LARC-160 as a Function of Cure Time and Temperature.

Cure Temp °C	Cure Time (min)	Nadic imide	Nadic acid-ester		Benzophenone tetracarboxylate acid-ester				BID amide	BID imide
		a	b	c	e	f	g	h	i	j
12	0	176.73	173.16	171.96	167.72	167.15	166.94	166.44		
105	60	176.70			167.64	167.12	166.87	166.39	166.26	166.22
105	120	176.64				167.04		166.40	166.32	166.19
125	15	176.74	173.12	171.94	167.65	167.11	166.88	166.38		
125	30	176.64			167.57	167.07	166.81	166.42	166.34	166.23
180	5	176.70			167.59	167.04		166.40	166.33	166.19
180	10	176.61							166.32	166.18
180	30	176.63								166.19

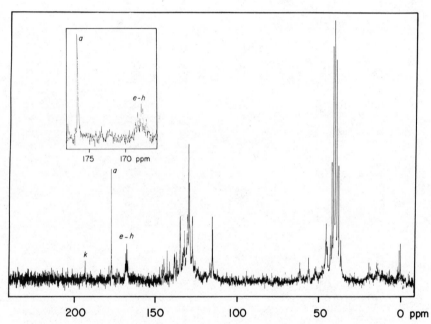

Figure 3. C^{13}-NMR spectrum of LARC-160 monomer mixture after storage for 14 months at 12°.

Table II.   Quantitative Analysis of Infrared Spectra of LARC-160
            Resin Percent Imidization versus Cure Time.

Cure Time (min.)	Cure Temp (°C)	% Imidization
0	70	0
10	70	-0.87
20	70	1.5
40	70	5.1
60	70	5.1
90	70	10
120	70	18
180	70	18
300	70	23
420	70	31
540	70	46
660	70	37
840	180	85
960	180	83
1020	180	83
1050	220	94
1110	220	89
1170	220	89
1290	220	99
1350	220	100

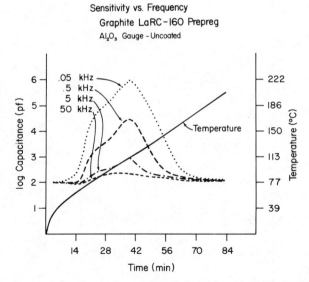

Figure 4.   Log capacitance versus time for imidizing LARC-160.

Table III. HPLC Analysis of Aging LARC-160 Monomer Mixture, Cool Mix*.

				Weight % after n days at 12°C							
Peak	Species	V_e (±0.1 ml)	n= 1	7	14	28	53	106	287	377	
1	NE	35.2	31	28	26	22	14	13	13	0	
2	Jeffamine, n=0	33.6	21	15	13	12	10	12	9	4‡	
3	Nadimide**	33.4	–	–	–	9	8	8	15	33‡	
3.5	Jeffamine, n=1	31.3	9	5	9	8	9	9	6	1‡	
4	BTDE	30.3	33	38	39	40	43	43	46	49	
5	imide	28.9	6	13	13	10	16	15	13	13	

* 0.2% wt/wt in THF; two 500A° + two 100 A° μ-styragel columns.

‡ Peaks 2, 3 and 3.5 are inseparable. Jeffamine peaks are estimated based on the theoretical amount of Jeffamine left, assuming no BTDE has reacted.

** In the unpolymerized mixture, both NI and BNI elute at this volume.

The difference in reactivity of NE and BTDE can lead to marked differences in the cure chemistry of the LARC resin with age. The longer the resin has been stored the more nadic groups have imidized. This leaves less end-capper to react with the polyimide chains forming during the cure process. The result is a cured resin having more longer chains plus a lot of the relatively small molecule, BNI.

Dynamic dielectric analysis (DDA) has been used to observe both the imidization and cross-linking reactions of LARC-160. Measurements have been made at frequencies from 5 to 5×10^6 Hz. This wide range of frequencies, not previously reported in DDA studies of curing polymer systems, increases the amount of information which can be obtained from DDA.

Figure 4 is a graph of the log of the capacitance vs. time for curing LARC-160. The solid line represents the temperature-time profile. The broken lines represent the capacitance at different frequencies. As lower frequencies are used, the magnitude of the capacitance increases, increasing the sensitivity of this measurement. The rise in capacitance with temperature is due to increasing rate of orientation of the dipoles and ions within the resin with decreasing viscosity as the resin melts. At the point of maximum capacitance, the viscosity starts to increase as the effects of the imidization reaction overcome the effects of increasing temperature on the viscosity (and therefore on the rate of dipole orientation). The capacitance continues to drop until a highly viscous point is reached when the dipolar components of the resin molecules can no longer follow the oscillating electrical field.

Figure 5 represents the dissipation or loss tangent (tan δ) of imidizing LARC-160 with time. The temperature ramp is given by the solid line. Two representative frequencies have been picked to show how the behavior observed during the reaction depends upon the frequency of the measurement. At 0.5 KHz the dissipation goes up as the dipole oscillation in the melting resin approaches the frequency of the external field, and decreases again as the viscosity continues to drop and the resonant frequency of the dipoles in the resin no longer match the external field frequency. A minimum in the dissipation is reached when the increasing viscosity due to the polymerization reaction begins to overcome the decreasing viscosity due to the temperature rise at that frequency. Again the dipoles go in and out of resonance with the external field as the reaction progresses, increasing the viscosity and slowing the molecular motion.

The behavior of the LARC-160 resin at 50 KHz is quite different from that at 500 Hz. At this frequency the dipoles never reach their resonant frequency before the increase in viscosity due to the polymerization reaction takes effect. The result is a single peak. As the frequency is further increased, a single peak is observed, at the same position on the graph, but with diminished amplitude as the observing frequency remains further away from the resonant condition.

There are two important reactions in the curing LARC system: the low-temperature imidization and the high-temperature cross-linking of the unsaturated nadic end-cappers. The previous two figures concerned the imidization reaction. The cross-linking reaction takes place above 250°C in a highly viscous medium.[4] Therefore, the frequency range available to us would not be expected to be low enough for observing dipolar resonance before and after this reaction. However, some observable changes in dipolar character would be expected during the reaction. Figure 6 isolates the dielectric behavior of the probe used for these measurements as a function of temperature and compares it to the dielectric behavior of cross-linking LARC-160 plus the probe at 5 KHz. The dissipation of the probe increases with temperature but remains constant at each temperature hold. The LARC+probe system, on the other hand, exhibits an increase in dissipation with temperature, but a gradual decrease at each temperature hold as some of the polymer slowly cross-links. Note that the dissipation ratio being measured is on the order of thousandths. We feel our instrumentation provides a very sensitive means of measuring dielectric changes in highly viscous media. Improvements in the probe should allow us to make these measurements at lower frequencies, and improve our sensitivity still further.

We have used our DDA technique to observe differences in resin composition. Samples that have a large amount of the lower melting NE should become less viscous sooner than samples relatively high in BTDE concentration. Furthermore, since NE begins to imidize at a lower temperature than does BTDE, the point at which the onset of the reaction affects dielectric measurements should occur sooner. This is what is observed. Figure 7 compares the dissipation at 500 Hz of a freshly prepared LARC resin rich in NE and that of LARC-160. The resin rich in NE has its preliminary peak at a lower temperature than does LARC-160. The dissipation minimum also occurs at a significantly lower temperature for NE-rich LARC than for LARC-160. The capacitance for these two resins is shown in Figure 8. The capacitance, which is sensitive to the fluidity of the system, both increases and decreases sooner for the resin high in [NE], reflecting the lower melt temperature and the earlier influence of imidization on viscosity caused by the additional nadic groups.

218

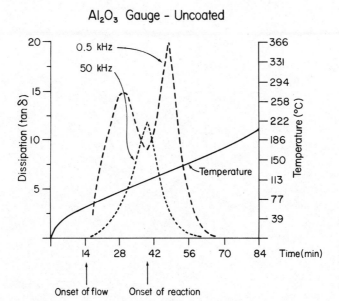

Figure 5.   Dissipation or loss tangent versus time for imidizing LARC-160.

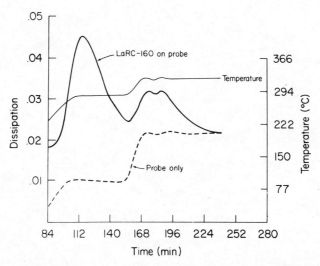

Figure 6.   Dissipation versus time for the cross-linking LARC-160.

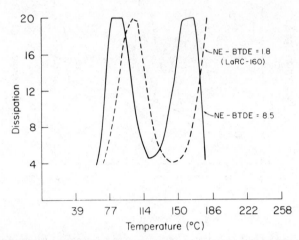

Figure 7.   Dissipation versus temperature for imidizing LARC-160 at two different compositions.

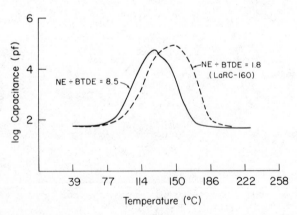

Figure 8.   Capacitance versus temperature for imidizing LARC-160 for two different compositions.

SUMMARY

The imidization of the LARC monomer mixture takes place in two steps: a rapid, low temperature reaction between NE and Jef., followed by a slower reaction between BTDE and Jef. which requires temperatures above 100° for initiation. Both reactions lead directly to the imide. Neither the nadic or benzophenone tetracarboxylic amide is present in significant quantities at any stage of the imidization reaction and is not a stable reaction intermediate.

The polymerization of LARC-160 has been monitored by DDA. The onset of melting, the onset and completion of the imidization reactions, and the onset of the cross-linking reaction have all been followed by this method. The effects of a change in resin composition on these aspects of the cure cycle have also been monitored dielectrically.

REFERENCES

1. S. E. Delos, R. K. Schellenberg, J. E. Smedley, and D. E. Kranbuehl, J. Appl. Polymer Sci., 27, 4295 (1982).
2. T. St. Clair and R. Jewell, Nat'l SAMPE Tech., Conf. Series, 8, 82 (1976); Nat'l SAMPE Symp. and Exhib. Proc., 23, 520 (1978).
3. C. N. R. Rao, "Chemical Applications of Infrared Spectroscopy," pp. 534-536, Academic Press, New York (1963).
4. A. C. Wong, A. N. Garroway, and W. M. Ritchey, Macromolecules, 14, 832 (1981).

$\eta = K M_w$

CHARACTERIZATION OF POLYIMIDES AND POLYAMIC ACIDS IN DILUTE SOLUTION

P. Metzger Cotts

IBM Research Laboratory
San Jose, California 95193

A fundamental understanding of the structure of the cured polyimide and the precursor polyamic acid from the condensation of pyromellitic dianhydride and 4,4'-diaminodiphenylether has been obtained through the use of several dilute solution techniques on samples prepared and cured under controlled conditions. The results obtained by these methods are compared with similar measurements on commercially available materials. Measurement of the concentrated solution viscosity as a means to identify changes in the molecular weight of the precursor polyamic acid is found to be inadequate because of its sensitivity to changes unrelated to molecular weight, i.e., strong polymer-polymer interactions. In this study we have augmented the concentrated solution viscosities with both light scattering and dilute solution viscometry to separate molecular weight changes from other effects. Unusual behavior in both dilute solution viscosity and light scattering measurements is attributable to polyelectrolyte effects arising from base impurities. We have also been able to measure directly the weight average molecular weight of the cured polyimide by low angle light scattering in concentrated sulfuric acid, and compare these molecular weights with those of the uncured precursor. These results have demonstrated that little change in molecular weight or chain dimensions occurs on curing, i.e., no evidence of branching or cross-linking is observed for this system.

223

INTRODUCTION

The successful development of polyimide materials for commercial use in electronic and other industries has produced much information in the literature about the properties of polyamic acid solutions and their relation to the final properties of the cured polyimide. Unfortunately, these studies were generally directed toward synthesis and cure of a wide variety of polyimide materials or toward a specific end use and thus did not go into depth in molecular characterization of any one material. In this study we have focused on one particular polyimide material; that prepared by the condensation of pyromellitic dianhydride (PMDA) with 4,4'-diaminodiphenylether (DAPE) to the polyamic acid, followed by final ring closure (cure) to the polyimide, shown in Figure 1. This is one of the most widely used polyimides and is commercially available as a concentrated solution of the polyamic acid. We have investigated both polyamic acids available as commercial products and samples prepared in our laboratory under controlled conditions[1]. Cured polyimide samples investigated include commercial materials (DuPont's Kapton®), commercial materials cured in our laboratory, and samples synthesized and cured in our laboratory. Isothermal cures at various temperatures and combination chemical/thermal cures were used.

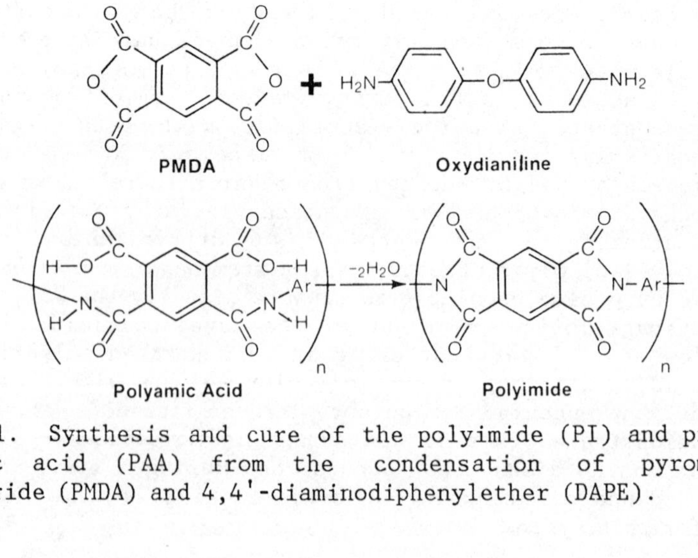

Figure 1. Synthesis and cure of the polyimide (PI) and precursor polyamic acid (PAA) from the condensation of pyromellitic dianhydride (PMDA) and 4,4'-diaminodiphenylether (DAPE).

Techniques used for this investigation included low angle light scattering (LALS), viscometry (η) in dilute and concentrated solutions, and membrane osmometry. The viscosity of concentrated solutions of polyamic acids has frequently been used in the literature as a measure of solution stability. Decreases have been attributed to degradation in molecular weight[2], and increases to a gelation phenomenon which is not well understood[3]. We have augmented the concentrated solution viscometry with dilute solution viscometry and low angle light scattering, both of which measurements can be expected to be insensitive to strong intermolecular effects which may be present in the more concentrated solution. Thus an assessment of the individual molecular size can be separated from the total solution viscosity, which reflects both molecular volume and intermolecular interactions. This allows recognition of polyelectrolyte or other effects which can contribute to the viscosity but do not contribute to the ultimate mechanical strength of the final cured material.

The direct measurement of the weight average molecular weight (M_w) by LALS on the cured polyimide dissolved in concentrated sulfuric acid has not been previously reported, although some [η] measurements in this corrosive solvent have been published[4]. As may be expected, certain precautions must be taken in working with acidic solvents, and additional criteria for data evaluation are necessary. Previous studies[4] have indicated a fairly rapid degradation of molecular weight in sulfuric acid, making accurate measurements nearly impossible. We have shown that this degradation is found only in polyimides which had been at least partially chemically cured, leading to the formation of the unstable isoimide, which then slowly degrades in the acid solvent.

EXPERIMENTAL

The polyamic acids used were either commercial materials or those synthesized in this laboratory[1]. Generally, the polyamic acids were obtained as concentrated solutions from the polymerization, and dilutions were made from this concentrated stock solution. Concentrations were checked gravimetrically by fully curing a known weight of solution to the polyimide. Organic solvents used (N-methylpyrrolidone (NMP), dimethylsulfoxide (DMSO), and dioxane) were all reagent grade, distilled in glass (Burdick and Jackson), and at times were redistilled before use. Lithium bromide (LiBr) used to repress polyelectrolyte effects was dried at 100 °C and 10 mm Hg to remove water before use. Cured samples were obtained by spin-coating the polyamic acid soluton (10-20%) onto glass slides to the desired thickness and then curing thermally on a controlled hot plate under a nitrogen atmosphere. The films were then peeled from the glass slides and dissolved in sulfuric acid. The sulfuric acid (97% reagent grade, MCB) was used without further treatment.

Viscosity measurements were made on both dilute and concentrated solutions of the polyamic acid and on dilute solutions of the polyimide. Concentrated solution measurements were made with a Brookfield viscometer equipped with a temperature controller and pressure transducer to allow the temperature and viscosity to be measured simultaneously. Ubbelohde suspended level capillary viscometers equipped with an automatic viscosity timer (Wescan) were used for the intrinsic viscosity ($[\eta]$) measurements. For the $[\eta]$ determinations, all solutions were filtered through 0.5 μm Fluoropore (Millipore Corp.) filters before measurement. In all dilute solution viscometry, solutions were maintained under dry nitrogen. Values for the intrinsic viscosity, $[\eta]$, were determined by extrapolation of the reduced and inherent viscosities to infinite dilution in accord with the Huggins and Kramers relations:

$$\eta_{sp}/c = [\eta] + k'[\eta]^2 c + \ldots \tag{1}$$

$$\ln(\eta_{rel})/c = [\eta] + (k'-1/2)[\eta]^2 c + \ldots \tag{2}$$

where η_{sp}/c is the reduced viscosity and $\ln(\eta_{rel})/c$ is the inherent viscosity. Linear plots were obtained except where polyelectrolyte effects were observed, which will be discussed further in the next section.

Membrane osmometry (Wescan) was used to measure the number average molecular weight (M_n) of two samples synthesized in this laboratory. The aromatic hydrocarbon solvent present in the commercial samples prevented use of osmometry for those samples. The regenerated cellulose membranes used were apparently highly swollen in NMP and thus the effective pore size was probably smaller than the nominal 0.005 μm. This resulted in both the undesirable effect of unusually long equilibration times and the desirable effect of extending the molecular weight range where measurements could be made down to 3000-4000 daltons. Due to the long equilibration times, measurement of a sample required 2-3 days; however, reasonable results were obtained and the virial coefficients were consistent with those measured by light scattering.

Low angle light scattering (LALS) measurements were made using a Chromatix KMX-6 light scattering photometer. This instrument measures light (632.8nm) scattered in a solid angle (~4°) about the incident beam. Both the sample volume (150 μl) and the scattering volume (5 μl) are much smaller than those in the classical wide angle photometers, and thus the problem of sample clarification is greatly reduced. Dust particles or other highly scattering particles are observed as spikes in the intensity output as they move through the small scattering volume, and are easily ignored. The design of the instrument to permit measurement at these very small angles also precludes measurement at wide angles and thus the information contained in the angular dependence (the radius of gyration, for sufficiently large molecules) is not obtained. Measurements of the Rayleigh factor, R_θ, were obtained for 4-5 concentrations and the weight average molecular weight was obtained through the relations:

$$Kc/R_\theta = M_w^{-1} + 2A_2c + 3A_3c^2 + \ldots \tag{3}$$

or

$$(Kc/R_\theta)^{1/2} = (M_w)^{-1/2} + A_2cM_w^{1/2} , \tag{4}$$

which are strictly valid only at $\theta=0°$ and are used here with $\theta = \sim4°$. Here, A_2 is the second virial coefficient, c is the concentration, and Equation (4) has been obtained by truncating Equation (3) after terms of order c^2 and making the substitution :

$$A_3 \sim M(A_2)^2/3 . \tag{5}$$

Linear plots are obtained over a larger range in c with the use of Equation (4), which has been used for most of the data presented here.

It should be noted that this instrument measures transmitted light as the incident beam rather than splitting off a portion of the beam prior to the sample as in some other light scattering instruments. Thus the Rayleigh factor is corrected for the absorption of the light by the sample, which is significant for the highly colored acid solutions. The extinction coefficient at 632.8 nm, which is far away from the maximum absorption, can be calculated from the ratio of the intensities of the transmitted beam through the solution and solvent. For the cured polyimide samples in sulfuric acid the absorption measured with this technique obeys Beer's law and yields $\varepsilon \sim 10$ cm^2/g with the concentration in g/ml and a pathlength of 1.5 cm. This is very small in comparison to other heterocyclic polymers yielding colored solutions in acidic solvents[6] and indicates that absorption of the 632.8 nm incident beam should not be a hindrance to light scattering for these materials. The possibility of a significant amount of fluorescence contributing to the observed scattering intensity from the acid solutions was eliminated by observing the scattering through a narrow bandpass filter in addition to the usual broadband red filter of larger transmittance. The Rayleigh factors obtained were the same with and without the bandpass filter indicating little or no inelastic scattering. All solutions for LALS measurements were clarified by filtration through 0.5 μm Fluoropore filters. These filters were found to be inert to concentrated sulfuric acid for the limited contact time required here. The observed solvent scattering was very small, similar to that observed from water or methanol, as expected. Use of other filters recommended for use with strong acids, such as Durapore (Millipore Corp.) or silver membrane filters (Selas Flotronics), produced enormous amounts of scattering with pure solvent, indicating degradation of the filter material. Use of the square root plot above (Equation (4)) yielded linear extrapolations except where polyelectrolyte effects were observed.

The differential refractive index increment, dn/dc, was measured using a Chromatix laser differential refractometer, at 25° C and 632.8 nm. Values for dn/dc were found to be independent of molecular weight over the range measured (2000 - 250,000), and yielded a value of 0.192 ml/g in distilled NMP. It should be noted here that measurement of dn/dc for commercial samples which are received as concentrated solutions is impossible by dilution of these with the solvent of interest. Although the mixed diluent used in most commercial materials may contribute little to the light scattered from a dilute solution, the contribution to the refractive index is substantial, on the order of that of the polymer itself. Here, we have used the dn/dc value measured on laboratory samples prepared in pure distilled NMP for all light scattering measurements in NMP. Precipitation of the commercial materials, followed by redissolution in distilled NMP yielded dn/dc of 0.192 in agreement with the laboratory synthesized samples. However, precipitation also resulted in a substantial loss of molecular weight due to hydrolysis, so only the dn/dc measurement was of interest.

RESULTS AND DISCUSSION

Molecular Parameters in Distilled NMP

Molecular weights (M_w and M_n), and second virial coefficients (A_2) measured by light scattering and membrane osmometry in distilled NMP are summarized in Table I for the samples indicated. Also given are the [η] values, in ml/g, for samples in distilled NMP and in the poorer mixed solvent NMP/dioxane (1:3 ratio); and the viscosities of the concentrated solutions (~10 wt%). These measurements were made in distilled NMP to avoid the polyelectrolyte effects observed in NMP used as received. A log-log plot of [η] versus M_w is shown in Figure 2. The data obey the Mark-Houwink relation:

$$[\eta] = k \, M^a, \tag{6}$$

yielding

$$[\eta] = 0.058 \, (M_w)^{0.74} \tag{7}$$

with [η] in ml/g in distilled NMP, using the weight average molecular weight obtained by light scattering. This relation is consistent with a linear flexible chain in a good solvent and

Table 1.   Molecular Parameters for Polyamic Acids

Sample	$10^{-3}\,M_w^{LS}$ daltons	$10^{-3}\,M_n^{OS}$ daltons	$10^4\,A_2^{LS}$ ml/g·dalton 25°C	$10^4\,A_2^{OS}$ ml/g·dalton 40°C	$[\eta]$ in NMP ml/g	$[\eta]$ in NMP/dioxone ml/g	η (cp) at ~10 wt %
PAA-1	4.5	—	43	—	28	—	20
PAA-2	9.0	—	40	—	50	—	70
PAA-3	37.0	—	29	—	132	—	700
PAA-4	77.0	—	21	—	245	—	4000
PAA-5	250.0	—	16	—	585	—	200(2%)
PAA-6	2.0	—	45	—	—	14	12
PAA-7	4.0	—	42	—	—	23	—
PAA-8	6.0	3.0	35	45	—	—	7500(32%)
PAA-9	9.1	—	38	—	48	—	—
PAA-10	9.8	—	37	—	49	—	—
PAA-11	10.4	—	48	—	50	44	—
PAA-12	16.0	—	35	—	72	61	—
PAA-13	16.0	—	30	—	74	—	—
PAA-14	22.0	—	31	—	87	—	—
PAA-15	29.0	15.0	25	32	100	—	10000(18%)
DuPont A	28.0	—	—	—	120	—	500
DuPont B	18.0	—	—	—	70	—	—

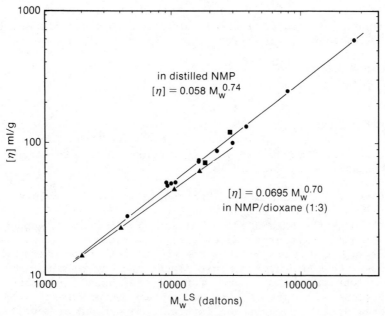

Figure 2.   Log $[\eta]$ versus log M_w for the polyamic acid in distilled N-methylpyrrolidone (NMP) (●,■) and in the mixed solvent NMP/dioxane (▲).

enables weight average molecular weights to be determined for a given sample using the easily available technique of capillary viscometry, provided the polydispersity is not too large. Here, no corrections have been made for polydispersity, since corrections would be minimal for samples with a most probable distribution in a good solvent[7]. For example, for a = 0.74, the viscosity average molecular weight, M_v, where

$$[\eta] = k \ (M_v)^a \tag{8}$$

is related to the weight average molecular weight, M_w, by

$$M_v/M_w = 0.93 \tag{9}$$

for a most probable distribution, which is within the uncertainty of the measurement. Experimentally determined molecular weights have been shown to be in agreement with those predicted from the monomer stoichiometry[1], which is evidence that the polymerization follows the expected chain condensation kinetics, without significant side reactions. In addition, membrane osmometry on two samples has yielded M_n values consistent with this distribution, Table I.

Equation (7) may be compared with other relations reported for the same polyamic acid in other solvents. In DMAc with 0.1 N LiBr Wallach[8] obtained:

$$[\eta] = 0.0185 \ (M_w)^{0.80} \tag{10}$$

and in pure DMAc and in a mixed solvent of DMAc and dioxane (1:2.5 ratio) Birshtein and co-workers[9] obtained

$$[\eta] = 0.0077 \ (M_w)^{0.80} \tag{11}$$

and

$$[\eta] = 0.20 \ (M_w)^{0.50} \tag{12}$$

respectively. The mixed solvent DMAc/dioxane was determined to be a θ-solvent by these workers[9].

These relations all give substantially smaller values for [η] for a given molecular weight than Equation (7) above for distilled NMP. Although the DMAc used by Wallach is also a very good solvent for PAA, and may be expected to give very similar results, the use of LiBr suggests that polyelectrolyte effects were observed. In this case the chain hydrodynamic dimensions may be more influenced by electrostatic interactions than by the usual polymer-solvent interactions. The use of LiBr to suppress these interactions would not necessarily reproduce the interactions present in a solution of an uncharged polymer. The data of Birshtein , which are also in DMAc, but without any LiBr, are inconsistent with both our data and that of Wallach. The measurements of [η] in the mixed solvent NMP/dioxane (1:3 ratio) from our laboratory (Table I and Figure 2) yield the relation:

$$[\eta] = 0.0695 \ (M_w)^{0.70}.\tag{13}$$

It was observed that addition of slightly more dioxane than the 1:3 ratio used caused precipitation of the polymer, which is consistent with the observation of θ-solvent behavior observed for a similar mixed solvent system used by Birshtein et. al., mentioned above. However, the exponent in Equation (12) above is larger than the 0.5 expected for poor solvents. These data have only been obtained for a limited number of low molecular weight samples, which are not expected to obey relations valid only for substantially longer chains. It is likely that as very low molecular weights (DP ≤10) are approached, the unusually long rigid backbone segment prevents the chain from coiling completely and attaining a random coil conformation.

Polyelectrolyte Effects

As mentioned above, polyelectrolyte effects were observed in dilute solutions prepared in NMP as received, without redistillation. Dilute solution viscometry yielded viscosities increasing with dilution, as shown in Figure 3 for sample PAA-4. Addition of a small amount of water to a solution in distilled NMP resulted in a lower intrinsic viscosity, but no polyelectrolyte effect. This viscosity effect has been observed previously[10] for polyamic acid solutions with a small amount of triethylamine (Et_3N) added. The behavior observed by these workers for a lower molecular weight sample is also shown in Figure 3 for comparison. Apparently, impurities in the NMP, such as amines, are basic enough to abstract a proton from the carboxilic acid groups, resulting in a partially charged macromolecule. In very dilute solution, the charges on different segments of the same molecule repel, greatly increasing the hydrodynamic size of the molecules, and thus their contribution to the viscosity.

Light scattering in the as received NMP also showed some unusual effects in very dilute solutions, as shown in Figure 4. The intensity of the light scattered from very dilute (< 2 mg/ml) solutions was reduced, producing curving plots which could not be extrapolated to infinite dilution to yield a molecular weight. This effect was more pronounced for higher molecular weights. In the case of the highest molecular weight sample studied (PAA-5), the effect was so strong at low concentrations that extreme negative dependence of the reciprocal Rayleigh factor was observed (see Figure 4). Addition of LiBr (0.005N) to the solutions in NMP with decreased scattering caused the scattering to increase somewhat, although not to its expected value. Larger amounts of LiBr caused the polymer to precipitate from solution. This effect on the

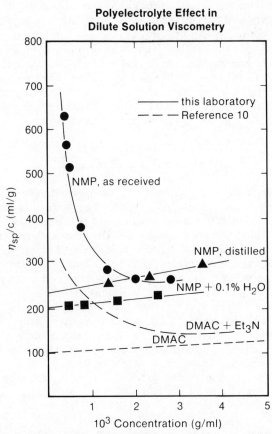

Polyelectrolyte Effect in Dilute Solution Viscometry

Figure 3. The reduced viscosity, η_{sp}/c as a function of concentration for PAA-4 in as-received and distilled NMP, showing the polyelectrolyte effect obtained in the as-received solvent. Also shown are similar data obtained by Bower and Frost[10] in DMAc and DMAc with triethylamine added.

scattered light intensity could not be caused by the intramolecular repulsion presumably responsible for the increased dilute solution viscosity discussed above. However, in this concentration range, for this molecular weight, it is likely that intermolecular repulsions are also present. These intermolecular forces can produce a non-random distribution of the scattering molecules, leading to a reduction in the observed intensity. For example, at ~5 mg/ml, sample PAA-4, with a M_w of 77,000 and an $[\eta]$ of 245 ml/g has its molecules essentially close-packed, and cannot space molecules farther apart to minimize electrostatic repulsion between chains. At lower concentrations, the molecules have the freedom to minimize this repulsive force by increasing the distance from their nearest neighbors, leading to a non-random distribution. Addition of a salt, such as LiBr, provides a population of charges which act to screen the repulsive forces so that the molecules then behave as uncharged species. This technique has been reported for this polymer using another amide solvent, dimethylacetamide (DMAc), which may be expected to contain similar base impurities as the NMP used here.

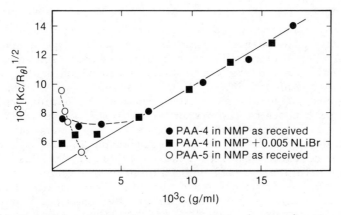

Figure 4. Reduced scattering intensity observed at very low concentrations ($[\eta]c < 1$) in as-received NMP, showing intermolecular repulsions leading to a degree of ordering in the system.

Cured Materials in Sulfuric Acid

Plots of $(Kc/R_\theta)^{1/2}$ versus concentration for two grades of a commercial polyimide film and for a commercial polyamic acid cured thermally in our laboratory with various cure cycles are shown in Figure 5. Although the uncertainty in the data is somewhat larger than is observed in organic solvents, the weight average molecular weight can be obtained. These molecular weights, which are tabulated in Table II, are the same within the experimental uncertainty, and are not substantially different from the precursor polyamic acids for which M_w is known. The $[\eta]$ values in sulfuric acid are also similar to those obtained for the precursor polyamic acid. The dilute solution viscosity is independent of solution age, in contrast to prior reports[4].

These data indicate that neither the chain dimensions nor the molecular weight are altered substantially during the curing process. It is important to emphasize that the flexibility of the cured chain as indicated by the intrinsic viscosity is an equilibrium measurement of the average hydrodynamic dimensions of the chain in sulfuric acid. It is not sensitive to local hindrances to motion which contribute, for example, to the extremely high glass transition temperature observed for the cured polyimide in comparison to the precursor polyamic acid. In addition, the chain is protonated in the strong acid solvent, and may assume somewhat different dimensions than in the bulk state. However, the general conclusions that no significant branching or cross-linking occurs, and that differences in the precursor molecular weights are carried over to the final cured material are valid.

Several other observations concerning various cure conditions are also noted in Table II. Samples cured with temperatures only up to 150 °C were found to be incompletely cured and degraded in sulfuric acid to produce a somewhat turbid black mixture. Samples cured between 200 and 300 °C were soluble in sulfuric acid, producing orange solutions which yielded molecular weights in agreement with those of the precursor material. Measurment of these solutions over a period of up to a week gave a constant molecular weight, indicating little or no degradation in the strong acid. This is in disagreement with results reported by Wallach[4], who observed a relatively rapid degradation in the viscosity of dilute solutions of the cured polyimide in sulfuric acid. Experiments carried out using samples cured first by chemical dehydration, using acetic anhydride and pyridine, followed by thermal cure at 200°, clarified this discrepancy. These materials were soluble in sulfuric acid, but produced solutions with an intense red color, as was observed by Wallach for samples also chemically cured. The red color diminished slowly with time and the samples were indistinquishable in color from thermally cured samples after ~24 hours. At this point, LALLS measurements yielded molecular weights of only ~7,000. It has been demonstrated[11] for this system that

Table II. Measurements of Cured Pl in 97% H_2SO_4

Sample	$10^{-3} M_w^{LS}$ PAA (daltons)	Cure Conditions	Film (μm) Thickness	Age of Solution	$10^{-3} M_w^{LS}$ PI (daltons)	$[\eta]$ in H_2SO_4 ml/g
Kapton®	—	—	7	5 hours	18	100
Kapton®	—	—	7	4 days	16	—
Kapton®	—	—	12	3 days	20	—
Kapton®	—	—	25	3 days	18	—
Kapton®	—	—	50	3 days	23	—
DuPont A	28	thermal, 150° C	8-10	degraded	—	—
DuPont A	28	thermal, 200° C	8-10	1 day	20	—
DuPont A	28	thermal, 300° C	8-10	1 day	22	—
DuPont A	28	thermal, 300° C	8-10	14 days	25	—
DuPont A	28	thermal, 300° C, 400° C	8-10	insoluble	—	—
DuPont A	28	chemical, 150° C	8-10	4 days	7	—
DuPont A	28	chemical, 200° C	8-10	4 days	9	—
PAA-11	10.4	thermal, 300° C	8-10	7 hours	9	50
PAA-11	10.4	thermal, 300° C	25	4 days	11	—
PAA-3	37.0	thermal, 300° C	8-10	1 day	30	—

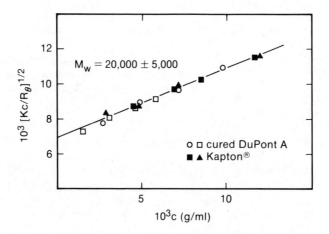

Figure 5. $[Kc/R_\theta]^{1/2}$ versus c for thermally cured samples of DuPont A PAA and for a sample of commercial PI film (Kapton®) in solution in concentrated sulfuric acid, showing that the same molecular weight is obtained for both. The strong concentration dependence indicates that the sulfuric acid is a good solvent for PI.

chemical curing produces a small percentage of the isoimide rather than the imide, which then is converted to the imide at high temperatures. The presence of a small residual amount of this isoimide moiety in the chemically cured samples may produce the red color upon protonation by the acid and then slowly degrade, leading to lower molecular weights and viscosities. An intense red color upon protonation has been observed for other heterocycles containing structures similar to the isoimide. The molecular weight of ~7,000 would indicate 1-2 scissions per chain, or ~1-2% residual isoimide after the 200° thermal treatment. Samples baked at 400°C for periods of 1 hour or more after a 300°C thermal cure were no longer soluble in sulfuric acid, and molecular weights could not be measured. Some further physical or chemical process occuring above 300°C leads to eventual insolubility, although more work is needed to identify this change. These observations have brought some consistency to the varied conclusions found in the literature concerning the solubility and stability of this polyimide in concentrated sulfuric acid.

We have shown that dilute solution measurements on polyamic acids and polyimides provide an understanding of molecular sructure which is not obtained by measurement of concentrated solution viscosity alone. Data obtained in amide solvents such as NMP, DMAc or DMF can be affected by trace impurities in the solvent, leading to polyelectrolyte effects. Measurement of both dilute solution viscosity and molecular weights by light scattering have shown that the cured polyimides retain essentially the same molecular weight and molecular conformation as their precursor polyamic acids, provided that the cure is extended to sufficiently high temperature. Samples which are cured chemically appear to require a high temperature post-bake to completely convert residual isoimide to the more stable imide.

REFERENCES

1. W. Volksen and P.M. Cotts, these proceedings, vol. 1, pp. 163-170.
2. L. W. Frost and I. Kesse, J. Appl. Polymer Sci., 8, 1039 (1964).
3. DuPont product literature, Technical Bulletin on Pyralin.
4. M. L. Wallach, J. Polymer Sci.:A-2,7,1995 (1969).
5. E. F. Casassa and G. C. Berry, in "Polymer Molecular Weights Part I", R. E. Slade, Editor, Marcel Dekker, New York, 1975.
6. G. C. Berry and T. G Fox, J. Macromol. Sci. Chem., A3,1125 (1969).
7. P. J. Flory, "Principles of Polymer Chemistry", Cornell University Press, Ithaca, New York, 1953.
8. M. L. Wallach, J. Polymer Sci.:A-2,5,653 (1967).
9. T. M. Birshtein et. al., European Polymer J.,13,375 (1977).
10. G. M. Bower and L. W. Frost, J. Polymer Sci.:A,1,3135 (1963).
11. P. Buchwalter and A.I. Baise, unpublished data.

POLYIMIDE CURE DETERMINATION

R. Ginsburg and J. R. Susko

IBM Corporation
General Technology Division
Endicott, New York 13760

A technique was developed for monitoring the conversion of polyamic acid to polyimide. The technique was used to characterize the cure reaction and to assist in understanding the relationship between degree of cure and etchability of films by a wet alkaline method. Interpretation of infrared spectra using a band ratio method gave consistent degree of cure determinations. Dynamic monitoring of films undergoing cure demonstrated that the rate of reaction is dependent on temperature and film thickness. Thicker films or thicker regions of nonuniform films were found to cure to a greater extent than thinner films or regions with identical thermal processing. Etchability was shown to have a high linear correlation with degree of cure. The study indicated that for optimal etch results film thickness uniformity and thermal treatment must be closely controlled.

EXPERIMENTAL AND RESULTS

The polyimide material investigated was a commercial pyromellitic dianhydride/4,4'-oxydianiline (PMDA/ODA) system. Polyamic acid, prepared by the polycondensation of PMDA and ODA in n-methyl pyrollidone (NMP), reacts by condensation cyclization to form an imide (Figure 1).

237

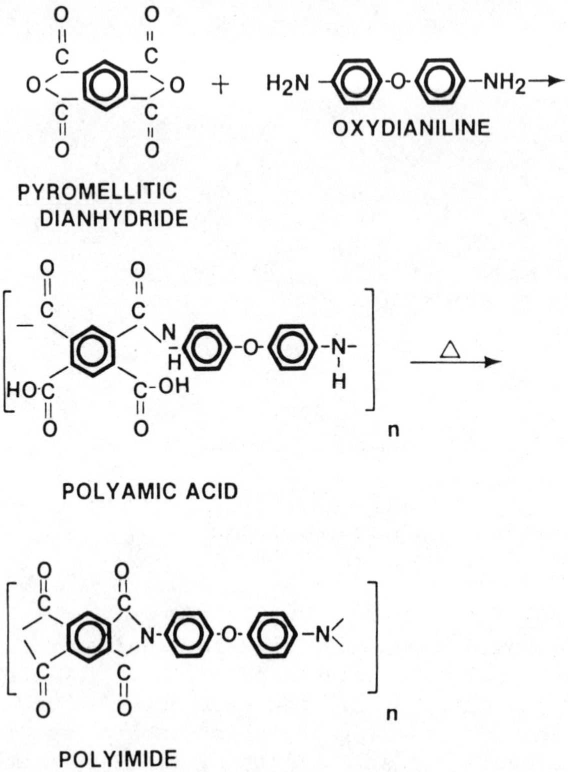

Figure 1. Pyromellitic dianhydride reacts with oxydianiline to form a polyamic acid which, when heated, undergoes a condensation cyclization to form the imide.

The technique employed for characterization of the polymer films was Fourier Transform Infrared (FTIR) spectroscopy. Spectra of polyamic acid/polyimide films were analyzed by a band ratio method.[1,2] The area of the symmetric carbonyl stretch at 1776 cm^{-1} [3] was ratioed with a reference aromatic vibration at 1012 cm^{-1}.[4] An absorbance spectrum of a partially cured film (Figure 2) includes the regions used for these measurements. Per cent imidization was calculated by normalizing a sample band ratio with that of a fully cured polyimide film. These absorbance bands were chosen because they were fairly well isolated from the rest of the spectrum, and they were relatively weak, i.e., their intensity could be quantified even with thick films. Spectral subtraction routines were used to further isolate the bands for more accurate measurement. A spectrum of uncured polyamic acid was subtracted to isolate the imide carbonyl absorbance from adjacent bands. To correct the standard band, a spectrum of the solvent, NMP, was interactively subtracted. There was an absorbance due to solvent at 980 cm^{-1} which interfered with the quantification of the standard band at 1012 cm^{-1}.

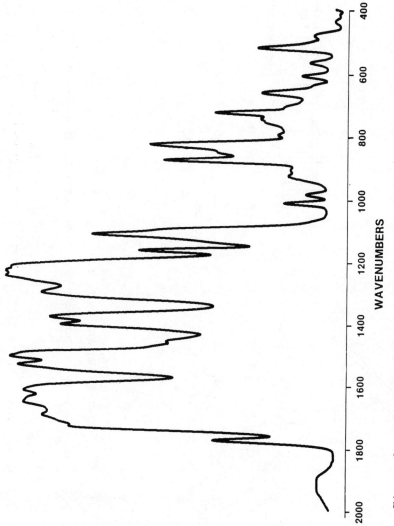

Figure 2. Infrared spectrum of partially cured polyimide. Carbonyl absorbance at 1776 cm^{-1} and aromatic vibration at 1012 cm^{-1} were used in quantifying cure.

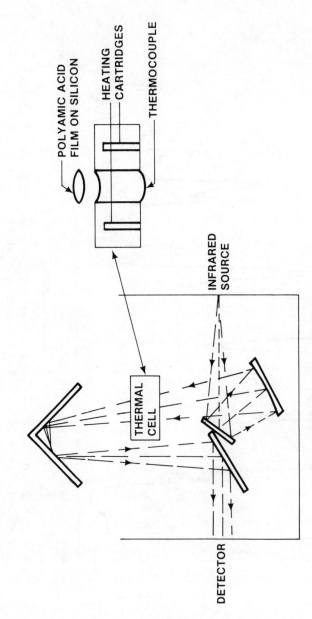

POLYAMIC ACID
FILM ON SILICON

HEATING
CARTRIDGES

THERMOCOUPLE

FTIR SPECTROPHOTOMETER
SAMPLE COMPARTMENT

INFRARED
SOURCE

THERMAL
CELL

DETECTOR

Figure 3. Accessory for FTIR spectrophotometer which permits
horizontal positioning of sample films in thermal cell.

240

The cure determination technique was first used in a set of dynamic experiments. Wet polyamic acid film samples of varying thickness were prepared on silicon substrates. Each sample was continuously scanned as the cure was advanced in a thermal cell in the spectrophotometer's sample compartment. A thermal cell was assembled which would permit the wet films to be positioned horizontally (Figure 3). The FTIR was programmed to collect scans for spectra at 4 minute intervals over a 40 minute period as the temperature was ramped and held at 110°, 130°, or 150° C.

The rate and extent of cure were found to be dependent on both temperature and film thickness. Plots of the experimental results are presented in Figures 4, 5, and 6. The slope of the plots indicates an initial rapid rate followed by a more gradual conversion. The "knee" in the curve has been observed previously[5] and is thought to result from the stiffening of the film which slows the reaction by reducing the availability to each other of the reacting groups. The more rapid reaction with thicker films is attributed to the increased retention of solvent permitting increased molecular chain motions.

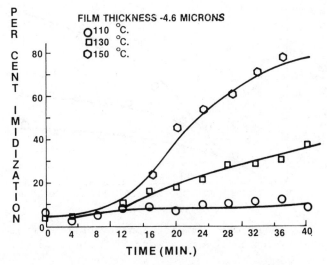

Figure 4. Per cent imidization versus time as determined in dynamic experiments by spectral band measurements. Experiment was run at each of three temperatures for thin (4.6 μm) films.

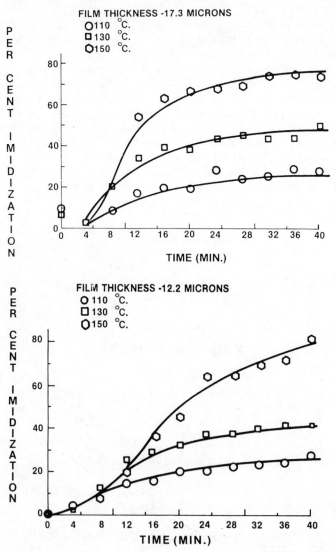

Figures 5 & 6.   Per cent imidization versus time from dynamic
experiments.   Film thicknesses were 12.2 and 17.3 μm.

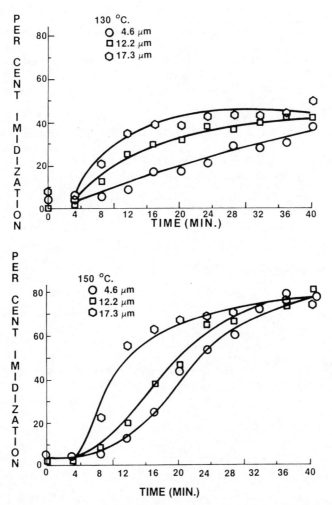

Figures 7 & 8. Per cent imidization versus time at 130° C and 150° C for three film thicknesses. Dynamic cure data was plotted to demonstrate the convergence of the degree of cure of films of varying thickness with time.

When cure values for films of varying thickness treated at the same temperatures are plotted on the same axes, there is initially a considerable divergence of the curves (Figures 7 & 8). However, with time, these values converge. This phenomenon has been demonstrated with processed films, i.e., imidization is uniform despite thickness differences if the time is sufficiently extended.

Solvent loss during cure can also be monitored by the FTIR technique. The solvent subtraction factors resulting from the 1012 cm^{-1} band correction described above were plotted to show the pattern of solvent loss (Figure 9). After an initial rapid desorption, the solvent remained constant at constant temperature.

When thickness differentials were present within a film, the cure was also seen to vary. By dipping silicon substrates into polyamic acid solution, films were prepared which had a thickness gradient. These films were baked for 5, 10, and 15 minutes at 130° C and measured for cure. The observed cure

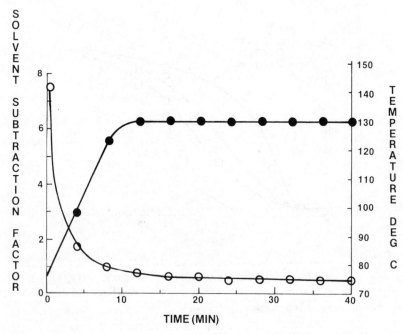

Figure 9. Solvent content and temperature of polyimide for 130° C run are plotted versus time. The FTIR subtraction factor (open circles) is a solvent concentration value. The film temperature (filled circles) was monitored by a temperature probe in contact with the silicon substrate.

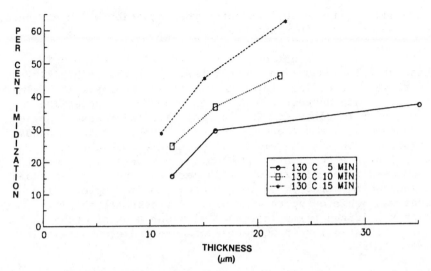

Figure 10. Per cent cure of thickness gradient films. Individual films were prepared in which thickness varied from about 10 to greater than 20 μm. After baking these films at 130° C, degree of cure was determined for thin, intermediate, and thick regions of each film.

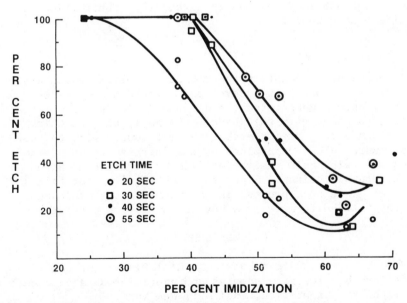

Figure 11. Percent etch versus percent imidization for a sample matrix with varying film thickness (7-30 μm) and degree of cure (25-70%). Complete film removal was not achieved as imidization increased. Etchant removed more film as etch time increased.

differentials (Figure 10) indicates the variation which can occur with films of nonuniform thickness.

FTIR cure determinations correlated well with the bulk etching of films on silicon substrates. Both the etchability and cure were determined for a matrix of 70 samples. The films were etched with aqueous KOH and then measured to determine per cent of film removed. The initial film thicknesses ranged from 7 to 30 μm, and the etch duration from 20 to 45 seconds. In Figure 11, bulk film removal is plotted versus degree of cure. With an etch time of 20 seconds, complete film removal was achieved with imidization of 25 per cent and then decreased with increasing cure. With longer etch times, complete film removal was achieved up to 42% cure. Per cent etchability of highly imidized films increased with duration of etch. This demonstrates that both degree of cure and etch parameters control etchability.

A cure/etch relationship was also demonstrated for replicate sets of partially cured 10 μm films. These samples were subjected to nonuniform thermal processing. While one subset was retained for cure measurement, small patterns were etched in the other subsets and the resultant dimensions were measured. With these samples, the degree of cure ranged from 36 to 90 per cent. A high linear correlation (.99+) was calculated for the etched pattern dimensions and the predetermined cure values. These data illustrate the usefulness of the cure determination technique for predicting etchability and the requirement of controlling cure for uniform etch results.

SUMMARY

A technique to quantify the degree of polyimide conversion using FTIR spectroscopy was developed. The dynamic measurement of thermal cure demonstrated the dependence of reaction rates on temperature and film thickness. The rapid cure of thick films accounts in part for etch rate differences with thicker layers. Cure values determined by the infrared technique correlated well with bulk removal of film by alkaline etch and etch pattern dimensions. Since the time required for film removal in a wet etch process is dependent on both thickness and cure, film uniformity is a key to controlled etching.

ACKNOWLEDGEMENTS

The authors wish to acknowledge Carl Diener for contributing to this work by preparing samples and providing data on etchability. He is with IBM Corporation, Endicott, New York.

REFERENCES

1. Chicago Society for Paint Technology, "Infrared Spectroscopy: Its Use in the Coatings Industry", pp. 52-62, Federation of Societies for Paint Technology, Philadelphia, 1969.
2. J. Hashem, H. A. Willis, and D. C. M. Squirrel, "Identification and Analysis of Plastics", pp. 328-329, Iliffe Books, London, 1972.
3. H. Ishida, S. T. Wellinghoff, E. Baer, and J. L. Koenig, Macromolecules, 13, 826 (1980).
4. A. N. Krasovskii, N. P. Antonov, M. M. Koton, K. K. Kalnin'sh, and V. V. Kudryavtsev, Polymer Science U.S.S.R., 21, 1038 (1980).
5. M. M. Koton, Vysokomol.soyed, A13, 1348 (1971).

IN-SITU MONITORING OF POLYAMIC ACID IMIDIZATION WITH

MICRODIELECTROMETRY

D. R. Day and S. D. Senturia

Department of Electrical Engineering and Computer
Science and Center for Materials and Engineering
Massachusetts Institute of Technology
Cambridge, Massachusetts 02139

The kinetics of the imidization reaction of various polyimides used in microelectronic applications
have previously been studied by infrared spectroscopy.
This paper reports the use of a new dielectric technique, called Microdielectrometry, to monitor the imidization reaction of PMDA-ODA polyamic acid films deposited on the surface of an integrated circuit. Dielectric
permittivity and loss factor in the frequency range
1-1000 Hz were measured with Microdielectrometry probes
during imidizations at temperatures between 100 and
160°C. At temperatures below 130°C, the permittivity
remains dispersive even after five hours; whereas at
higher temperatures, the permittivity becomes constant
in less than two hours. The frequency dependence of the
loss factor indicates that the film exhibits a weakly
dispersive bulk conductivity that varies between 3×10^{-10}
and 5×10^{-14} (Ohm cm)$^{-1}$, increasing with temperature but
decreasing with degree of imidization. At temperatures
below 130°C, the conductivity remains relatively high
even after five hours; whereas above 130°C, the conductivity decreases much more rapidly to low values,
although conductivity changes are still detectable in
the films at 2.5 hours. Results show a significant
change in rate of reaction between 120 and 140°C, but
with clearly observable changes still taking place well
after the nominal one hour cure time.

INTRODUCTION

The kinetics of the imidization reaction of various polyimides used in microelectronic applications have been previously studied by infrared spectroscopy[1,2,3]. It has been generally observed that for PMDA-ODA, the temperature range where significant imidization occurs is from 100 to 160°C, and that the degree of achievable imidization increases markedly through this range. The present paper examines this imidization reaction using a new dielectric technique, called Microdielectrometry[4], in which a small integrated circuit probe with a pair of interdigitated electrodes and added amplification circuitry serves to monitor both the low frequency dielectric permittivity and loss factor of films placed on the surface of the integrated curcuit. The loss factor can, in turn, be interpreted in terms either of conductivity or dipole orientation effects, depending on the observed frequency dependence. As will be seen below, conductivity effects dominate the loss factor for the polyamic acid imidization reaction studied in this work.

Because the Microdielectrometer probe (or "chip") is itself a silicon integrated circuit, its geometry is ideal for the in-situ study of films intended for microelectronic applications. A micro-photograph of a Microdielectrometer chip is shown in Figure 1. The interdigitated electrode pattern is the portion of the device where contact with the sample under study is important. A typical cross section through the electrode portion is illustrated in Figure 2. One of the interdigitated electrodes is driven with a sine wave,

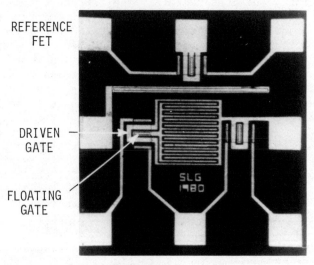

Figure 1. Photograph of microdielectrometer chip (chip dimensions are .075 x .075 inches).

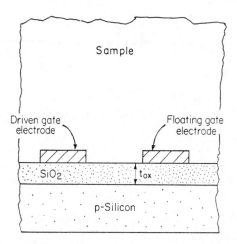

Figure 2. Cross section of the interdigitated electrode section of the microdielectrometer chip.

and the other serves as the input of a differential amplifier, the details of which have been published elsewhere[5]. The sinusoidal steady state transfer function between the driven electrode and the response observed at the amplifier output can be interpreted through a calibration table to yield the permittivity and loss factor at the frequency of the driving sinusoid. This calibration is independent of film thickness, provided that the film is thicker than the spacing between the electrode fingers on the device, which, for the sensor used in this work is 12 μm. If the film is thinner than 12 μm, one can still obtain an accurate measurement of the sheet conductance of the film[5], which is the bulk conductance multiplied by the film thickness plus any surface conductance that may be present. However, determination of the permittivity of thin films is much less precise than for thick films, even though relative changes within a sample as a function of time, or between samples of the same thickness, can be reliably interpreted in a semi-quantitative way.

EXPERIMENTAL

Experiments were carried out with a PMDA-ODA polyamic acid precursor/NMP solution (14% solids, supplied by Du Pont) prefiltered to 0.2 μm. Both thin films (1 μm thickness) and thick films (>12 μm thickness) were studied. Thin films were spin deposited on A-1100 (γ-aminopropyltriethoxysilane) treated microdielectrometer sensor chips where the sensor surface consisted of SiO_2 and aluminum. Thick films were formed by dipping A-1100 treated sensors into the polyamic acid solution allowing the excess to run off. Both thick and thin films gave qualitatively similar results.

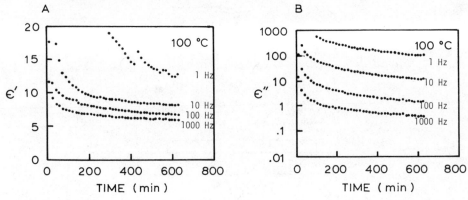

Figure 3. Plot of A) ε' and B) ε" as a function of reaction time of PMDA-ODA at 100°C.

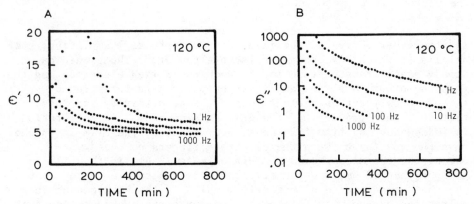

Figure 4. Plot of A) ε' and B) ε" as a function of reaction time of PMDA-ODA at 120°C.

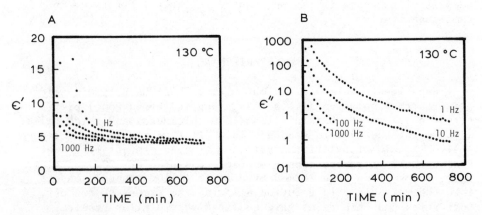

Figure 5. Plot of A) ε' and B) ε" as a function of reaction time of PMDA-ODA at 130°C.

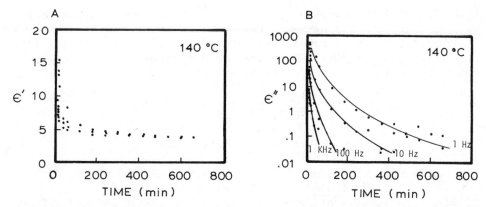

Figure 6. Plot of A) ε' and B) ε'' as a function of reaction time of PMDA-ODA at 140°C.

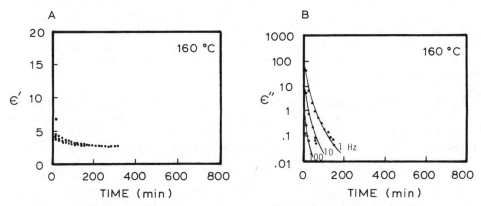

Figure 7. Plot of A) ε' and B) ε'' as a function of reaction time of PMDA-ODA at 160°C.

Because of the more accurate calibration available for the thicker films, only the thick film results are reported here. Following deposition of the films from solution, the film coated devices were placed in vacuum for 24 hours at room temperature to remove unbound solvent. The devices were then subjected to various cure temperatures, and the dielectric permittivity ε' and loss factor ε'' were measured as functions of time and temperature for frequencies in the range of 1-1000 Hz. Automated data logging equipment was used to record the results and control the sequential changes of frequency during each imidization experiment.

At present, the calibration of the measurement is well defined for loss factors greater than about 0.1 and less than about 1000. The contribution of a bulk conductivity to loss factor is proportional to the conductivity divided by the angular frequency. For the loss factor range cited above, and for frequencies in the range 1-1000 Hz, the corresponding range of conductivities covered by the measurement is $5 \times 10^{-7} - 5 \times 10^{-14}$ (Ohm cm)$^{-1}$. Using the best available calibration table, we estimate the accuracy of the values of ε' and ε'' obtained from the thick films to be about 10%, except at low loss factors approaching 0.1, where the present sensor and measurement apparatus still have some residual systematic errors.

RESULTS AND DISCUSSION

Microdielectrometry Data

Cures were conducted at 100, 120, 130, 140, and 160°C, during which ε' and ε'' were monitored at frequencies of 1, 10, 100, and 1000 Hz. The results are shown in Figures 3-7, with the (A) portion of each figure being the permittivity versus time and the (B) portion being the loss factor. There are several noteworthy features in the data. First, at 160°C (Figure 7), the value of ε' drops to a non-dispersive (i.e., frequency independent) value within 100 min., while the corresponding loss factor reaches a stable value at about 150 min. Taking these stable values as representing the behavior of fully imidized films, it is seen that at all other temperatures, there are still observable changes in properties occurring at three hours and beyond.

The apparent frequency dispersion in ε', particularly in the partially imidized films, is somewhat surprising. The origin of this dispersion is believed to be the combined effect of an ionic conduction mechanism in the film together with blocking electrodes. When the drive signal is applied, ions drift through the bulk of the sample toward the electrodes, but the blocked electrodes do not support redox reactions. As a result, ionic charge accumulates near the electrodes until the phase of the driven signal reverses, at which point the drift direction reverses. As measured externally from the electrodes, this phenomenon is indistinguishable from a large effective permittivity in the medium in which the charge would be polarization charge rather than ionic charge. An analysis of this effect has been made by MacDonald[6,7], and observations of this effect in Microdielectrometry measurements on epoxy-amine systems have been reported[8]. An alternative explanation of the dispersion in ε' could be a hindered dipole orientation mechanism (see Reference 8 for a discussion). However, the frequency dependence of ε'' does not support such an explanation, as discussed below.

A good test of the appropriate model is to examine the frequency dependence of ε'' at various temperatures and for various degrees of imidization. If the material has a non-dispersive bulk conductivity, then a log-log of ε'' versus frequency should yield a straight line with a slope of -1, whereas for a dipole orientation mechanism, peaks in the ε'' versus frequency are seen at that frequency which is equal to the reciprocal of the dipole relaxation time. Figure 8 shows selected data for various temperatures and states of imidization. The open points represent the behavior of ε'' versus frequency at successive times during the cure at 120°C. The solid points show the data for the sample initially cured at 100°C for 300 min., then rapidly heated (32°C/min) first to 120°C, then to 140°C. A total time of less than 60 sec. was spent at each temperature. Due to rapid ramping of temperature and short measurement times, little reaction occurred in the 100°C samples while at the elevated temperatures. In all cases, the data support the conductivity rather than the dipole orientation model, but the slope is less than –1. In fact, there is a characteristic shape to the dispersion in ε''. From the data for the 100°C cure, it is seen that as temperature increases, the ε'' curve shifts both toward higher values and higher frequencies. From the data for the 120°C cure, it is seen that as cure increases, the curve shifts toward lower values and lower frequencies. Therefore there are two effects; an increase in conductivity with increasing temperature as shown by the measurement on a short time scale with a given cure state at elevated temperatures and a decrease in conduction with time at a given temperature as imidization proceeds. Further

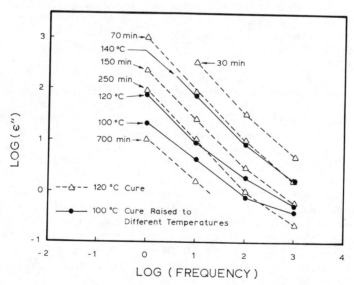

Figure 8. ε'' as a function of various temperatures and states of imidization.

experiments will be required to better document the shape of this dispersive conductivity, including extension of the measurement frequency both to lower and higher values. In addition, an assessment of the relative contribution to the ionic conductivity of unimidized acid groups, water released during imidization, residual solvent, and unreacted endgroups would be of great interest.

Further examination of the behavior of the ε" data with time in Figures 3-7 shows that at each temperature, there is an initial rapid drop followed by a long period of relatively slow change. For example, at 120°C (Figure 4), ε" changes by an order of magnitude in the first hour and by a second order of magnitude over the next six hours. When attempting to control the properties of a material, such as during prebake prior to wet etching, one would generally prefer to carry out process steps in regions where the relative rate of change of the property of interest is slow. This suggests that one could use the ε" data as a guide to achieving a reproducible degree of imidization by processing for successively longer times at successively lower temperatures. When trying to relate electrical properties to degree of imidization at various temperatures, however, one must correct for the intrinsic temperature dependence of the electrical properties. For example, in heating a film from 100 to 120°C, the ε" values increased by a factor of 6. Similar factors were measured between 100, 120, 130, 140, and 160°C to produce scaled ε" values ('equivalent 160°C' values) versus cure time using the 1 Hz data obtained at each cure temperature. These values are plotted in Figure 9, with the axis converted to a conductivity scale. Recalling that a decrease in conductivity corresponds to increased cure, the data in Figure 9 are similar in appearance to imidization-conversion curves determined from IR[1,2,3], where changes occur rapidly at first and then level off after a period of time.

Note in Figure 9 that at a reaction time of zero, all samples should have an equivalent cure state and, therefore, the scaled conductivities should all reach the same equivalent 160°C conductivity value. In fact, even though the scaling factors used to develop Figure 9 were obtained from samples that had undergone several hours of cure, when the data are scaled, all curves do appear to start from a common point. Whereas the IR data, usually presented on a linear conversion scale, level off after 60 min.[1,3], the 160°C equivalent conduction data (plotted on a log scale) show that changes still occur at all temperatures even after 10 hours at a given temperature. In spite of the fact that a precise correspondence between the conductivity data and degree of imidization cannot be made, the equivalent 160°C conductivity data nevertheless do provide an indication of the state of cure of the material and the rate at which that state changes.

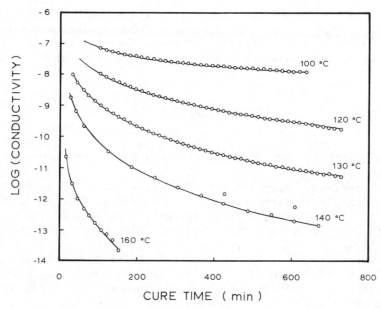

Figure 9. Equivalent conductivity at 160°C plotted as a function of reaction time at various temperatures (1 Hz data only).

SUMMARY

Microdielectrometry has been shown to be a good technique for monitoring dielectric properties during the imidization of polyamic acid films. The results indicate that above 130°C the dielectric constant rapidly converges to values of 3.5 at all frequencies, and conduction becomes low; whereas below this temperature, the dielectric constant remains dispersive with frequency and the the conduction remains relatively high. In contrast with infrared data, which suggest that reactions level off after approximately one hour, results presented here indicate that significant changes are still occurring in the film well after 10 hours at a given cure temperature.

ACKNOWLEDGEMENT

This work was supported in part by the E. I. du Pont de Nemours & Company, and used Microdielectrometry sensors and apparatus from a program supported in part by the Office of Naval Research, with additional equipment provided by the National Science Foundation under Grant ENG-7717219. Polyimide materials were obtained from du Pont. The authors wish to acknowledge the contributions of Norman F. Sheppard, Jr., and Huan L. Lee of MIT for their development of the Microdielectrometry instrumentation and the corresponding calibration curves. Thanks are due to Ed Yuan of du Pont for his participation in technical discussions of this work.

REFERENCES

1. Y. K. Lee and J. D. Craig, "Polyimide for Microelectronic Applications", ACS Organics Coatings and Plastics Chemistry Preprints, $\underline{43}$, 451 (Aug. 1980).
2. G. Samuelson, "Polyimide for Multilevel VLSI and Alpha Protection", from lecture notes for "Polyimide Coatings for Microelectronics with Applications", Continued Education in Engineering, University Extension, Univ. of Cal., Berkeley, (1981).
3. P. van Pelt, D. J. Belton and S. R. DiIorio, "Dissolution Behavior of Polyimide Films on a Silicon Substrate", paper presented at the 160th Meeting of the Electrochemical Society, Denver, CO, Oct. 11-16 (1981).
4. N. F. Sheppard, D. R. Day, H. L. Lee, and S. D. Senturia, "Microdielectrometry", Sensors and Actuators, $\underline{2}$, 263 (1982).
5. S. L. Garverick and S. D. Senturia, "An MOS Device for AC Measurement of Surface Impedance", IEEE Trans. Elec. Dev., $\underline{ED-29}$, 90 (1982).
6. J. R. MacDonald, Phys. Rev., $\underline{92}$, 4 (1953).
7. J. R. MacDonald, Phys. Rev., $\underline{91}$, 412 (1953).
8. H. L. Lee, S. M. Thesis, Massachusetts Institute of Technology, August 1982, unpublished.

258

STUDIES ON THERMAL CYCLIZATION OF POLYAMIC ACIDS

Shun-ichi Numata, Koji Fujisaki, and Noriyuki Kinjo

Hitachi Research Laboratory
Hitachi Co., Ltd.
Hitachi-shi, Ibaraki-ken, Japan

The imidization reaction of various polyamic acids having different chemical structures has been followed by measuring the weight losses that occurred during dehydro-cyclization. From these studies it was found that when polyamic acids were heated rapidly to a given temperature, the imidization reaction proceeded during the temperature rise but slowed down very markedly after the given temperature was reached. The temperature at which the imidization reaction ended was closely related to the glass transition temperature of the resulting polyimide. Based on these observations, it is concluded that the imidization reaction slows down markedly because the glass transition temperature of the polymer rises as the reaction proceeds, and molecular motion is frozen. In other words, the free rotation of amide bonds in the main chain is frozen. As a result, suitable conformation for imidization cannot take place any more.

INTRODUCTION

Among the organic polymers, aromatic polyimides have various advantages, such as high thermal decomposition temperature, high glass transition temperature(Tg), high mechanical strength, etc.[1~3] Accordingly, they are being studied with keen interest. Possible applications of polyimides have been extensively researched, especially with respect to insulation films for electrical equipment and electronic devices, such as tape carriers[4], flexible printed circuit boards[5], and multilevel metallization of LSIs[6,7].

259

It has been known for some time that polyimides are synthesized through a dehydro-cyclization reaction of polyamic acids. Therefore, the kinetics of the imidization reaction has been rather thoroughly researched. The results of this research indicate that in an imidization reaction of a polyamic acid brought about by heating, the reaction rate slows down as the reaction proceeds.[8~10] A decrease in entropy resulting from a decrease in the amount of solvent,[8] a decrease in the number of functional groups,[9] and a change in the skeleton motion[10] have been cited as the causes for this slowdown. Laius and others[10] followed the imidization reaction in the temperature range of 150°C to 200°C and confirmed that it almost stopped when the softening temperature of the partially imidized polyamic acid approached the reaction temperature. Also, Lavrov and others[11,12] followed the imidization reaction of amide acid in solvents and reported on the relationships between the reaction rate and the electron density on nitrogen atoms in the amide group and the basicity of solvents.

However, in all these investigations, the percent conversion to imidization were determined quantitatively using the infrared spectroscopy method and, therefore, the precision of the measurements is in doubt. Considering that the imidization reaction is a condensation reaction, the authors attempted to follow it by measuring the weight losses that occur during the dehydro-condensation of polyamic acid.

EXPERIMENTS

Specimen

The polyamic acids used in our experiments were obtained by a reaction between diamine and tetracarboxylic dianhydride, as shown in Table 1. The process of polyamic acid synthesis and imidization is shown in expression (1) below.

$$H_2N-R_1-NH_2 + O\overset{CO}{\underset{CO}{\diamondsuit}}R_2\overset{CO}{\underset{CO}{\diamondsuit}}O \xrightarrow[\text{at } 5-50°C]{\text{in NMP}} \xrightarrow[\text{at } 80-85°C]{\text{cooking}} -R_1-\underset{HOCO}{NHCO}-R_2-\underset{COOH}{CONH}-$$

Diamine   Dianhydride   Polyamic acid
                                                                      varnish

$$\xrightarrow{\text{Coating on a Glass Plate}} \xrightarrow{\text{Cleaning in Water}} \xrightarrow{\text{Vacuum Drying}}$$

$$\text{Polyamic acid Film} \xrightarrow{\text{Heating}} -R_1-N\overset{CO}{\underset{CO}{\diamondsuit}}R_2\overset{CO}{\underset{CO}{\diamondsuit}}N- + H_2O \quad ----(1)$$

Polyimide

260

Table 1. Synthesis of polyamic acids and their properties.

No.	Diamine	Dianhydride	Calculated water loss (wt%)	Glass transition temperature of polyimide (°C)
1.	4,4'-diamino diphenylether	PMDA*	8.6	385
2.	4,4'-diamino diphenylether	BTDA+	6.9	295
3.	2,2-bis 4-(p-amino phenoxy)phenyl propane	PMDA*	5.7	350
4.	2,2-bis 4-(p-amino phenoxy)phenyl propane	BTDA+	4.9	240
5.	2,2-bis 4-(p-amino phenoxy)phenyl hexafluoropropane	PMDA*	5.7	365
6.	2,2-bis 4-(p-amino phenoxy)phenyl hexafluoropropane	BTDA+	4.3	260

* Pyromellitic dianhydride
\+ Benzophenone tetracarboxylic dianhydride
Solvent; N-Methyl-2-pyrrolidone (solid content, 15 wt%)

When the diamine and tetracaboxylic dianhydride are made to react in N-methyl-2-pyrrolidone (abbreviated as NMP) in the temperature range of 5°C to 30°C, the varnish viscosity gradually increases, and in 2 to 5 hours stirring is no longer possible. The varnish is then held at 80-85°C until the viscosity drops to 10 Pa.s at 25°C, which results in an ultimate polyamic acid varnish concentration of 15%.

Specimen preparation

As is clear from Expression (1), the amount of imide cyclization and dehydro-condensation are stoichiometrically equal. In this paper, the imidization reaction is followed by measuring the weight loss that accompanies the dehydro-condensation reaction. Accordingly, there should be no weight losses attributable to other causes during the measurements. Conditions must be chosen so as to eliminate all factors that might contribute to weight loss, such as the varnish solvent, the absorption of water vapor from the air and the production of unwanted products during thermal decomposition.

Kreuz and others[8] attempted to follow the imidization reaction using a thermobalance but dropped the idea since they could not eliminate the contribution due to the solvent. After studying various methods, we have established a method of eliminating the solvent by cleaning the polyamic acid film in distilled water. That is, a thin coat(about 0.3mm in thickness) of the polyamic acid varnish was applied on a glass plate, and the plate was then dipped in distilled water to precipitate the polyamic acid in brittle and porous film form. The film was cleaned 5 times in distilled water with an ultrasonic cleaner, for 30 minutes each time. After being vacuum dried for about 2 hours at a temperature of 80°C and a pressure no higher than 0.5mmHg, the amount of solvent that remained was determined using thermal decomposition gas chromatography. The results obtained indicated that the NMP content was below the barely detectable level of 0.01%.

The infrared spectrum for polyamic acid, which was obtained in this way, indicated extremely low levels of absorption for imide rings. The levels observed were almost negligible when compared with the level of absorption observed for amide acid.

The water vapor absorbed by the polyamic acid film was easily removed before measurements were made by vacuum drying for 15 minutes at a temperature of 80°C and a pressure no higher than 0.5 mmHg with the film set on a thermobalance. It was also confirmed from infrared measurements that the imidization reaction progressed very little under these conditions.

The amount of gas produced by side-reactions of the imidization reaction was measured concurrently with the residual solvent characteristics described above. That is, a heat treatment similar to the imidization reaction was applied and the amount of gas generated was determined from thermal decomposition chromatography. The result was that decomposed products, having a somewhat low boiling point, were generated. However, the amount was quite negligible when compared with the dehydration accompanying the imidization reaction.

Korshak and others[13] reported that when polyamic acid was thermally decomposed in the temperature range 200 to 400°C, extremely small amounts of gases like carbon dioxide were produced and, otherwise, the only product was water.

Following of Imidization Reaction

The progress of the imidization reaction was followed by measuring the weight loss caused by dehydro-condensation with a precision thermobalance (TGD 3000, made by Shinku Riko). A polyamic acid sample (50 to 100 mg) was set on a thermobalance and

262

heated for 30 minutes at 80°C in vacuo (at a pressure no higher than 0.5 mmHg) in order to remove the absorbed water vapor.

Two imidization conditions were adopted; one was heating at a constant rate, and the other was a two-step isothermal heating. In the case of the constant rate heating method, the temperature was raised at a rate of 5 °C/min. In the case of the isothermal heating method, the temperature was raised to the primary reaction temperature at a rate of 200°C/min and the specimen was then left at that temperature for 60 minutes. After that, the temperature was raised to the secondary reaction temperature at the same rate to complete the imidization reaction. The secondary reaction temperature was determined, through trial and error, to be the highest temperature within the range at which thermal decomposition would not develop (the weight loss being less than 0.2 % during a period of 15 minutes). The secondary reaction temperature was 450°C for specimens No. 1 and 2, and 400°C for specimens No.3, 4, 5, and 6 (cf. Table I). As will be described later, the total measured weight loss agreed with the calculated values from the chemical formula for polyamic acid. These imidization reactions were carried out in a nitrogen atmosphere.

Measurement of Infrared Spectrum

The polyamic acids and imidized specimens were powdered and measured in accordance with the KBr tablet method using an infrared spectrometer (Type 260-50, made by Hitachi).

Glass Transition Temperature (Tg)

The specimens used were films prepared by the solvent casting method, each of which was 50 μm thick, 7 mm wide, and 65 mm long (between chucks). Tg was obtained from dimension changes with a thermo-mechanical analyzer (Type TMA-1500, made by Shinku Riko), when the temperature was raised at the rate of 5°C/min.

RESULTS

Infrared Spectra of Specimens

Figure 1 shows the infrared spectra for polyamic acids prepared with the method described above and the resulting polyimides. The characteristic absorption bands are 1660 cm^{-1} for amide acid, and 1780, 1380 and 725 cm^{-1} for imide rings. A trace of absorption caused by the imide ring was observed for each polyamic acid specimen. This absorption apparently occurred because the specimens had been cooked and vacuum dried at 80°C.

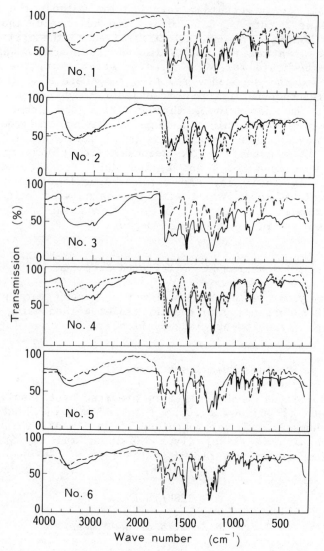

Figure 1. Infrared spectra for polyamic acids (——) and polyimides (----).

Imidization with the Constant Rate Heating Method

Figure 2 shows the measured weight loss caused by the imidization reaction, when the polyamic acids (specimens No.1 and 3) were heated at a constant rate. These results show that weight loss occurred in two steps. Weight loss first occurred in the temperature range of 150 to 250°C, and next above 400°C. First-step weight losses almost agreed with the calculated water losses. Therefore, these weight losses were assumed to be caused by the imidization reaction, while the second-step weight losses were assumed to be caused by thermal decomposition. However, the end points of the imidization reaction were not obvious.

Imidization with the Two-step Isothermal Heating Method

Figures 3 and 4 show the measured weight loss that occurred during the imidization reaction. Heating was applied in two steps, the first of which is indicated by the solid lines in the figures and the second by the broken lines.

Attention should be paid to the weight loss that occurs during first-step heating. In all cases, most of the weight loss occurred during the temperature rise to the primary reaction temperature, and after that temperature was reached, the weight loss became smaller. In other words, while the temperature was rising at the fast rate of 200 °C/min, dehydro-condensation proceeded quickly, but the reaction almost stopped when the primary reaction temperature was reached.

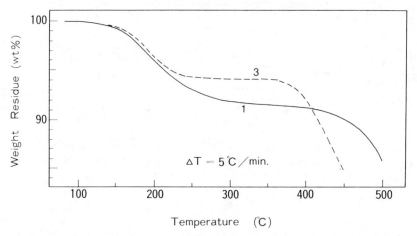

Figure 2. Weight loss during heating with constant rate.

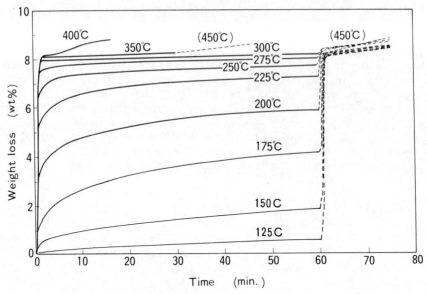

Figure 3. Weight loss of specimen No.1 during imidization reaction.

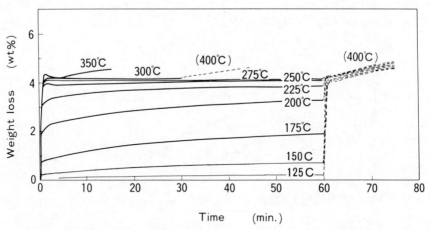

Figure 4. Weight loss of specimen No.6 during imidization reaction.

In the case of second-step weight loss, most of the loss occurred during the temperature rise too. There was very little weight loss after the secondary reaction temperature was reached. As described above, the secondary reaction temperature was determined by trial and error. Thermal decomposition occurs slightly at this temperature but the degree of decomposition is small enough to be ignored. In other words, it is assumed that the weight loss that occurs during the temperature rise is caused by the imidization reaction and that the weight loss that occurs after the secondary reaction temperature has been reached is caused by thermal decomposition.

"Total weight loss", which corresponds to the amount of water discharged as a byproduct of imidization is obtained by adding the weight loss that occurs during the first step and the loss that occurs during the temperature rise up to the secondary reaction temperature. The calculated water loss values for various polyamic acids at the time of imidization (shown in Table 1) and measured water loss values are in good agreement for almost all specimens. For example, the calculated water loss for specimen No.1 (Figure 3) is 8.6% and the total measured weight losses were in the range of 7.9 to 8.2%. The total weight losses are somewhat smaller than the calculated loss. This difference apparently occurred because the specimens were subjected to a slight degree of imidization, as confirmed by the infrared spectra described above.

These results indicate that, even if the imidization and thermal decomposition reactions should occur concurrently, as in the two-step isothermal heating method, the weight loss of each reaction can be determined separately.

Figure 5 shows the primary reaction temperatures for all specimens and their percent conversion of imidization after reaction at those temperatures. The percent conversion of imidization is defined as the ratio of the weight loss that occurs up to completion of the first step and the total weight loss that occurs up to the secondary reaction temperature. Figure 5 shows that most of the imidization reaction occurs in the temperature range of 150 to 250°C but that the reaction is completed at temperatures higher than 250°C. The figure shows that the temperature at which the reaction is completed varies according to the type of polymer. For example, specimen No.2 has the lowest reaction rate at all temperatures and the temperature at which the imidization reaction is completed is the highest.

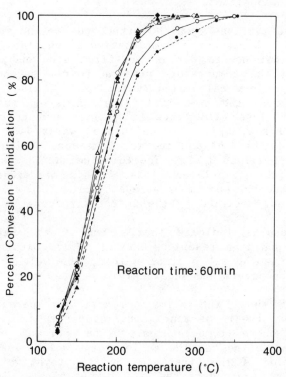

Figure 5. Relationship between percent conversion and imidization temperature, ○: No.1, ●: No.2, ◇: No.3, ◆: No.4, △: No.5, ▲: No.6.

DISCUSSION

As described in the preceding section, the imidization reaction is very rapid during the rise in temperature of specimens but slows down once a predetermined reaction temperature is arrived at. However, if the specimens are heated again, the reaction starts again, indicating that it was not completed under the initial conditions. Such behavior, which is not observed in reactions in solution systems, can be explained as follows.

It is assumed that the imidization reaction proceeds in a state in which molecular chains can rotate freely but slows down in the glassy state in which molecular motion is frozen. Polymer molecular motion is comparatively rapid in the early stage of the imidization reaction, but as the reaction proceeds and imide rings are formed in the molecular chains, molecular chain rigidity increases and the transition temperature of the reacting polymer shifts toward a higher temperature. When the glass transition temperature of the polymer reaches the reaction temperature, polymer molecular motion is frozen and the imidization reaction becomes inactive.

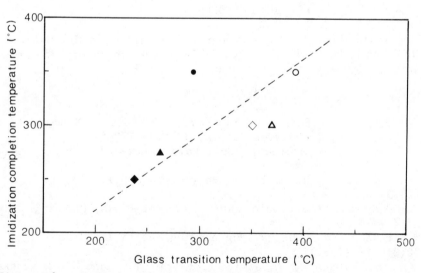

Figure 6.   Relation between Imidization completion temperature and glass transition temperature.

If the above assumption is correct, the temperature at which the imidization reaction is complete should be the same as the glass transition temperature of the polymer. The glass transition temperature of polyimide as obtained from thermal expansion behavior, is shown in Table I . Figure 6 shows the relationship between the glass transition temperature and the temperature at which the imidization reaction is complete. Figure 6 shows the correlation between the two temperatures.

CONCLUSION

The progress of the imidization reaction was followed by using various polyamic acid film specimens and by measuring weight losses that resulted from the dehydro-condensation of polyamic acid with a thermobalance. When the polyamic acid films were heated quickly and then kept at a specified temperature, most of the imidization reaction occurred during the temperature rise in all polyamic acid specimens and the reaction became very slow after the specified temperature was reached. The temperature at which the imidization reaction is completed is closely related to the glass transition temperature of the resulting polyimide. The reaction slows down after a specified temperature has been reached because imide rings form in the molecular chains as the reaction progresses and the molecular chains then become rigid and the glass transition temperature of the polymer rises. When the glass transition temperature reaches the reaction temperature, free rotation of the molecular chains becomes possible.

ACKNOWLEDGEMENTS

We would like to express our appreciation to Mr. Mikio Sato and Dr. Junji Mukai of the Hitachi Research laboratory who gave us valuable guidance in this research.

REFFERENCES

1. C.E.Sroog, A.L.Endrey, S.V.Abramo, C.E.Berr, W.M.Edwards, and K.L.Olivier, J. Polymer Sci.:A, 3, 1373 (1965)
2. C.S.Marvel, J. Macromol. Sci.-Revs. Macromol. Chem., C13(2), 219 (1975)
3. J.K.Gillham and H.C.Gillham, Polymer Eng. Sci., 13, 447 (1973)
4. G.Takahashi, Electronic Parts and Materials (Japan), 20(6), 27 (1981)
5. K.Harasawa and I.Ohsawa, Engineering Materials (Japan), 30(3), 27 (1982)
6. K.Sato, S.Harada, A.Saiki, T.Kimura, T.Ohkubo, and K.Mukai, IEEE Trans. Parts, Hybrids and Packag., PHP-9, 176 (1973)
7. A.Saiki, H.Umesaki, and K.Sato, Electronic Parts and Materials, 20(11), 22 (1981)

8. J.A.Kreuz, A.L.Endrey, F.P.Gay, and C.E.Sroog, J. Polymer Sci.: A-1, $\underline{4}$, 2607 (1966)
9. L.A.Laius, M.I,Bessonov, Ye.V.Kallistova, N.A.Adrova and F.S.Flinskii, Polymer Sci. U.S.S.R., $\underline{9}$, 2470 (1967)
10. L.A.Laius, M.I.Bessonov, F.S.Florinskii, Polymer Sci. U.S.S.R., $\underline{13}$, 2257 (1971)
11. S.V.Lavrov, A.Ya.Ardashnikov, I.Ye.Kardash, and A.N.Pravednikov, Polymer Sci. U.S.S.R., $\underline{19}$, 1212 (1977)
12. S.V.Lavrov, I.Ye.Kardash and A.N.Pravednikov, Polymer Sci. U.S.S.R., $\underline{19}$, 2727 (1977)
13. V.V.Korshak, Yu.Ye.Doroshenko, V.A.Khomutov, and L.M.Mochalova, Polymer Sci. U.S.S.R., $\underline{16}$, 2515 (1974)

Kinetics

THE INTERPLAY BETWEEN SOLVENT LOSS AND THERMAL CYCLIZATION IN LARC-TPI

P. D. Frayer

Rogers Corporation
One Technology Drive
Rogers, CT 06263

Isothermal reaction kinetics are reported for the thermal cyclization of the poly(amide-acid) based on benzophenone tetra-carboxylic acid dianhydride reacted with 3,3'-diaminobenzophenone. Isothermal solvent loss kinetics are also reported. The data suggest a strong interdependence between the rate of imidation and the rate of solvent loss. Additional reactions may occur for isothermal reaction temperatures above about 180°C. Infrared analysis showed the presence of moisture even at elevated temperatures; hence, hydrolysis is one possible side reaction. Infrared spectra of films exposed to elevated temperatures further suggest that complete solvent removal is difficult. The infrared band assignments are proposed. The kinetic data were fitted to a modified absolute reaction rate theory equation where the time dependence of the reaction rate constant may be explained by entropic changes which follow a continuous function of time. The reaction kinetics yielded reasonable values for the frequency of vibration of the activated complex transition state and the activation energy. Low values for the activation entropy may be explained by solvent effects. The solvent loss kinetic model yields an estimate of the polymer-solvent interactions.

INTRODUCTION

This paper concerns the thermal cyclization of the poly(amide-acid) (PAA) formed by the nearly stoichiometric condensation of benzophenone tetracarboxylic acid dianhydride (BTDA) with 3,3' diaminobenzophenone (DAPB). More specifically, we discuss the interplay between the reaction and the solvent loss kinetics.

Progar, Bell, and St. Clair of NASA patented the method for preparing the PAA in ether solvents which do not complex with the PAA as do the highly polar, aprotic solvents[1]. They speculated that a lower extent of imidation relative to that in more polar solvents may be obtained and that superior adhesive bonding results perhaps because the unreacted poly(amide-acid) groups may be responsible for adhesion.

Bell of NASA patented a process for making film[2]. The method is time consuming, but a thermoplastic polyimide film was claimed. Subsequently, St. Clair et al.[3] made thermoplastic films by a shorter process cycle, 60 minutes each at 100°C, 200°C, and 300°C. LARC-TPI is an acronym for these materials.

Laius et al.[4] studied by IR the reaction of 4,4'-diaminodiphenyl ether with pyromellitic dianhydride. They used a preliminary thermal treatment for placing their films in an initial standard state. A final standard state was established by using thermal conditions where no further changes could be detected by IR. Thus, the extent of cyclization could be determined between these limits. They used the 1380 and 720 wave number imide bands for their kinetics. Kreuz et al. reported a similar kinetic study[5].

Although the 1380 imide band appears in the LARC-TPI spectrum, this region has a high background absorbance which reduces measurement accuracy. Thus, we used the 1780 imide band instead. In LARC-TPI a doublet occurs near 720 cm^{-1}. We followed changes in both bands, but the 710 was used for kinetics.

Whereas Laius et al.[4] and Kreuz et al.[5] followed only the increase in imide band intensity, we also tracked the decrease in two amide bands near 3325 and 1540 cm^{-1}. Thus, our kinetic equations are based on the average of two data sets for the two different types of chemical changes which should be in a 1:1 correspondence if no side reactions occur, e.g., amide loss by hydrolysis rather than cyclization to imide.

The pre-treatment used by Laius et al.[4] made the initial standard state contain about 10% imide. A significant observation was made when films with and without prior thermal treatment were compared. The pre-treated, reacted film contained less imide. This difference in conversion may be a plasticizer effect or may be a polymer-solvent complex effect. The latter possibility has been proposed by others who have reported strong interactions between the PAA form and the more polar solvents[5-7]. Hodgkin[8] and Kumar[9] have found evidence for chemically bound solvent residues.

If ether solvents do not complex with the PAA form and if they can be volatilized at lower extents of cyclization, then the rate of thermal conversion may vary with the rate of solvent loss. Thus, we have looked for this effect in LARC-TPI.

EXPERIMENTAL

Gulf Research & Development Company supplied the samples of LARC-TPI as solutions of the poly(amide-acid) form at 16% solids by weight either in 75% diglyme (diethylene glycol dimethyl ether) plus 9% ethanol, or in 39% diglyme and 36% tetrahydrofuran (THF) plus 9% ethanol. The alcohol is added subsequent to the polymerization to the poly(amide-acid) form and provides solution stability.

Thin films (about 1×10^{-5}m "dry" thickness) of the poly (amide-acid) form were made by casting the solution on a substrate using a constant speed (2 cm/s) doctor blade on a flat bed. We used 30 minutes in a circulating air oven at 85°C to place the free films in their initial standard state. Further solvent removal and conversion to polyimide were tracked using a dual beam Perkin Elmer Model 283 infrared spectrophotomer. Both single beam (transmission and absorbance) and dual beam (differential analysis) experiments were used.

Isothermal kinetics were studied by employing a multiple sample approach. Thus, for each time and temperature a different starting film was used. However, to minimize side reactions, the final reference state film was first heated to about 183°C for 60 minutes followed by 370°C for 15 minutes.

Non-isothermal kinetics were studied by heating each sample at a constant rate between 2 to 140°C/min. in a temperature controlled infrared cell while recording a specific band intensity continuously. Each sample began in the initial standard state and heated from room temperature to 300°C then cooled to near room temperature. The IR absorbance was followed both on heating and cooling. The lower frequency bands were found to be temperature sensitive.

Negligible cyclization occurs under the drying conditions used as judged by differential spectra of dried films and KBr plate smears; only solvent content differences were observed. However, the as-received samples contained a small amount of imide formed either during the preparation of the poly(amide-acid) or during storage, although no change occurred in the as-received samples in the time frame of our experiments. Two lots of poly(amide-acid) were used; one lot was used only for the isothermal kinetics ($\sim$5% initial imide content) and the other ($\sim$2% initial imide content) for the non-isothermal kinetics.

RESULTS AND DISCUSSION

Probable Infrared Band Assignments

For kinetics by the IR absorbance ratio method, one must find an absorption band whose intensity depends only on film thickness and not on solvent content, degree of imidation, or side reactions. Such a reference band is likely to be assignable to an aromatic ring vibration which is absent in the solvents but present to a constant extent in both the poly(amide-acid) and the polyimide forms. Such a band appears to occur in LARC-TPI at 995 wave numbers.

In general, comprehensive functional group band assignments have not been reported for polyimides. One exception is the paper by Reimschuessel et al.[13] who reported specific functional group vibrations for most of the IR absorptions of one novel polyimide with supportive evidence by x-ray, NMR, and deuteration experiments. However, neither is this polyimide aromatic nor its precursor a poly(amide-acid); it is formed by the thermal polymerization of β-carboxymethyl caprolactam.

Since a detailed analysis of LARC-TPI infrared spectra apparently has not been previously reported, we attempted to assign each band to the most probable chemical groups that are likely to be the major components of each observed absorption. Our assignments are not intended to be comprehensive and were obtained by making correlations between the chemical structures and the available infrared spectra of many polyimides[10-24]. Our proposed assignments appear in Figures 1-3.

Figure 1 contains several transmittance infrared spectra in the high wave number region. Spectrum A was made by smearing a small amount of LARC poly(amide-acid) solution onto a KBr plate. Samples B through F2 were all cast as free films. Samples C through F2 were all dried at 85°C for 30 minutes prior to exposure to the indicated temperatures and times. Sample F2 was also exposed to 183°C for 60 minutes prior to high temperature exposure. These samples contained THF and diglyme solvents as described previously.

Except for the apparent intensity changes caused by changes in the overlapping band intensities (N-H amide and C-H solvents), the aromatic C-H stretching vibrations should be unchanged by solvent

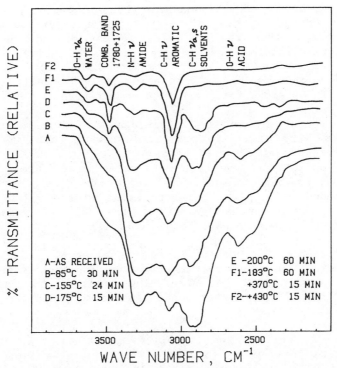

Figure 1. Transmittance infrared spectra in the high wave number region. (F2=F1 Conditions + 430°C for 15 min.).

or amide loss. Although this aromatic absorption cannot be used for quantitative analysis, its intensity relative to the N-H and the solvent C-H stretching vibrations as a function of thermal history suggests several qualitative observations.

1. the drying conditions are effective in removing the THF without causing reaction (compare curves A, B, and C between 2800 and 3350 cm-1);

2. below about 175°C the conversion of amide, presumably to imide, occurs at a rate about equal to the rate of

diglyme loss for the time shown;

3. from 175 to 200°C the rate of amide loss exceeds the rate of solvent (diglyme) loss;

4. residual or unreacted amide is observed in samples F1 and F2; hence, complete intramolecular cyclization is difficult to achieve under these conditions;

5. apparently, the imidation reaction occurs so rapidly at 200°C that the polyimide Tg may exceed the reaction temperature, thereby depressing the solvent diffusion rate even though the reaction temperature exceeds the normal diglyme boiling point by 38°C (Bell[2] reported a Tg of 216°C for a reaction temperature of 200°C).

At a recent course on "Polyimides for Microelectronics"[25], mass spectrometry data were presented showing that two types of water are present at elevated temperature. This first form, called "free" or mobile water, was found to exist in a "cured" polyimide at temperatures below about 250°C while the second form, called "tightly bound" water, was found to exist in the polyimide up to 450°C. Therefore, our assignment of the 3630 cm^{-1} band to OH stretching vibrations appears to be justified.

The N-deuteration work of Reimschuessel et al.[13] suggests that the band near 3500 wave number is a combination band of the symmetrical and asymmetrical C=O stretching bands of the imide ring, i.e., 1790 + 1722 wave number bands combine to yield the 3502 wave number band.

Figure 2 contains two transmittance IR spectra. The dashed line spectrum corresponds to sample B of Figure 1 (i.e., dried 30 minutes at 85°C) while the solid line spectrum corresponds to sample E (i.e., dried plus 60 minutes at 200°C). The probable band assignments are indicated.

The absorption near 1852 cm^{-1} appears to be characteristic of the LARC polymers. We have not seen it in any of the approximately 75 spectra of converted imide-containing polymers[23], including those based on BTDA. St. Clair and Progar suggested that this band may be related to residual anhydride[26]. We observed that the band is not present in the poly(amide-acid), but rather becomes apparent with conversion. However, the intensity does not increase at the same rate as the imide bands, but increases slowly and then decreases at elevated temperatures. Kudryavtsev et al. suggested for another system that a band in this region may be due to the isoimide structure which can thermally convert to imide[27]. Koton et al. have suggested that a band in this region may be attributed to intermediate products[28].

Essentially all of the reported imide-containing polymer spectra exhibit the 1780 and 1725 bands, the former always being

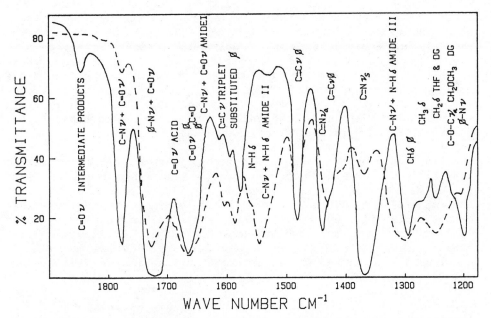

Figure 2.   Transmittance IR before (---) and after (———) imidation at 200°C.

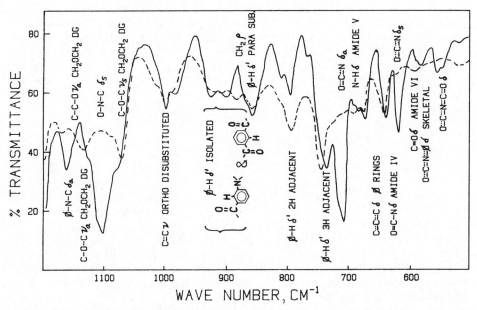

Figure 3.   LARC IR Spectrum before (---) and after (———) imidation at 200°C.

lower in intensity. We propose mixed mode assignments for these bands analogous to those in polyamides. However, in this case the Amide I band equivalent is two bands and both occur at higher frequency perhaps because of the rigidity of the imide ring. Some authors[13,14] have proposed that the occurrence of two bands is due to symmetric and asymmetric "imide ring" or "carbonyl" vibrations. We propose that the 1725 band is a mixed mode whose intensity is enhanced on imidation by the addition of a second carbonyl to the phenyl-nitrogen pair. Either of these could be correct.

The apparently isolated aromatic carbon-carbon stretching vibration at about 1483 cm^{-1} is not sufficiently isolated to be useful as the internal reference band. The reader will note that the Amide II band intensity controls the maximum transmittance near a wave number of 1500. Consequently, the aromatic band has a changing baseline from positive to negative. Residual diglyme solvent bands can be seen in Figure 2. The 1100 to 1400 region is quite complex. The majority of the transmittance maxima are below about 50% transmittance.

Figure 3 reveals a less complex region. We believe that the triplet located between 950 and 1025 cm^{-1} arises from the ortho-disubstituted aromatic carbons that participate in the 5-membered imide ring system as well as in the poly(amide-acid) form, i.e., in the BTDA aromatic rings. Thus, we used the middle band of the triplet at 995 cm^{-1} as our internal reference for absorbance ratios. Results by differential transmission IR support the proposal that the 995 can be used for quantitative kinetic studies, i.e., the intensity is proportional to sample thickness and independent of reaction conditions or degree of imidation. In Figure 3 one also notices a strong imide band doublet at 700 to 725 cm^{-1}. The 710 cm^{-1} band was used for the kinetics since the 720 band intensity reached a limiting value before imidation was complete. More work is needed to understand this behavior.

Isothermal Reaction Kinetics

The kinetics of imidation are complex. The calculated reaction rate "constant" is variable. The most common explanation is that the material passes through a change of physical state due to solvent loss and cyclization, both of which reduce polymer chain flexibility[30,33]. Thus, the rate constant is observed to fall to zero as the polymer passes into its glassy state. A Tg above the reaction temperature results from both imidation and solvent loss, the latter continuing even though reaction may cease. Measurement of an intermediate Tg is possible because the rate of further reaction is slow compared to the measurement time[28,30].

By this hypothesis, a graph of the intermediate glass transition temperature versus the specific imidation temperature should intersect a reference line of slope unity, i.e., Tg = reaction temperature, when the ultimate Tg is reached. From literature values for LARC-TPI[1-3,26] we made such a plot in Figure 4, but we discovered that above about 235°C, the intersection point, the reaction temperature always <u>exceeds</u> the Tg. Thus, a phase change to the glassy state cannot occur in this region. Also, the apparent ultimate Tg occurs above the intersection point.

We speculated that this deviation from the above hypothesis may be a consequence of side reactions which reduce the ultimate Tg by preventing complete cyclization. If these side reactions are not crosslinking reactions, then the partially converted polyimide chain should have greater flexibility than the fully converted polyimide. Perhaps the unusual (for aromatic polyimides) high extent of softening above the Tg observed by Gillham <u>et al.</u>[29] for the LARC-TPI based on 4,4'-DABP is due <u>in part</u> to this effect in addition to the existence of two polymer backbone carbonyl "hinge" groups which are important in chain flexibility and mechanical properties[30].

We believe our isothermal reaction kinetics data, which follows, show that some type of side reactions do occur when films dried at 85°C for 30 minutes are rapidly heated to temperatures above about 180°C. It should be emphasized that our initial standard state differs slightly from that used by others for LARC-TPI. In particular, these results are for the THF/diglyme solvent system where the initial standard state contains essentially no THF. The diglyme only system discussed later has a higher diglyme content in the initial standard state.

Following other researchers we assumed first order kinetics. Initially we used an Arrhenius relation between the rate constant and temperature. However, since the rate "constant" decreases with time, one obtains a family of Arrhenius curves as shown in Figure 5, each curve corresponding to a constant time. If the rate constant were not time dependent, all four curves shown would merge into one.

Figure 5 also shows that the data cannot be fit with one <u>set</u> of straight lines which one would expect if a single reaction mechanism were involved. Thus, we concluded that between 175°C and 200°C another reaction or mechanism predominates.

Although Laius <u>et al.</u>[4] showed that for thin films the half-life for diffusion of water is several orders of magnitude faster than the reaction rate constant, our infrared spectra suggested that not all of the water diffuses out of the film. Thus, we considered a diffusion limited reaction model, but we did not arrive

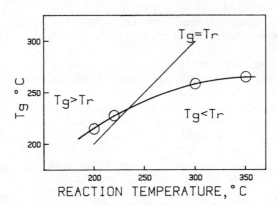

Figure 4.   Glass transition vs reaction temperature.

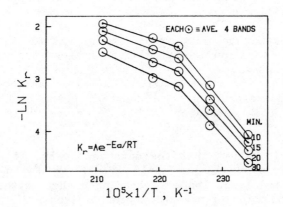

Figure 5.   Arrhenius plot for reaction kinetics.

at the proper time dependence. Johnson discussed hydrolysis as a side[7] reaction especially in preparing thick films or solid moldings.

We also considered a more complex kinetic analysis with two rate constants, one for imidation and one for hydrolysis, in an equilibrium reaction process similar to that used in polyamide chemistry[12]. However, again we were unsuccessful in fitting the observed time dependence. We then turned to reaction rate theory.

According to the absolute reaction rate theory, the pre-exponential frequency factor is associated with the dissociation of an activated complex transition state to the final products where the vibrational energy, hv, depends on the thermal energy, kT. When the restoring force to maintain the activated complex structure approaches zero, the final product forms. In the above, k is the Boltzmann constant, and v is frequency. If the imidation reaction reduces molecular mobility, then the decreasing rate constant may be explained by a time dependence of the entropy.

A multiple linear regression fit for data points using natural logarithmic transformed variables gave a high correlation coefficient of R = 0.9986. Our regression equation may be written as follows:

$$\ln K_r = A_o + A_1 \ln t + A_2/T \tag{1}$$

where
$$A_o = \ln (60\ ekT/h) = 31.2113,\ \ln\ min^{-1}$$
$$A_1 = \Delta S^{\ddagger}/R\ln t = -0.7555,\ \ln\ min^{-1}$$
$$A_2 = -E_a/R = -14.25 \times 10^3,\ K$$
$$R = 8.3143\ J/mol\ K = gas\ constant$$
$$E_a = 118.5\ kJ/mol = activation\ energy$$
$$t = time,\ min$$
$$\Delta S^{\ddagger} = change\ of\ molar\ entropy\ of\ activation$$
$$K_r = reaction\ rate\ constant,\ min$$
$$T = absolute\ temperature,\ K$$

The reaction rate theory equation is

$$K_r = (60\ ekT/h)\ e^{\Delta S^{\ddagger}/R}\ e^{-E_a/RT} \qquad\qquad or$$

$$\ln K_r = \ln (60\ ekT/h) + \Delta S^{\ddagger}/R - E_a/RT \tag{2}$$

where the first term is the frequency factor, the second the entropy of activation energy factor. (The 60 converts s^{-1} to min^{-1}.)

If one equates the vibrational energy to the thermal energy, one calculates for a gaseous state reaction a frequency of $2.48 \times 10^{13}\ s^{-1}$. Calculating the frequency from our regression coefficient yields a value of $6 \times 10^{11}\ s^{-1}$. Since imidation is an intramolecular cyclization, it involves the reaction of two species with restricted mobility, hence a lower frequency.

Anufriyeve et al.[31] reported rotary diffusion coefficients for a solution of a PAA with one ether "hinge" group per polymer chain repeat corresponding to a frequency of only 4×18^8 s^{-1}. The exact nature of the activated complex is unknown, but since the carbonyl "hinge" groups may participate in resonance states with the benzophenone phenyl groups, the above calculations suggest that the activated complex frequency is independent of rotational freedom about the two carbonyl "hinge" groups in the polymer chain. This conclusion is expected and further supports our experimental frequency factor which is between 10^8 and 10^{13} s^{-1}.

Kreuz et al.[5] reported an activation energy of 108.8 kJ/mol (as did Kydryavtsev et al.[17] more recently) for poly(pyromellitimide) at reaction temperatures from 160°C to 190°C. They divided the reaction into two steps, fast and slow, but found no significant difference between the two experimental activation energies. Our suggestion of a constant activation energy seems justified. Laius et al. used an Arrhenius equation and suggested that both the activation energy (range of 96 to 125 kJ/mol) and the frequency factor increase as the reaction proceeds. However, we believe the rate theory approach is more appropriate where one assumes that the time dependence is related to entropic changes.

It should be noted that Koton[30] and Laius et al.[32] later proposed the phase change hypothesis and suggested the polymer chain should shrink in length by 10% on ring closure to imide. Thus, solvent loss and degree of imidation not only affect intramolecular conformational freedom, but also solvent loss should decrease intermolecular chain slippage which may restrict ring closure. Thus, the activated complex entropy may be linked to solvent content.

Kreuz et al.[5] reported activation entropies of -41.8 J/mol K for the initial fast reaction and -100.4 J/mol K for the slow reaction. They compared this to the tri-n-butylamine/PAA salt whose relative rate constant was 5.4 times greater and whose entropy of activation was only -22.2 J/mol K. Our calculated activation entropies vary with the natural logarithm of time and are surprisingly low, ranging from -20 to -30 J/mol K for the time scale of our experiments. The low negative values may be a reflection of the lower degree of imidation in LARC-TPI under the reaction conditions. Johnson reported that the maximum rate of imidation for more polar solvents generally occurs between 120 and 140°C[7]. As shown later, for LARC-TPI the maximum rate is near 175°C. This may explain the higher entropy for the activated complex transition state, i.e., a lower negative value.

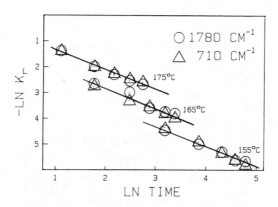

Figure 6. Absolute reaction rate theory plot for the imide bands showing that the regression equation accounts well for the effect of time on the reaction rate constant.

Figures 6 and 7 show the fit of the experimental data with the regression equation for 155°C, 165°C, and 175°C reaction temperatures. For clarity, we separated the data points for the two imide bands and two amide bands. The data above about 175°C do not fit the regression equation. This fact can be shown better in Figure 8.

Note in Figure 8 that up to about 175°C all the data now fall on a single straight line. At elevated temperatures less imide forms than the amount of amide lost. This difference may be due to amide group degradation. Oxidation and hydrolysis are known to occur in polyamides as the temperature approaches 200°C[12]. Qualitative measurement support this hypothesis; as the isothermal reaction temperature increases the films became more brittle and darker yellow in color.

If the diglyme were to react with the polymer and to prevent complete cyclization, then the polymer chain entropy would be increased in two ways: 1) greater intra-molecular flexibility and

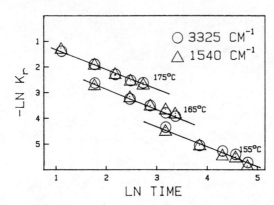

Figure 7.   Time effect rate theory plot - amide bands.

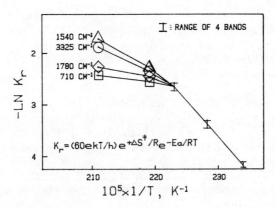

$$K_r = (60ekT/h)\,e^{+\Delta S^{\ddagger}/R}\,e^{-Ea/RT}$$

Figure 8.   Reaction rate theory plot for all data.

2) easier inter-molecular chain slippage. Since others have observed solvent bound residues in other systems[8,9], we wondered whether a kinetic study of solvent loss might reflect some interplay with the reaction kinetics.

Isothermal Solvent Loss Kinetics

The solvent diffusion process is complex. The PAA form is soluble in the solvent; the PI form is not. The solvent undergoes a phase change (liquid to vapor) in the temperature range of interest. Although diglyme normally boils at 162°C, one expects some boiling point elevation due to PAA-solvent interactions. However, the number of PAA sites for solvent interaction decrease as imidation proceeds. Not only are the solvent content and physical state changing, but also the polymer composition and physical state (viscous liquid to glassy) are changing during the thermal cyclization process.

Chartoff and Chiu[34] used diffusion techniques in combination with viscoelastic measurements to probe polyimides (including the LARC-TPI type) for crosslinked and branched structures. However, their samples presumably were neither changing in chemical structure nor their diffusing "solvent" (DMF) molecules undergoing a phase change during their experimental sorption measurements at 23°C. On the other hand, their results suggest not only that their imidation conditions can create crosslinked and branched structures, but also that the diffusion of the "solvent" (probe) molecules is sensitive to imidation conditions.

Early in this century statistical and quantum mechanics were used to theoretically justify the formerly empirical Arrhenius equation for chemical reactions[39]. However, the principles which lead to the concept of an activation energy are not limited to thermal energies which induce chemical changes[39-41]. Thus, the Arrhenius type function has also been used to describe the temperature dependence of physical processes such as viscosity[42], conformational energetics[36], solubility and diffusion[43]. Although interpretation of the activation energy may be difficult, solvent loss from a changing polymer composition is likely to be a thermally activated process following an Arrhenius type relation.

Activation energies for diffusion in polymers depend on polymer-solvent interactions and polymer chain mobility[43]. Just above the polymer Tg, the former are believed to predominate yielding activation energies around 25 to 50 kJ/mol. Just below Tg, the latter is believed to predominate yielding activation energies from 40 to as high as 200 kJ/mol[37,38,43]. Theoretical equations for diffusivities incorporate intermolecular potential energy functions[42,44]. Thus, we expect the solvent loss activation energy will contain several contributions which may not be easily identified individually.

Polymer-solvent interactions are especially difficult to predict for hydrogen bonding species with individually different donor/acceptor forces. Hydrogen bonding in aprotic ethers such as diglyme may not be strong alone, but interaction with acid or amide hydrogens may result in stronger interactions. Epley and Drago[45] reported calorimetric and spectroscopic measurements of hydrogen bonding for various "acid-base adducts" and showed that diethyl ether strongly interacts with a phenolic hydrogen with a bond energy of 22 to 23 kJ/mol, even though diethyl ether alone is only weakly hydrogen bonding, perhaps only 6kJ/mol.

For diglyme loss from a reacting poly(amide-acid), polar and dispersion forces are also important. However, it is particularly significant that hydrogen bonds must be oriented to exist which is not true for polar and dispersion forces. Thus, when the hydrogen bonds are thermally disoriented, we would expect a relatively abrupt loss in polymer-solvent interactions. Abrupt means within the limits set by the distribution of quantum energies according to Planck's thermal radiation law and the population of activated species according to the Maxwell-Boltzmann law[39]. thus, we would anticipate a temperature dependence for the activation energy in the present case since the temperatures used span the boiling temperature range expected.

Although dispersion forces are not inherently temperature dependent for the temperatures of interest, as are polar and hydrogen bonding forces[36], dispersion forces are dependent on intermolecular distance to the inverse sixth power, as are polar forces. Thermal expansion occurs mostly by the net increase in intermolecular distance as a result of increased vibrational displacement. Hence, all intermolecular forces are temperature dependent in that sense.

The transition state theory has been applied to physical processes such as viscous flow and diffusion[42,43]. Thus, we again used a multiple linear regression of transformed variables just as for the reaction kinetics. If strong interactions exist between the solvent and the activated complex for the reaction, then one might expect that the coefficients would again have significance in terms of frequency, entropy, and activation energy. However, we discovered that the regression coefficients are temperature dependent. These are given in Table I with the calculated activation energies.

We noted a striking similarity in the values where an abrupt change occurs near 446K with the values obtained for the single regression equation for the reaction kinetics, i.e., A_0 = 31.2113, $-A_1$ = 0.755, and $-A_2$ = 14250. Is this result fortuitous or does it reveal significant interplay between solvent loss and reaction kinetics?

288

Table I. Solvent Loss and Regression Equation Coefficients, and Calculated Activation Energies.

Temp., K	A_o	$-A_1$	$-A_2$	E_a (kJ/Mol)
428	43.2587	0.7835	19827	165
438	43.2587	.7835	19827	165
443	42.8078	.7833	19625	163
446	32.8971	.7782	15185	126
450	19.2051	.7712	9051	75
455	16.1360	.7696	7676	64
460	12.2540	.7676	5937	49
473	11.9962	.7675	5821	48

To explore this phenomenon further we plotted the reaction rate equation in terms of the partial derivative of conversion, C, with time on one ordinate axis and the solvent loss activation energies on the second ordinate axis both versus temperature as shown in Figure 9. Dashed lines above 180°C (453K) indicate the end of validity of the reaction kinetics equation. Although not shown in Figure 9 for clarity, we also plotted the analogous equations for solvent loss and discovered that the point at which the rate constants for reaction and solvent loss are equal (447K) also corresponds to the same maximum point in the time rate of change functions, i.e., where the regression equation coefficients are about equal.

We interpret the critical point as the temperature where the expected abrupt change in hydrogen bonding occurs. Thus, the difference between the upper limiting value of the solvent loss activation energy (165 kJ/mol) and that value at 447K (112 kJ/mol) should be a reasonable estimate of the hydrogen bond energies. Accounting for the 3 ether oxygens per molecule in diglyme yields 53/3 or 17.7 kJ/mol. This compares favorably with the values found in Epley and Drago[45] discussed earlier, i.e., 22-23 kJ/mol for diethyl ether/phenolic hydrogen interactions.

The continued decrease in activation evergy above 447K is not easily explained. Although the polar forces are temperature sensitive, we do not believe this alone accounts for the continued drop in activation energy for solvent loss at the same rate as for disruption of the hydrogen bonds. Since the temperatures above 447K are substantially above the normal boiling point of diglyme (435K), we speculate that the solvent induces greater "thermal" or volume expansion than would have occurred if the solvent phase transition (liquid to vapor) were not occurring. As mentioned earlier, both the polar and dispersion forces are strongly dependent on inter-

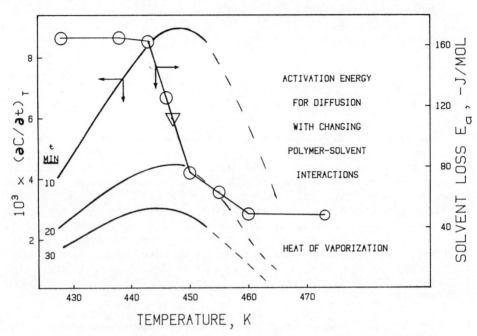

Figure 9. Partial derivative of conversion with time and solvent
loss activation energies vs. absolute temperature.
O = calc. activation energies; ∇ = solvent loss
activation energy at max. reaction rate.

molecular distance. Thus, the continued drop may be a result of a reduction in these forces caused by additional volume expansion when the relatively short hydrogen bond lengths are replaced by a larger average intermolecular distance.

We also have difficulty explaining the change in slope of the activation energy at 450K. It might be a reflection of the onset of side reactions which affect solvent diffusion. For example, if hydrolysis of unreacted amide groups occurred, then the average polar interactions with the solvent might increase. If solvent molecules actually reacted with the polymer, then the activation energy for solvent loss might not drop as quickly. Either or both phenomena could happen.

At temperatures where we are fairly certain side reactions are significant (above 460K), the activation energy levels off to about 48 kJ/mol. We suspect that for these isothermal reaction conditions, the reaction temperatures and the rate of heating are such that side reactions prevent complete cyclization or reduce the molecular weight. Thus, in this region we think the reaction temperature exceeds the achievable Tg. Hence, the dominant factor in the resistance to diffusion is not polymer chain segment mobility, but rather the remaining polymer-solvent interactions. It is interesting that the normal heat of vaporization for diglyme is 46.4 kJ/mol[34]. Since this value includes the hydrogen bonding forces that normally exist in diglyme (which should be lower than that which exists between diglyme and the PAA), the observed value of 48 kJ/mol suggests that the polymer segmental motions may make some contribution to this remaining term of the overall "activation energy for diffusion".

Note that below the critical temperature for hydrogen bond disorientation (447K), the activation energy of 165 kJ/mol verifies that polymer segmental mobility is a dominant factor[43]. However, the separation of polymer-solvent interactions from polymer segmental mobility is difficult in the region of the abrupt change of the solvent loss activation energy because of the phase changes that may occur.

We conclude from the solvent loss kinetics that a strong interplay exists between the imidation reaction and solvent loss rates. The solvent-PAA hydrogen bonds may play a role in the reaction mechanism. We are not sure how to interpret the temperature dependent frequency and "activation entropy" factors for the solvent kinetics. However, the near matching of terms at the maximum rate of change of imidation seems to be more than fortuitous.

Nonisothermal Kinetics

Although we have not completed our quantitative analysis of the nonisothermal kinetics, we would like to describe some interesting qualitative conclusions. First, we observed that films cast from the THF/diglyme solvent mixture contain less diglyme in the initial standard state than films cast from diglyme only as the solvent as expected from the volatility differences. Second, by studying a broad range of heating rates (2 to 140°C/min), we found that at slow heating rates the reaction rates were the same while at faster rates the films based only on diglyme reacted faster. We have subsequently confirmed this by isothermal studies. Our result is analogous to that of Laius et al.[4] mentioned earlier. Whether or not the solvent can act as a plasticizer appears to depend on the relative rates of reaction versus solvent loss.

The above conclusion is not surprising. However, we would like to point out that the constant heating rate required to take advantage of the plasticizing effect in LARC-TPI in diglyme may result in side reactions. We noted that if the heating rate is too fast relative to the reaction rate, then the films darken in color and become brittle.

PRACTICAL IMPLICATIONS

The literature gives "cure" schedules for imidation and solvent removal. However, they are rather disappointing. Limiting the schedule to three or four constant temperature stages with 50 to 100°C steps without controlling the rate between steps does not appear to be the best approach. Our work shows that at least three rates are important: 1) heating, 2) reaction, and 3) solvent loss. Our kinetics data allowed us to derive a more reasonable schedule with more stages and controlled rates. The result was a reduction in the total time to achieve a specified degree of conversion.

Our interest is in LARC-TPI as a high performance adhesive[46,47]. Unfortunately, adhesion depends on many other parameters. Consequently, we have not yet been able to arrive at definite conclusions concerning adhesive bonding and degree of imidation. However, we feel that further kinetic studies beyond the simple isothermal approach will be important. We hope that other researchers will include solvent loss kinetics in their future studies of reaction kinetics of the imidation process.

REFERENCES

1. D. J. Progar, V. L. Bell and T. L. St.Clair, U.S. Patent 4,065,345 (Dec. 27, 1977).
2. V. L. Bell, U. S. Patent 4,094,862 (June 1978).
3. A. K. St.Clair, W. S. Slemp and T. L. St.Clair, Adhesives Age, 35 (Jan. 1979).
4. L. A. Laius, M. I. Bessenov, Y. V. Kallistova, N. A. Androva and F. S. Florinskii, Vysokomal. soyed. A9(10), 2185 (1967); translated in Polymer Sci. USSR, 9(10), 2470 (1967).
5. J. A. Kreuz, A. L. Endrey, F. P. Gay and C. E. Sroog, J. Polymer Sci., A-1, 4, 2607 (1966).
6. G. Bower and L. Frost, J. Polymer Sci., A, 1, 3135 (1963).
7. E. Johnson, J. Appl. Polymer Sci., 15, 2825 (1971).
8. J. Hodgkin, J. Appl. Polymer Sci., 20, 2339 (1976).
9. D. Kumar, J. Polymer Sci., Chem. Ed., 18, 1375 (1980).
10. H. F. Mark, N. G. Gaylord, N. M. Biikales, Editors, "Encyclopedia of Polymer Science and Technology", Vol. 10, p. 682, Interscience Publisher, New York, 1969; also see Vol. 11 on Polyimides.
11. B. Vollmert, "Polymer Chemistry", p. 227, Springer-Verlag, New York, 1973.
12. M. I. Kohan, Editor, "Nylon Plastics", pp. 32-40, John Wiley and Sons, Inc., New York, 1973.
13. H. K. Reimschuessel, L. G. Roldan and J. P. Sibilia, J. Polymer Sci., Part A-2, 6, 559 (1968).
14. H. Lee, D. Stoffey and K. Neville, "New Linear Polymers", McGraw-Hill Book Co., New York, 1967.
15. C. L. Segal, Editor, "High-Temperature Polymers", Marcel Dekker, Inc., New York, 1968.
16. A. H. Frazer, "High Temperature Resistant Polymers", John Wiley and Sons, Inc., New York, 1968.
17. W. B. Alston, in "Proc. 12th National SAMPE Technical Conference-Materials 1980", p. 121.
18. R. W. Lauver, J. Polymer Sci., Polymer Chem. Ed., 19, 451 (1981).
19. M. M. Koton, Polymer Sci. USSR, 21, 2756 (1980).
20. A. V. Galanti, J. Polymer Sci., Polymer Chem. Ed., 19, 451 (1981).
21. J. H. Hodgkin, J. Polymer Sci., Polymer Chem. Ed., 14, 409 (1976).
22. C. J. Pouchert, Editor, "The Aldrich Library of Infrared Spectra", 2nd Edition, Aldrich Chemical Co., Milwaukee, WI, 1975.
23. D. O. Hummel and F. Scholl, "Infrared Analysis of Polymers, Resins and Additives: An Atlas", Vol. I and II, Wiley-Interscience, New York 1971.
24. J. C. Henniker, "Infrared Spectrometry of Industrial Polymers", Academic Press, New York, 1967.

25. G. Samuelson (Motorola, Phoenix, AZ), Course Notes for "Polyimides for Microelectronics", held in Palo Alto, CA (Aug. 4-5, 1981).
26. T. L. St.Clair and D. J. Progar, Polymer Preprints, 16 (1), 538 (1975).
27. V. V. Kudryavtsev, M. M. Koton, T. K. Meleshko and V. P. Sklizkova, Polymer Sci. USSR, 17 (8), 2029 (1975).
28. M. M. Koton, V. V. Kudryavtsev, N. A. Adrova, K. K. Kalnin'sh, A. M. Dobnova and V. M. Svetlichnyi, Polymer Sci. USSR, 16 (9), 2411 (1974).
29. J. K. Gillham, K. D. Hallcock and S. J. Stadnicki, J. Appl. Polymer Sci., 16, 2595 (1972).
30. M. M. Koton, Polymer Sci. USSR, 13 (6), 1513 (1971).
31. Y. V. Anufriyeva, Y. Y. Gotlib, V. D. Pautov, Y. Y. Svetlov, F. S. Florinskii and T. V. Sheveleva, Polymer Sci. USSR, 17 (12), 3220 (1975).
32. L. A. Laius, M. I. Bessonov, and F. S. Florinskii, Polymer Sci. USSR, 13 (9), 2257 (1971).
33. J. K. Gillham and H. C. Gillham, Polymer Eng. Sci., 13 (6), 447 (1973).
34. R. P. Chartoff and T. W. Chiu, Polymer Eng. Sci., 20 (4), 244 (1980).
35. R. C. Weast, Editor, "CRC Handbook of Chemistry and Physics", 58th Edition, C-738, CRC Press Inc., Cleveland, OH 1977.
36. D. H. Kaelble, "Physical Chemistry of Adhesion", pp. 65-67, Wiley-Interscience, New York 1971.
37. J. A. Barrie, M. J. L. Williams and K. Munday, Polymer Eng. Sci., 20 (1), 20 (1980).
38. S. P. Chen and J. A. D. Edin, Polymer Eng. Sci., 20 (1), 40 (1980).
39. R. C. Tolman, J. Am. Chem. Soc., 47, 1524 (1925).
40. F. H. Getman and F. Daniels, "Outlines of Theoretical Chemistry", p. 336, John Wiley & Sons, New York, 1937.
41. R. B. Lindsay and H. Margenau, "Foundations of Physics", pp. 227-269, John Wiley & Sons, New York, 1936.
42. R. B. Bird, W. E. Stewart and E. N. Lightfoot, "Transport Phenomena", p. 26, John Wiley & Sons, New York, 1966.
43. C. E. Rogers, in "Engineering Design for Plastics", E. Baer, Editor, Chapter 9, p. 644, Reinhold Book Corp., New York, 1968.
44. C. J. Geankoplis, "Transport Processes and Unit Operations", p. 277, Allyn and Bacon, Inc., Boston, 1978.
45. T. D. Epley and R. S. Drago, Paint Technol., 41 (536), 500 (1969).
46. K. L. Mittal, Editor, "Adhesion Aspects of Polymeric Coatings", Plenum Press, New York, 1983.
47. K. L. Mittal, Editor, "Adhesive Joints: Formation, Characteristics and Testing", Plenum Press, New York, 1984.

KINETICS AND MECHANISM OF THERMAL CYCLIZATION OF POLYAMIC ACIDS

L. A. Laius and M. I. Tsapovetsky

Institute of High Molecular Compounds
Academy of Sciences
Leningrad 199004, USSR

Polyamic acids with different chemical structure
and molecular weights have been used to analyze the
kinetic relationships of their transformation into
polyimides. Models describing the cyclization of polyamic acids under both isothermal and non-stationary
conditions are developed. Cyclization is regarded as a
two-stage physico-chemical process in which both the
molecular mobility of the polymer matrix and the
physical state of intrachain groups transformed into
the imide rings play an important role. The nature of
non-equivalent physical states of amic acid groups is
discussed.

INTRODUCTION

The transformation of polyamic acids (PAA) into polyimide (PI) should be regarded as a first-order reaction because it proceeds between two functional groups belonging to the same amic acid fragment. However, in the coordinate system of the first-order reaction the cyclization isotherms are not straight lines, rather are curves with a gradually decreasing slope.[1,2] In the early investigations dealing with the PAA cyclization[1,2] the following main features of the kinetics of this process were established:

1. When the reaction is carried out under isothermal conditions, two stages may be distinguished: the fast initial stage and the slow second stage. In the initial stage the rate constant calculated according to the expression for the first-order reaction remains virtually invariable and in the second stage it decreases continuously and can drop almost to zero long before the reactive species are consumed. This stage is often called the kinetic interruption, and reactions exhibiting such kinetics are called polychromatic.

2. With increasing experimental temperature, the contribution due to the first stage increases, and as a result the degree of cyclization corresponding to the kinetic interruption of the reaction becomes higher. A family of cyclization isotherms typical of polyimides is shown in Figure 1 for a polyimide (PM) obtained from pyromellitic dianhydride and 4,4'-diamino-diphenyl ether.

These kinetic relationships are manifestations of the cyclization mechanism of PAA, and, probably, for this reason, such relationships have attracted considerable attention of researchers. There are different opinions about what causes the retardation of cyclization. The following possibilities are considered:

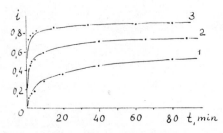

Figure 1. Cyclization isotherms of PAA PM at temperatures 1) 413, 2) 433 and 3) 453 K. Points represent experimental values; solid curves are calculated according to Eq. (4).

1) In PAA, the amic acid groups (AAG) are in the non-equivalent kinetic states. Both more and less active groups exist. As the cyclization proceeds, the active groups are consumed and this leads to a decrease in the rate constant[2,3].

2) The activity of AAG decreases during the reaction owing to the change in the properties of the polymer matrix itself[2,3].

3) Apart from the main reaction, i.e., transformation of PAA into PI, an important contribution is provided by a side reaction: the decomposition of AAG at amide bonds with the formation of anhydride and amine groups

$$D_1 + D_2 \underset{k_2}{\overset{k_3}{\rightleftarrows}} A \overset{k_1}{\rightarrow} I \tag{1}$$

where A are the amic acid groups, I are the imide rings and D_1 and D_2 are the anhydride and amine end groups, respectively.

As a result of the side reaction, amic acid groups are partly excluded from the process for a certain time; however as this reaction is reversible, so during further process the amic acid groups are slowly reformed and then cyclized. It is assumed that the regeneration rate is lower than that for cyclization and this leads to a drastic change in the slope of cyclization isotherms[4].

RESULTS AND DISCUSSION

To elucidate the true mechanism determining the transformation of PAA into PI, two models were developed and with their aid the analytical expressions for the degree of cyclization i as functions of experimental time t and temperature T were obtained. It was impossible to distinguish between them on the basis of the analysis of kinetic data alone.

The first model[5] is based on the assumption that amic acid groups (AAG) in PAA are kinetically non-equivalent and differ in their "preparedness" for the closure of the imide ring and the extent of this preparedness may be unequivocally related to the parameter j (state parameter). AAG are distributed along this parameter with a density $\rho(j,t)$ (Figure 2a). The reaction scheme has the following form

$$I \overset{k_c}{\leftarrow} P \overset{k(j)}{\underset{k(j)}{\leftarrow}} j_1, j_2 \cdots j_m \tag{2}$$

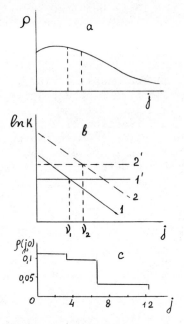

Figure 2. a) Scheme of AAG distribution according to the states and b) dependence of logarithm of constants k(j) (1,2) and k_c (1',2') on state parameter j at temperatures T_1 (1,1') and T_2 (2,2') $(T_2 > T_1)$; ν_1 and ν_2 are shell boundaries at temperatures T_1 and T_2, respectively; c) stepwise distribution for PAA determined experimentally.

The most favorable state for cyclization is P (the pre-start state) and the least favorable state is j_m. For the passage of an AAG from a certain state j into the imide state I (for chemical transformation) it should initially pass on to the most favorable state for cyclization, P. The rate constant for this transition k(j) depends on the initial state of AAG. This dependence may be given by

$$k(j) = A_o e^{-j} e^{-\frac{U}{RT}} \tag{3}$$

where A_o and U are constants. The reverse transition at the same rate constant is also possible.

The rate constant of chemical transformation will be given in the usual form

$$k_c = B_o e^{-\frac{E}{RT}}$$

where B_o and E are constants.

In accordance with Eq. (3) the dependence of $\ln k(j)$ versus j will be represented by a straight line with a negative slope (Figure 2b) and the dependence of $\ln k_c$ versus j will be described by a straight line parallel to the j axis. At a point $j = \nu$ these straight lines intersect. The point of intersection divides the whole plot into two regions in which AAG behave differently. It is known that if the process occurs in several stages, the resulting rate is determined by the slowest stage limiting the rate of the entire process. Hence, the cyclization rate of AAG, being in the states $j < \nu$, is limited by the constant k_c regardless of the value of j. All these AAG can be regarded as though they were in the same state and the totality of these states $0 < j < \nu$ will be termed "shell".

Free exchange of AAG is possible between the shell states because here $k(j) > k_c$. It should be noted that the shell is not a physical volume, rather a combination of such states in which the transition of AAG to the imide state $(j < 0)$ is more probable than on leaving the shell (transition to the states $j > \nu$). The shell also includes the pre-start state. For AAG existing in the states $j > \nu$ the limiting constant of the entire process is $k(j)$.

It is easy to explain qualitatively the above features of PAA cyclization with the aid of the plot shown in Figure 2b. In fact, it follows from this diagram that during cyclization the intra-shell states are exhausted first at a rate determined by k_c. Subsequently AAG in the states $j > \nu$ are cyclized and among them those in the states adjoining $j = \nu$ are cyclized at the fastest rate. Each next state (with a higher j) has a lower rate constant than the preceding state. This leads to a decrease in the overall cyclization rate and to a kinetic interruption. Hence, it is clear that during the reaction its limiting stages vary.

When the experimental temperature is raised, the constants k_c and $k(j)$ change in different ways in accordance with the values of activation energy U and E. If $U > E$ (as will be shown below, this condition is fulfilled), then when the temperature increases from T_1 to T_2, $k(j)$ increases faster than k_c and the shell boundary is displaced to the right, towards $\nu_2 > \nu_1$. This means that the shell volume increases. Correspondingly, the degree of cyclization at which the kinetic interruption occurs also increases.

Let us calculate the overall rate of AAG transition from all the amic acid states to the imide states, i.e. the cyclization rate. The following symbols will be used: $i(t)$ is the fraction of the cyclized AAG at moment t, $a(j,t)$ is the fraction of AAG in all the amic acid states with the coordinate varying from 0 to j (integral distribution function), $\rho(j,t) = da(j,t)/dj$ is the density of AAG distribution according to the states, $a_\nu(t) = a(\nu,t) = \int_0^\nu \rho(j,t)\, dj$ is the number of AAG in the shell at moment t, and m

is the coordinate of the most distant of the states populated with AAG (maximum value of j). The equation $i(t) + a(m,t)=1$ is always fulfilled.

The main factor determining the change in shell population with time is (di/dt), the flow of AAG into the imide state at a rate constant k_c. The exchange between the shell and the extra-shell states may be neglected because at $j > \nu$ we have $k(j) < k_c$ and this inequality increases with increasing j. Consequently

$$\left(\frac{di}{dt}\right)_{shell} = - \frac{da_\nu(t)}{dt} = k_c a_\nu(t)$$

which leads to

$$\frac{da_\nu(t)}{a_\nu} = - k_c dt; \quad a_\nu(t) = a_\nu(0)e^{-k_c t}$$

and

$$\left(\frac{di}{dt}\right)_{shell} = k_c a_\nu(0)e^{-k_c t}$$

For an AAG passing from the extra-shell state into the pre-start state the transition to the imide state is more probable than the return to the preceding state because at $j > \nu$ we have $k_c > k(j)$. Hence, the back flow of AAG from the pre-start state into the extra-shell states may be neglected. Then the number of AAG $\rho(j,t)$ dj, contained in the range of states dj in the interval $\nu < j < m$ may be expressed by the equation

$$- \frac{d}{dt}[\rho(j,t)dj] = k(j)\rho(j,t)dj$$

to give $\rho(j,t) = \rho(j,0)e^{-k(j)t}$ where $\rho(j,o)$ is the initial distribution density. The flow of AAG from all extra-shell states to the imide state $(di/dt)_{ex}$ is equal to the integral

$$\left(\frac{di}{dt}\right)_{ex} = \int_\nu^m k(j)\rho(j,t)dj$$

The overal cyclization rate is given by

$$\frac{di}{dt} = \left(\frac{di}{dt}\right)_{shell} + \left(\frac{di}{dt}\right)_{ex} = k_c a_\nu(0)e^{-k_c t} + \int_\nu^m k(j)\rho(j,0)e^{-k(j)t}dj$$

300

The degree of cyclization is expressed by the equation

$$i = \int_0^t \frac{di}{dt} dt = a_\nu(0)(1-e^{-k_c t}) + \int_0^t \int_\nu^m k(j)\rho(j,0)e^{-k(j)t} dj.dt \qquad (4)$$

All the paremeters contained in Equation (4) including the distribution $\rho(j,0)$ may be obtained from a comparison of these equations with the experimental cyclization isotherms, e.g. by using the procedures described in Ref. 5. The family of isotherms shown in Figure 1 is in reasonable agreement with the following values: $E = 81.7$ kJ/mol, $U = 280.0$ kJ/mol, $\ln B_o = 19.4$ and $\ln A_o = 77.8$. The values of A_o and B_o are expressed in seconds. The position of the shell boundary $\nu = \ln(A_o/B_o) - (U-E)/RT$ is found from the condition that at $j = \nu$, $k(j) = k_\nu = k_c$. The distribution $\rho(j,0)$ is shown in Figure 2 .

Cyclization isotherms calculated according to Equation (4) with the above values of parameters are in good agreement with experimental data (Figure 1).

The other model[6] is based on the reaction scheme analogous to scheme (2)

$$I \xleftarrow{k_c} B \begin{array}{c} k_1(i) \\ \xleftarrow{} \\ \xrightarrow{} \\ k_2(i) \end{array} A$$

but differing from this scheme by the fact that instead of a series of extra-shell states distributed along the parameter j only one state common to all AAG contained in this state exists. The states B and I correspond to the shell and the imide state respectively as in the previous model. AAG exchange may occur between the A and B states. In this case it is important to point out that the constants k_1 and k_2 are assumed to be dependent on the degree of cyclization. They are given in the form $k_1(i) = f(i)e^{-U_1/RT}$ and $k_2 = f(i)e^{-U_2/RT}$ where $f(i)$ is the decreasing function of i. Hence, this model assumes that the activity of AAG depends on the properties of the polymer matrix itself, i.e., both the spatial (states A and B) and the time $(k(i)=k[(i(t)])$ kinetic non-equivalence of AAG occurs.

The determination of the dependence $i = i(t)$ in this model reduces to the solution of the second-order differential equation

$$\frac{d^2 i}{dt^2} + [k_1(i)(1+\beta)+k_c]\frac{di}{dt} - k_1(i)k_c(1-i) = 0 \qquad (5)$$

where $\beta = k_1(i)/k_2(i)$ is the equilibrium constant.

301

All the parameters contained in Equation 5 can be obtained experimentally, and the differential equation can be solved in the numerical form. Agreement with experimental data is approximately the same as in the preceding model. It should be emphasized that in both cases the comparison of the theoretical and experimental data was carried out as follows. All the required parameters of the model were calculated on the basis of the analysis of limited parts of two or three cyclization isotherms. Then the complete isotherms were obtained for any temperature according to Eqs. (4) and (5) and compared with the experimental results.

According to concepts of the models, cyclization is a physico-chemical process consisting of two stages:

a) the physical stage, i.e., the diffusion of AAG into various states and

b) the chemical stage, the closure of the imide ring.

At different moments of time, the process may be limited either by stage a) or by stage b). Initially the rate of the entire process is limited by the chemical transformation, and in the final stages it is limited by the AAG flow from the states unfavorable for cyclization to those favorable for it. Since these models can describe all the kinetic effects observed in the cyclization of PAA in the solid state in good agreement with experimental data, it might be assumed that they reflect correctly the cyclization mechanism.

However, the nature of kinetic non-equivalence of AAG on which both models are based is still open to question. If this problem is not solved, the models remain devoid of physical meaning, although all the parameters contained in them have a real physical meaning. Moreover, the relationship between these models should be eluci-dated.

The amic acid groups broken down at the amide bonds according to Scheme (1), i.e. the pair of amine and anhydride end groups, should be considered to be the states kinetically unfavorable for cyclization. The distance between these groups will serve as the parameter of kinetic non-equivalence. Evidently, these broken-down groups can affect the cyclization kinetics only if their number in the system is relatively large (comparable to the total amount of AAG). Specially designed experiments have been carried out to elucidate the significance of this factor[7]. For this purpose, the extent of anhydride end groups formed in various cyclization stages was monitored by the 1860 cm^{-1} IR band.

The anhydride content corresponds to that of the broken amide bonds D_1. Figure 3 shows the change in the D_1 value in the PAA-PM polymer during stepwise heating. It can be seen from Figure 3 that D_1 does not exceed 2% of the total number of amide bonds capable of breaking down. The degree of degradation depends on the chemical structure of the polymer. The greatest degree is observed in polyamic acids having no pin-joint bonds in the chain. In this case degradation may be 10%. In solutions, the extent of degradation may be much greater. However, in this case we are only interested in the cyclization in bulk. As calculations have shown, this extent of degradation cannot account for the phenomenon of kinetic interruption of cyclization. Hence, the decisive role played by the side reaction is not confirmed experimentally.

The problem of the nature of the kinetic non-equivalence has been considered[8] from the viewpoint of possible conformational situations. Two of the possible planar conformations are shown below:

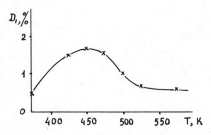

a   b

Figure 3. Temperature dependence of relative concentration of anhydride groups D_1 upon stepwise heating of PAA PM.

Conformation a) is more and conformation b) is less favorable for cyclization. The transition of conformation b) into the state favorable for cyclization is possible only when the CONH group turns together with the adjoining part of the polymer molecules. If this is actually the case then the transition of the reaction into the slow stage should depend on the chain length. This assumption has been confirmed experimentally[9]. Figure 4 shows cyclization isotherms of oligomeric PAA-PM analogues ranging from a monomer with a molecular weight of 406 to a polymer with M = 20000. In order to rule out the possible effect of the phase behavior and the interaction of reacting groups on the cyclization kinetics, all compounds were introduced at the same low molar concentration into a neutral amorphous polymer matrix. The properties of this matrix did not vary upon any thermal treatment carried out. The matrix consisted of a polyamide with a softening point of 483K. The samples were prepared by mixing the solutions of polyamide and a model compound in dimethylformamide. The concentration of model compounds in the matrix was always 10 mol%.

It was found that only monomer cyclization proceeds at a rate which is invariable from the beginning to the end. For other samples, the kinetic interruption state is observed and becomes increasingly pronounced with increasing chain length.

Hence, it is possible to say with confidence that kinetic non-equivalence is, to a certain extent, due to conformational differences in AAG. However, it cannot be ruled out that this non-equivalence is also affected by specific conditions of the medium surrounding the various groups.

If AAG can really exist in different states, they must be able to pass from one state to another under certain conditions. Consequently, it should be possible to activate the cyclization process which has attained the kinetic interruption stage with the

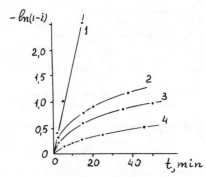

Figure 4. Dependence of −ln(1−i) on cyclization time for oligomeric PAA PM analogues in a neutral polymer matrix at 433 K. Molecular weight M: 1) 406, 2) 914, 3) 4180 and 4) 20000.

aid of a treatment that can redistribute AAG between the states and fill up more active states exhausted in the initial cyclization stage. Both models described above predict this effect.

In practice, this effect may be attained by the dissolution of partly imidized PAA film and its repeated preparation as well as with the aid of a short thermal pulse[10]. Figure 5 shows the change in the degree of imidization of PAA with the following structure

HOOC — O — COOH / ~HNOC — CONH — — C(CH_3) — (phenyl)

This polymer is soluble in both the imide and amic acid forms . The experimental conditions were as follows. The temperature was 413 K; and at the moment t_1 during the slow reaction stage the heating was interrupted and the film was dissolved in dimethylformamide and recast after which the cyclization experiment was continued at the same temperature. It can be seen in Figure 5 that as a result of this procedure the cyclization rate increased markedly.

Figure 6 shows the results of an experiment carried out to activate cyclization by a thermal pulse for polyamic acid, PAA PM, at a cyclization temperature of 433 K. The experimental procedure was similar to that in the preceding case except that at the moment t_1 the sample was not dissolved but abruptly heated up to 503 K and rapidly cooled to the previous temperature.

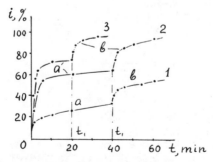

Figure 5. Degree of cyclization i vs heating time t under conditions of dissolution and repeated casting of the film of soluble polyimide during cyclization. a - cyclization before dissolution; b - cyclization after dissolution. Cyclization temperatures, T :
1) 393, 2) 413 and 3) 433K.

In this case also the cyclization rate increased. If the parameters of the models are known, it is possible to calculate the change in i(t) after activation. Solid line in Figure 6c is the calculated line which is in good agreement with the experimental data points.

The experimental results and calculations can be interpreted only on the basis of the assumption that AAG are distributed in the kinetically non-equivalent states. However, it cannot be assumed that the kinetics of PAA cyclization is completely governed by this factor alone.

It has been shown[11] that the transition from the fast cycliza-tion stage to the slow stage is, to a considerable extent, due to a decrease in molecular mobility in the polymer. With increasing imide ring content in the chain, the softening temperature of the

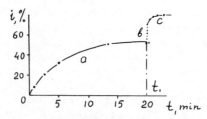

Figure 6. Degree of imidization i of PAA PM vs heating time t under conditions of thermal pulse. a and c - before and after the thermal pulse, T=433 K; b - thermal pulse, T = 503 K, t = 15s.

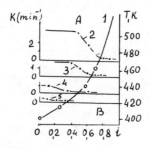

Figure 7. Softening temperature (curve 1) and cyclization rate constants (curves 2-5) for PAA DPhO vs degree of imidization. Imidization temperatures: 2) 473, 3) 453, 4) 433 and 5) 423K. Ranges above (A) and below (B) of curve 1 correspond to the softened and glassy states, respectively.

polymer increases[12]. Hence with increasing degree of cyclization T is a constant after a certain value of $i = i_g$ is attained, PAA passes from the softened state into the glassy state. It was found that near this transition the most drastic decrease in the rate constant is observed (Figure 7). In Figure 7 the experimental results for the DPhO polymer are shown.

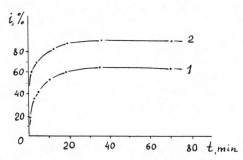

This polymer exhibits a distinct softening point in both the amic acid and the imide form. The intersection of the curve T_g = f(i) with the straight line T = const representing the experimental temperature corresponds to the transition of the polymer from the softened state to the glassy state. In prepolymers of some polyheteroarylenes, for example in polyamic hydrazides at $T < T_g$, the reaction does not proceed at all[13]. In polyamic acids the reaction proceeds even in the glassy state but at a lower rate.

Hence, the kinetics of PAA cyclization is determined by both the kinetic non-equivalence of amic acid groups and the physical state of the polymer. The relative roles played by these factors are clearly seen if PAA cyclization isotherms in a rigid polymer matrix (Figure 8, curve 1) are compared to those for pure PAA (curve 2). Although in the former case the state of the matrix remains invariable throughout the experiment, the kinetic interruption stage is distinctly seen in the isotherm. As already mentioned, in this case this phenomenon should be ascribed to the kinetic non-equivalence. The cyclization isotherm of pure PAA differs from that for the PAA in the matrix: in the former case the

Figure 8. Degree of imidization of i of PAA PM vs heating time t at 453 K in a polymer matrix (curve 1) and pure PAA (curve 2).

initial part is longer. Moreover, up to $i = i_g = 0.25$ at a given experimental temperature the polymer is in the softened state. At a higher i value, the polymer is transformed to the glassy state and then the reaction proceeds in the solid phase under the conditions analogous to the previous case. If curve 2 is displaced by $i = i_g$, it coincides reasonably well with curve 1. Hence in the glassy state further change in molecular mobility upon cyclization no longer appreciably affects the kinetics of the process, and the longer initial part of curve 2 is due to the cyclization of the polymer in the softened state.

Hence, the change in the type of molecular mobility plays the part of a "switch" limiting cyclization stages. In the softened state, molecular mobility is high and the transitions between kinetically non-equivalent states are facilitated. These transitions cannot limit the process; the limiting factor is the rate of chemical transformation. Hence, the rate is high and the parameters of the process are weakly dependent on conversion and are close to those of the reaction occurring in solution. After the polymer passes from the softened into the glassy state, the transitions between kinetically non-equivalent states become difficult and begin to limit the process rate. The rate constant decreases sharply and becomes dependent on conversion. The amic acid groups are distributed according to reactivities and this distribution determines the further kinetics of the process.

This appears to be the overall mechanism of cyclization of polyamic acids. As to the relationship between the two models considered, the following observations might be pointed out. In the first cyclization state – from the start of the process to the moment of glass transition of the polymer – the second model is in better agreement with the physical meaning of the phenomenon because it takes into account, in an explicit form, the dependence of the rate constant on conversion, and this dependence is responsible for the transition of the polymer from the softened into the glassy state, which greatly affects the process rate and leads to the change in the limiting reaction stages. In contrast, in the final cyclization stage the shell model correctly reflects the essence of the phenomenon: the kinetics is determined by the shape of distribution formed in the polymer during its hardening. The conversion level at which the change in the limiting stages occurs is taken into account in this model by the shell volume.

SUMMARY

Cyclization of polyamic acids is a complex physico-chemical process. It consists of the stage in which the amic acid groups change their physical state and the second (principal) stage – chemical transformation of amic acid groups into the imide rings.

The overall rate of the process in the initial period is usually limited by the chemical stage and that in the final period is determined by the first stage. The transition from one stage to the other leads to a drastic change in cyclization rate and the moment of this transition is determined, to a considerable extent, by the physical state of the polymer. The kinetic non-equivalence of amic acid groups is due to their conformation differences, molecular weight distribution and different interaction with the medium. This appears to be the mechanism for the formation of polyimides. Presumably it is a general mechanism for all thermally stable polymers formed as a result of intrachain cyclization of pre-polymers.

REFERENCES

1. J. A. Kreuz, A. L. Endrey, T. P. Gay and C. E. Sroog, J. Polym. Sci., A-1, 4, 2607 (1966).
2. L. A. Laius, M. I. Bessonov, E. V. Kallistova, N. A. Adrova and F. S. Florinsky, Vysokomol. Soedin., 9A, 2185 (1967).
3. N. A. Adrova, M. I. Bessonov, L. A. Laius and P. A. Rudakov, Poliimidy-novyi klass termostoikich polimerov, Leningrad, Ed. "Nauka", 1968.
4. E. N. Kamzolkina, P. P. Nechaev, S. V. Markin, Ya. S. Vygodsky, T. V. Grigoriev and G. E. Zaikov, Dokl. Akad. Nauk SSSR, 219, No. 3, 650 (1974).
5. L. A. Laius and M. I. Tsapovetsky, Vysokomol. Seodin., 22A, No. 10, 2256 (1980).
6. M. I. Tsapovetsky and L. A. Laius, Vysokomol. Soedin., 24A, No. 5, 979 (1982).
7. M. I. Tsapovetsky, L. A. Laius, M. I. Bessonov and M. M. Koton, Dokl. Acad. Nauk SSSR, 290, No. 1, 132 (1976).
8. I. S. Milevskaya, N. V. Lukasheva and A. M. Eliashevich, Vysokomol. Soedin., 21A, No. 6, 1302 (1979).
9. M. I. Tsapovetsky, L. A. Laius, M. I. Bessonov and M. M. Koton, Dokl. Akad. Nauk SSSR, 256, No. 4, 912 (1981).
10. M. I. Tsapovetsky, L. A. Laius, M. I. Bessonov and M. M. Koton, Dokl. Akad. Nauk SSSR, 243, No. 6, 1503 (1978).
11. L. A. Laius, M. I. Bessonov and F. S. Florinsky, Vysokomol. Soedin., 13A, No. 9, 2006 (1971).
12. L. A. Laius and M. I. Bessonov, Proceeding of the 15th Scientific Conference of the Institute of Macromolecular Compounds, the Academy of Sciences of the USSR, Leningrad, Ed. "Nauka", pp. 139-146, 1968.
13. L. M. Bronshtein, B. I. Zhisdiuk and A. S. Chegolya, Vysokomol. Soedin., 20B, No. 6, 443 (1978).

THE CURE RHEOLOGY OF PMR POLYIMIDE RESINS

P.J. Dynes, T.T. Liao, C.L. Hammermesh, and E.F. Witucki[*]

Rockwell International Science Center
Thousand Oaks, California 91360

The dynamic viscoelastic properties of PMR-15
polyimide resin during thermal curing have been studied
by the parallel-plate technique. The stoichiometry
and molecular weight distribution of the imidized
prepolymer are shown to have a strong influence on
flow behavior during cure. Hydrolysis of polyamide
acid intermediates and preferential reactivity of
monomeric ingredients are discussed as factors contributing to the broad molecular weight distribution
of the imidized prepolymer. Low molecular weight
oligomers were prepared, and their effect on cure
rheology studied. The presence of the lowest molecular weight oligomer was found to be critical to
PMR-15 flow properties during cure.

[*]Rocketdyne Division of Rockwell International, Canoga Park,
California

INTRODUCTION

Polyimides based on the PMR (polymerization of monomeric reactants) approach show excellent potential for use as high temperature
(600°F) resistant polymer matrix materials. In these systems, the
starting monomers are dissolved in a low-boiling solvent for impreg-
nating the reinforcing fibers. This solvent, together with the
condensation volatiles, is removed at low temperatures during processing. Final cure occurs at higher temperature through an addition-
type reaction without release of further volatiles. The two most
widely utilized of these type resins are the NASA developed PMR-15[1]
and LARC-160[2] systems. A key factor contributing to their success
is the satisfactory flow behavior they display at elevated temperature, which allows for easy processability. It is generally accepted
that this flow is the result of the low molecular weight of the PMR
prepolymer product. Recently, it has been shown that the flow is
sensitive to the initial stoichiometry[3] and to addition of reactive
oligomers[4]. However, the relationship between the rheology and the
molecular weight distribution of the prepolymer has not been clearly
shown. The objective of this study was to determine the causes of
the broad distribution observed in PMR prepolymers and its conse-
quences on flow during cure.

EXPERIMENTAL

The three starting ingredients used to prepare PMR-15 resin
were commercially available 3,3', 4,4'-benzophenone-tetracarboxylic
dianhydride (BTDA), 5-norbornene-2,3-dicarboxylic anhydride (NA),
and 4,4'-methylene dianiline (MDA). NE (the monomethyl ester of
NA) and BTDE (the dimethylester of BTDA) were prepared by refluxing
the corresponding anhydrides in methanol for 2 hours subsequent
to their dissolution, to yield a 50 weight percent solution.
MDA, dissolved in methanol, was added to the ester mixture to give
a solution containing a molar ratio of BTDE/MDA/NE of N/N+1/2,
respectively, where N=2.087 for PMR-15. The imidized prepolymer
was prepared by first reducing the PMR-15 monomer solution to ~10%
solvent content by heating at 121°C in an air circulating oven.
Imidization was completed by heating for 2 hrs. at 200° in an
air circulating oven.

Flow measurements were made with a Rheometrics Visco-Elastic
Tester in the parallel-plate mode of operation and using a shear
frequency of 10 rad/s with a strain of 10%. The imidized prepolymers were pressed into 13 mm diameter discs, and flow measurements
began at a temperature when a pellet was reduced to a thickness
of 0.5 mm between the 25 mm diameter parallel plates. Rheological
data, consisting of the shear storage modulus G' and shear loss
modulus G" together with the dynamic viscosity defined as

312

$$\eta* = (G'^2 + G''^2)^{1/2}/\omega,$$

where ω is the applied frequency, were recorded as a function of temperature or time.

Molecular weight characterization of the imidized prepolymers was made by gel permeation chromatography (GPC) analysis, using 2(100Å) and 1(500Å) μ-Styragel columns from Waters Associates. The samples were dissolved in a 25/75 dimethylformamide/tetrahydrofuran solvent mixture and the eluting solvent was tetrahydrofuran at a flow rate of 1 ml/min. Ultraviolet detection at a wavelength of 254 nm was used.

Partition liquid chromatographic separations were carried out using a reverse-phase technique. A Spectra-Physics ODS 10 column with a 10-90% water/acetonitrile solvent gradient over a 50 min. time period was used. The aqueous portion of the mobile phase contained a 0.01M KH_2PO_4 buffer at pH=3.00. Sample solvents and detection were the same as for GPC.

DISCUSSION

PMR Cure Behavior

The polymerization sequence for PMR-15 is generally described as shown in Figure 1. The first step involves the formation of an end-capped polyamide-acid from the three starting monomer ingredients. The polyamide-acid then undergoes imidization at higher temperatures. Finally, crosslinking occurs via, it is believed, a reverse Diels-Alder type mechanism.[5-7]

The formulated molecular weight of the resin is controlled by the stoichiometric proportion of BTDE, MDA, and NE. For PMR-15, it is N=2.087 mol of BTDE with N+1 mol of MDA and 2 mol of NE to yield an imidized prepolymer with a molecular weight of 1500. The GPC separations of imidized PMR prepolymers corresponding to stoichiometries of N = 1, 2.087, and 3 are shown in Figure 2. The three resins show a broad distribution of molecular weights, with oligomers ranging from n = 0 to four or higher. One previously described aspect of this type of system is the apparently greater reactivity of NE with MDA compared to that of BTDE.[8-10] This results in the preferential formation of the n = 0 oligomer, which depletes the system of end caps and thus forces the formation of higher molecular weight species

The question arose as to whether the observed distribution of products was entirely a consequence of the statistical considerations outlined above, or whether other mechanisms were also involved. A second area of interest was how the processing rheology of a monodisperse resin would differ from the mixed product obtained by the commercial synthesis.

313

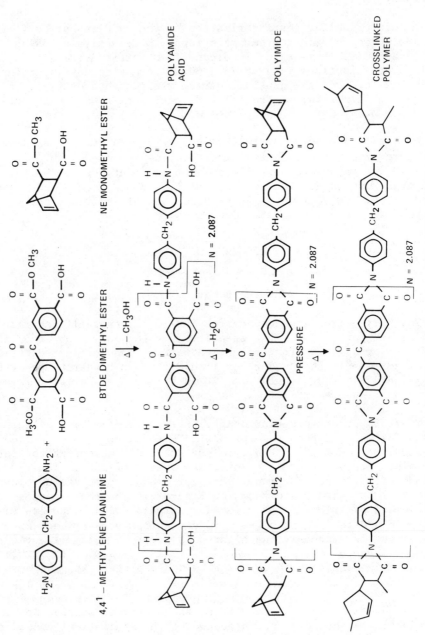

Figure 1. Idealized Cure Sequence for PMR-15 Resin.

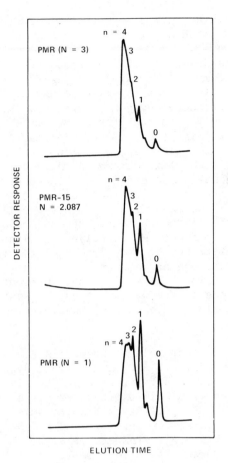

Figure 2. GPC separation of PMR prepolymer resins of varying formulated molecular weight.

Oligomer Synthesis

The question of an ideal oligomer versus the commercial product was investigated by synthesizing directly the simplest PMR oligomer containing the three monomer ingredients (the $n = 1$ compound) and comparing its properties to a commercially produced PMR with a stoichiometry formulated to $N = 1$. The $n = 1$ polyamide-acid, $E A_a B_a A_a E$, where E is the endcap, A the diamine, and B the dianhydride with the subscript a denoting the amide-acid linkage, was prepared by reacting two moles of $E A_a$, the (mononadamide of MDA), with one mole of the dianhydride form of B in tetrahydrofuran. The $E A_a$ was prepared by adding E dropwise to a solution of A in ethyl ether at a 10:1 A/E molar ratio. $E A_a$ precipitates under these conditions and was filtered and washed with ether. The $E A_a B_a A_a E$ product was precipitated with ethyl

315

ether, then filtered and dried.   Partition liquid chromatography and GPC indicated a pure product.

This polyamide-acid was imidized under the same conditions
described for the commercial resins, to yield ideally, $E_iA_iB_iA_iE$,
where the subscript $\underline{i}$ denotes the imide linkage.   The GPC
separation of the product, however, showed a wide distribution
of products very similar to that for commercially prepared
N = 1 PMR.   The most likely cause for this rearrangement is
hydrolysis of the amide linkages, a common problem in the synthesis of thermoplastic polyimides.   The same phenomenon was observed
in attempts to thermally convert E_aA to E_iA, where in addition to
E_iA, substantial amounts of E_iA_iE and some free A were detected
by partition liquid chromatography.   It is difficult to assess
the contribution of this mechanism to the molecular weight distribution observed in imidized PMR prepolymers, although it appears
that if a narrow distribution of polyamide-acids is formed in
the first step of the synthesis, a rearrangement would occur to
broaden the distribution.   It would be of interest to examine
the PMR system at the polyamide-acid stage of synthesis, but it
is difficult to complete this step without significant imidization also occurring.

Because it is not possible to prepare $E_iA_iB_iA_iE$ directly from
thermal imidization of $E_aA_aB_aA_aE$, a different synthesis route was
necessary.   Rather than start with the amide-acid E_aA, the imide
form E_iA was reacted with B in dimethylformamide at 120° C under
nitrogen atmosphere.   First, the half imide/amide acid $E_iA_iB_aA_aE$
is formed, then the three-quarter imidized $E_iA_iB_iA_aE$, and finally
the fully imidized $E_iA_iB_iA_iE$ product. E_iA, the mononadimide of
MDA, was prepared by reacting a 10:1 molar excess of A with E in
DMF at 120° C for 2 hrs under nitrogen atmosphere.   The product
contained a large amount of A and some E_iA_iE, and was purified by
separation on a preparative scale alumina chromatographic column.
The key to the success of this approach is that the thermal imidization of the initial $E_iA_aB_aA_iE$ intermediate cannot lead to hydrolysis products capable of forming higher molecular weight oligomers.
The molecular weight separation of $E_iA_iB_iA_iE$ prepared by this
technique and purified by passing through an alumina column is
shown in Figure 3.   The elemental analysis of the three imide
prepolymers prepared is summarized in Table I.

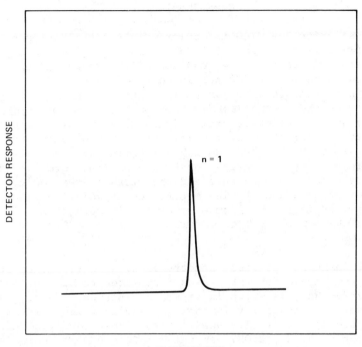

Figure 3. GPC Separation of n = 1 polyimide prepolymer.

Table I. Elemental Analysis of PMR Polyimide Precursors.

	Nitrogen		Carbon		Hydrogen	
	Theo.	Exp.	Theo.	Exp.	Theo.	Exp.
E_iA	8.14	7.72	76.8	75.34	5.86	5.76
E_iA_iE	5.72	6.00	76.0	75.74	5.35	5.62
$E_iA_iB_iA_iE$	5.75	5.82	74.0	74.61	4.35	4.33

Cure Rheology

Considering the wide molecular weight distribution of imidized PMR-15 prepolymer, it was of interest to determine how important this feature is to the flow behavior of the resin during cure. A critical aspect of a PMR polyimide resin when used as a matrix for a fiber reinforced composite is that it exhibits sufficient flow between the temperatures where immidization is complete ($\sim 200°$C) and crosslinking begins ($\sim 275°$C) to allow for consolidation into a coherent void-free laminate. The dynamic viscoelastic properties of PMR-15 resin during a $1°$C/min cure cycle are shown in Figure 4. Key parameters for this experiment are the minimum viscosity reached during cure and the width of the low viscosity region or "flow window." The occurrence of gelation can also be determined from the dynamic viscoelastic data, and it corresponds to that temperature or time at which the storage (G') and loss (G") moduli are equal.[11]

The influence of stoichiometry on dynamic viscosity during a $1°$C/min cure cycle for PMR resins formulated to stoichiometries of N = 1,2, and 3 is shown in Figure 5. Although these resins do not vary substantially in their molecular weight distributions (see Figure 2), their flow behavior is significantly affected. The selection of N = 2.087, on which NASA's PMR-15 resin is based, appears to represent a compromise between a large value for N, which provides a cured resin with superior thermal stability, and a small N, which allows for enhanced flow during cure. In addition to the value of N (which controls the formulated molecular weight), the distribution of oligomer molecular weights is also critical. The flow behavior of PMR formulated to N = 1 is compared to pure $E_iA_iB_iA_iE$ (n = 1) as well as to E_iA_iE (n = 0) in Figure 6. The commercially synthesized resin has a much wider "flow window" than either of the individual oligomers. The broader molecular weight distribution of the commercial product thus may be beneficial by providing a eutectic composition having a reduced melting temperature and enhanced flow behavior.

The trend observed for the flow of the pure n = 0 and n = 1 oligomers in Figure 6 indicates that oligomers of n > 1 are unlikely to exhibit significant flow prior to the onset of the crosslinking reaction. It was, therefore, of interest to determine the effect that the addition of these two oligomers would have on the cure rheology of PMR-15 resin. The first set of experiments involved adding various molar weight percentages of E_iA_iE and $E_iA_iB_iA_iE$ to the imidized PMR-15 prepolymer. This was accomplished experimentally by grinding the two solids thoroughly together with a mortar and pestle. The influence E_iA_iE on cure rheology is shown in Figure 7. A large reduction in the width of the flow window occurred as a result of only a small addition of E_iA_iE.

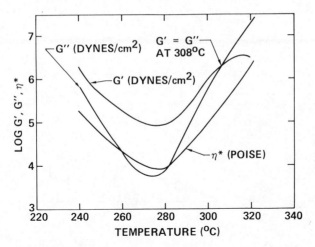

Figure 4. Dynamic viscoelastic properties of preimidized PMR-15 during 1°C/min cure cycle.

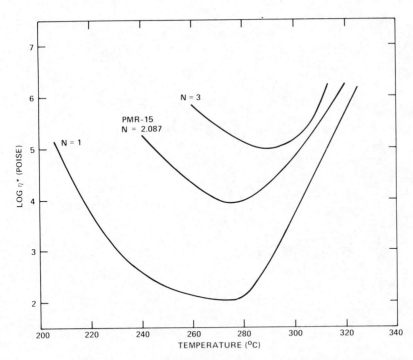

Figure 5. Effect of stoichiometry on the dynamic viscosity of PMR resins during a 1°C/min cure cycle.

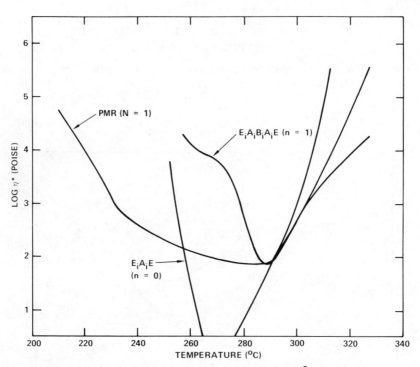

Figure 6. The dynamic viscosity during a 2°C/min cure cycle for PMR (N = 1) and n = 0 and 1 oligomers.

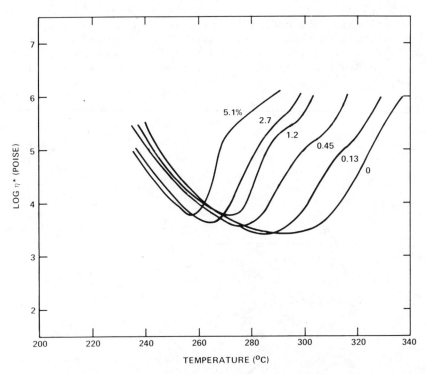

Figure 7. The dynamic viscosity during cure of PMR-15 resin with varying molar weight percentages of E_iA_iE mechanically blended into the resin.

The minimum viscosity reached during cure, however, increased only slightly. A similar effect is shown in Figure 8 for the addition of $E_iA_iB_iA_iE$ to PMR-15. The flow window was again narrowed, but much less than in the case of E_iA_iE.

Rather than mechanically blend E_iA_iE and $E_iA_iB_iA_iE$ with the imidized PMR-15 prepolymer, a series of PMR-15 resins were prepared in which the two oligomers were added during the initial BTDA and NA esterification step. In this way, the two additives were present in solution in the imidized prepolymers. The changes in PMR-15 cure rheology with changing concentrations of E_iA_iE and $E_iA_iB_iA_iE$ added by this technique are shown in Figures 9 and 10, respectively. The effect is opposite to that observed for physical mixing, in that the flow window is unchanged but the minimum viscosity is reduced with increasing concentration of E_iA_iE or $E_iA_iB_iA_iE$. The addition of N-phenylnadimide as a flow

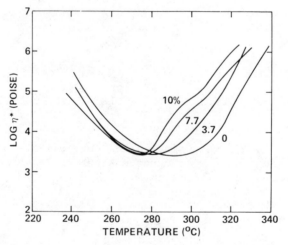

Figure 8. The dynamic viscosity during cure of PMR-15 resin with varying molar weight percentages of $E_iA_iB_iA_iE$ mechanically blended into the resin.

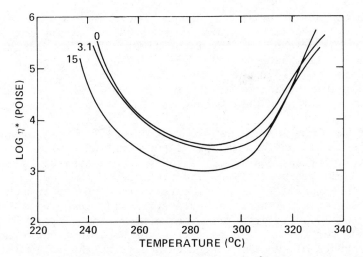

Figure 9. The dynamic viscosity during a 2°C/min cure cycle for PMR-15 prepolymer resin with varying molar weight percentages of E_1A_1E in solution in the resin.

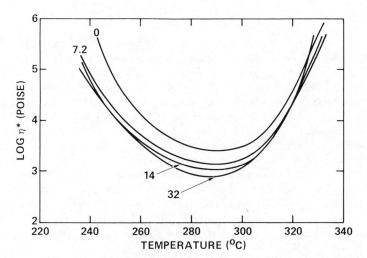

Figure 10. The dynamic viscosity during a 2°C/min cure cycle for PMR-15 prepolymer resin with varying molar weight percentages of $E_1A_1B_1A_1E$ in solution in the resin.

enhancer to PMR-15 resin was recently shown to produce a similar effect.[4] The results obtained by the physical mixing method may be due to the presence of the E_iA_iE or $E_iA_iB_iA_iE$ as finely dispersed particles which melt and act as a separate phase and tend to dominate the rheological properties. When the oligomers are present in solid solution in the PMR-15 prepolymer, this mechanism is not applicable and they behave as reactive plasticizers.

The cure rheology of PMR-15 resin was further elucidated by flow experiments on imidized prepolymer extracted with tetrahydrofuran and methylene chloride. The GPC separations of the soluble and insoluble resin fractions are compared to imidized PMR-15 prepolymer in Figure 11. Both solvents extract the lower molecular weight oligomers, with the methylene chloride removing slightly more of the higher oligomers. The complementary insoluble fractions show very similar molecular weight distributions with significant amounts of the two lowest molecular weight oligomers removed. The flow behavior of these materials in relation to imidized PMR-15 prepolymer is shown in Figure 12. The two insoluble fractions show very little flow over the entire temperature range. The soluble counterparts, however, display a very wide flow window and low minimum viscosity compared to PMR-15. These results demonstrate the sensitivity of the flow properties of PMR-15 to the presence of the low molecular weight oligomers and to the seemingly fortuitious circumstances resulting in their presence.

CONCLUSIONS

The results of this study demonstrate several important aspects of the relationship between the chemistry and cure rheology of PMR-15 resin. Commercially prepared PMR-15 prepolymer was found to be much broader in molecular weight distribution than expected from purely statistical considerations. One mechanism responsible for this effect was determined to be the hydrolysis of polyamide-acid intermediates during the imidization stage of PMR-15 processing. Another mechanism was the preferential reaction of endcapper with diamine to form E_iA_iE, the bis-nadimide of methylene dianiline. Rheologically, both these factors were found to be critical to obtaining the flow necessary for successful autoclave processing of composite laminates. It was shown that neither pure PMR-15-type oligomers nor imidized PMR-15 with low molecular weight components removed possessed adequate flow properties during cure.

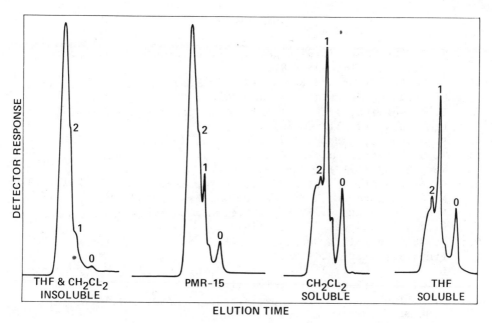

Figure 11.   GPC separation of PMR-15 and its methylene chloride and
tetrahydrofuran soluble and insoluble fractions.

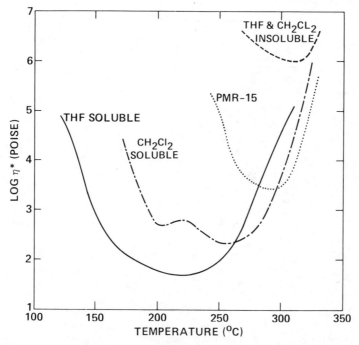

Figure 12.   Comparison of the dynamic viscosity during a 2°C/min
cure cycle for PMR-15 and its methylene chloride and
tetrahydrofuran soluble and insoluble fractions.

REFERENCES

1. T. T. Serafini, P. Delvigs, and G. R. Lightsey, J. Appl. Polym. Sci., 16, 905 (1976).
2. T. L. St. Clair and R.A.Jewell, 8th National SAMPE Technical Conference, Preprints, 8, 82 (1976).
3. F. I. Hurwitz, 12th National SAMPE Technical Conference, Preprints, 12, 517 (1980).
4. R. H. Pater, SAMPE J., 17 (Nov./Dec. 1981).
5. T. T. Serafini and P. Delvigs, Appl. Polym. Symposium No. 22, 89 (1973).
6. R. W. Lauver, J. Polym. Sci., Polym, Chem. Ed., 17, 2529 (1979)
7. N. G. Gaylord and M.Martan, Polym. Preprints, 22(1), 11 (1981).
8. P. R. Young and G. F. Skyes, Organic Coatings and Plastics Chem., Preprints, 40, 935 (1979).
9. P. J. Dynes, 12th National SAMPE Technical Conference, Preprints, 12, 402 (1980).
10. D. E. Kranbuehl, J. E. Smedley, S. Delos and J. M. Buchanan, Polym. Preprints, 22(2), 107 (1981).
11. C. M. Tung and P. J. Dynes, J. Appl. Polym. Sci., 27, 569 (1982).

RHEOLOGICAL CHARACTERIZATION OF ADDITION POLYIMIDE MATRIX RESINS

AND PREPREGS

M. G. Maximovich and R. M. Galeos

Lockheed Missiles and Space Company (LMSC)
P.O. Box 504
Sunnyvale, CA 94086

Graphite reinforced polyimide matrix composite
materials are attractive because of their outstanding
specific strengths, specific stiffnesses, and thermal
oxidative stability. Processing problems have, however, curtailed their application in aerospace hardware.
The key to better understanding and control of the processing of such materials is their rheological behavior.
Addition type polyimide resins and prepregs were
obtained or synthesized and characterized. Rheological
studies were carried out on neat resin systems.
Problems were encountered because of outgassing during
cure. A pressure cell transducer failed to solve these
problems. A staging technique was developed to successfully handle polyimide samples. Techniques were developed
to generate rheological curves on the graphite reinforced
prepreg. Rheological prepreg data correlated well with
the processing characteristics of these materials. Several commercial graphite/polyimide systems were studied
and compared, including PMR 15, LARC 160 (AP-22), LARC
160 (Curithane 103), and V378A.

INTRODUCTION

Graphite reinforced polyimide matrix advanced composites are attractive candidates for many aerospace applications. They exhibit high mechanical properties at elevated temperature, low density, and excellent thermal-oxidative stability. However, their actual application on production programs has been extremely limited to date because of processing problems and lack of control during fabrication of graphite/polyimide composites.

Good control is essential in processing polyimide composites, as several physical and chemical changes can occur simultaneously during cure. For example, modern addition polyimides such as PMR 15 or LARC 160 are based on the Ciba-Geigy P13N chemistry[1]. Figure 1 shows the temperature ranges for several different phenomena that occur during P13N processing. If the viscosity or flow characteristics of the prepreg during cure can be similarly determined, then the processing window can be accurately defined. Variables such as heat up rates, hold and cure temperatures, and pressure application points can be optimized for autoclave, press and vacuum bag processing. The purpose of the work described below was to identify, define, and develop the rheological techniques required to study polyimide resins and composite prepregs.

EXPERIMENTAL

Synthesis of LARC 160 Resins

Samples of LARC 160 (AP-22) and LARC 160 (Curithane 103) were synthesized according to the usual procedure in the literature[2].

A one-to-one substitution of Curithane 103 for Jeffamine AP-22 was made. Curithane 103 was considerably more difficult to soften or dissolve than Jeffamine AP-22. This behavior seems consistent with the isomer concentrations listed in Table 1. Curithane 103 has the highest concentration of 4.4' MDA and as such it is expected to have a more crystalline nature.

In order to verify chemical structures, IR spectroscopy and liquid chromatography were employed. Samples of both LARC 160 batches appear to show no significant differences in IR absorption peaks.

As synthesized, the volatile content of each LARC 160 batch, primarily water and alcohol, was running approximately 48 percent. This was determined by inserting a resin sample into a preheated oven at 316°C for 30 minutes and then calculating the sample weight

328

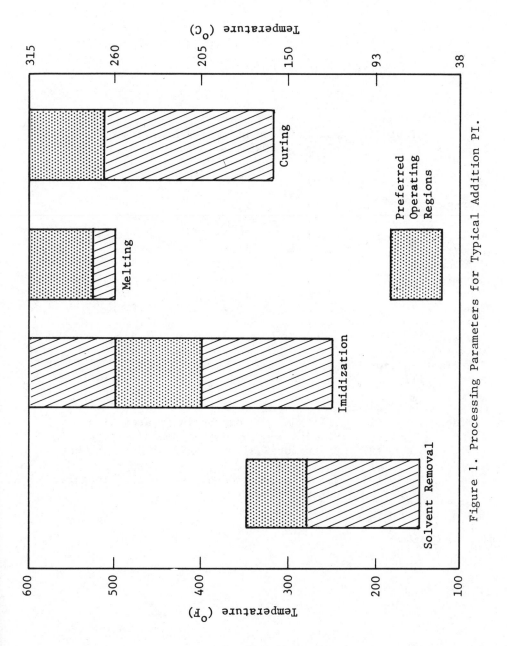

Figure 1. Processing Parameters for Typical Addition PI.

Table I. Results of Quantitative Analysis on
Polyamine Mixtures (Reference 4).

Percent Methylenedianiline

SAMPLE	2,2'-[1]	2,4'	4,4'-	Σ
AP-22	14.18	20.97	48.73	83.88
Tonox	4.24	14.80	61.21	80.25
Curithane-103	1.62	2.78	73.89	78.29
Ancamine DL	5.63	23.36	42.35	71.34
Tonox LC[2]	8.01	11.15	51.27	70.43
Tonox 22	0.78	3.98	54.75	59.51
Tonox JB	2.67	9.80	36.97	49.44
Tonox 60/40	1.60	8.71	37.85	48.16
Poly MDA	—	—	4.36[3]	4.36
2,4-Bis (p-amino-benzyl)aniline	—	—	1.83[3]	1.83

Footnotes: (1) 2,2'-MDA peak may be integrated with 4,4'-diamino-3-methyldiphenylmethane peak.
(2) Peaks for 2,2'-MDA and 2,4'-MDA were not well resolved.
(3) Value is questionable because of long retention time.

loss. This figure includes both solvent and condensation by-products. In order to be more representative of resin found in a prepreg, it was decided that a realistic volatile to resin ratio should be determined and also that the as-synthesized resin should be dried to this level. The dry resin content (after 30 minutes at 316°C) and volatile content (same conditions) of fabric LARC 160 prepreg used at LMSC runs approximately 36 percent and 12 percent, respectively. This gives a ratio of 1:3 or in other words 25 percent of the net resin weight is volatile. Drying conditions of 50°C for 2 to 3 hours in a vacuum oven seem to be appropriate and IR data show no side reaction products.

HPLC Evaluation

An HPLC procedure for the quality assurance of the synthesized polyimide resins was evaluated.[3] In our case the aqueous portion of the gradient mobile phase was not buffered with KH_2PO_4. However, the change did not appear to affect the quality of the component separations.

Liquid Chromatography Test Conditions:
µBONDAPAK C_{18} Column (Waters Assoc.)
U.V. at 200 nm, 0.5AUFS, Band Width 16nm, Time Constant at Normal
   (Varian VARICHROM)
U.V. at 254nm, 0.2AUFS (Waters Assoc. 440 Detector)
10%   50% CH_3CN (Waters Assoc.) vs. H_2O (J. T. Baker)
15 min. Linear Gradient, 10 min. Hold at Final, 15 min. Equilibration Delay
Flow Rate at 1.0 ml/min.

The necessity of the buffer to obtain "good" separations seems to depend on the sum total effects of all test parameters as well as the desired end use of the resultant chromatogram. The use of a buffer in polyimide analysis appears not to be universal with all investigators so the decision was made to first analyze the synthesized resins with pure solvents.

For comparison of the responses at 200nm the chromatograms for the blank and neat resins are shown in Figures 2 through 5. No attempt was made to optimize integration so the percent Peak Area results are not included since integration points were not consistent. The differences between the responses at 200nm and 254 nm are not substantial. Note that using these test conditions nadic anhydride (5-norborene -2, 3-dicarboxylic acid anhydride) is not detected.

To determine the effects of staging on the separated peaks, staged resins were analyzed by HPLC. The chromatogram comparison is shown in Figure 6. Keep in mind that the resin staged 1 hour

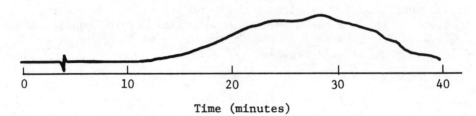

Time (minutes)

Figure 2. Chromatogram of Blank (CH$_3$CN).

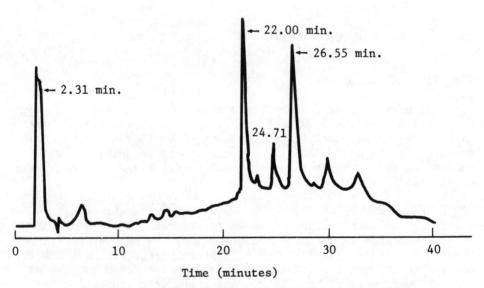

← 22.00 min.

← 26.55 min.

← 2.31 min.

24.71

Time (minutes)

Figure 3. Chromatogram of LARC 160 with AP-22.

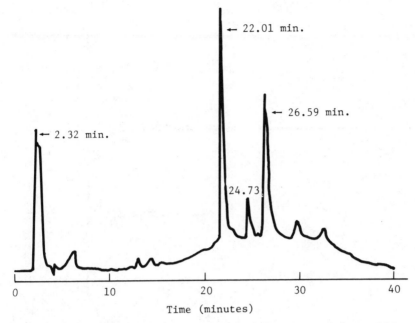

Figure 4. Chromatogram of LARC 160 with Curithane 103.

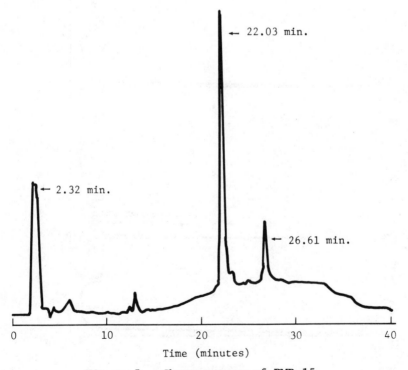

Figure 5. Chromatogram of PMR-15.

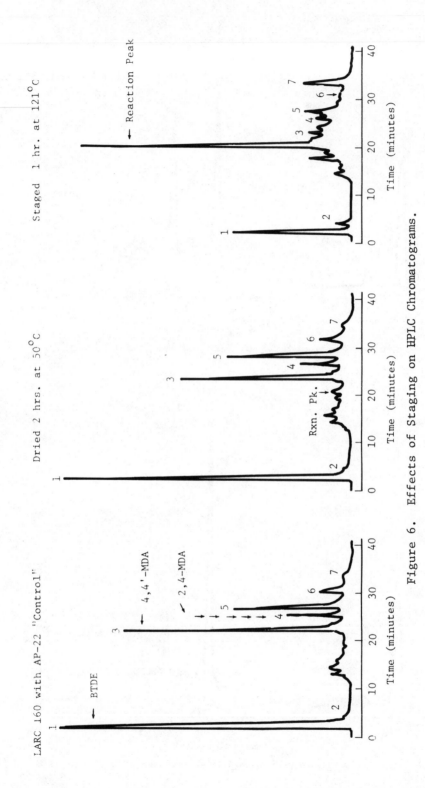

Figure 6. Effects of Staging on HPLC Chromatograms.

334

at 121°C was not completely soluble in acetonitrile. Thus, the curve represents resin that was soluble and prefiltered through a 0.45μ fluoromembrane filter. As can be seen, there are obvious effects on all peaks as a result of staging. Slight changes are evident even when the resin is dried for 2 hours at 50°C. The significance of these differences is not known at this time. Radical variations are seen in the extreme case of 1 hour staging at 121°C. The main reaction peak predominates in this curve.

General identification of specific peaks is made in a few instances. Lack of chromatogram complexity as compared to that obtained by Rockwell[4] may be attributed to two factors. First, our syntheses were carried out in fairly pure ethanol whereas Rockwell's esterification was conducted using an industrial alcohol mixture. Second, our resins were prepared under dilute solvent conditions. According to the 9th Quarterly Report (June–September 1980) from Rockwell (Ref. 5), high viscosity during the co-esterification of the starting anhydride materials results in incomplete conversion of BTDA to the BTDE diester and, thus, the presence of the BTDA tetra-acid (BTA) and BTDE monoester will be evident. The infrared spectra of the three synthesized resins are shown in Figures 7 through 9.

DSC Characterization

DSC studies were carried out on both unstaged resin and resin staged ½ hour at 120°C. Only minor differences were noted between the two LARC 160 versions. A set of curves was also run on a PMR-15 batch prepared at LMSC and is included for comparative purposes. Figures 10 through 15 show these data.

LARC 160 Resin Rheological Studies

After the initial batches of LARC 160 were prepared, a study was conducted to evaluate the effect of volatile content on viscosity. Results are shown in Figures 16 and 17 where η^* and Tan δ are shown as a function of time. The resin is LARC 160 with Curithane 103 and the heating rate was 2°C/min. As synthesized, the initial viscosity at RT was 62 poise and the minimum viscosity at 87°C was 9.2 poise. The volatile content was 48 percent. After 2 hours of drying at 50°C resulted in a volatile content of 27 percent, the RT viscosity increased to 3400 poise and the minimum viscosity at 85°C was 39.7 poise. Note, too, the appearance of an initial peak in Tan δ which indicates resin softening. Measurement of Tan δ is expected to be useful when correlating viscosity measurements to dielectric measurements for the purpose of cure monitoring. The second peak in Tan δ, in the vicinity of 131°C (55 min.), can be associated with resin hardening caused by imidization. The scatter in both curves is the result of bubble formation

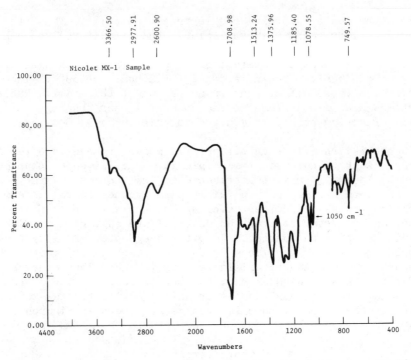

Figure 7.   IR of Synthesized LARC 160 with Jeffamine AP-22.

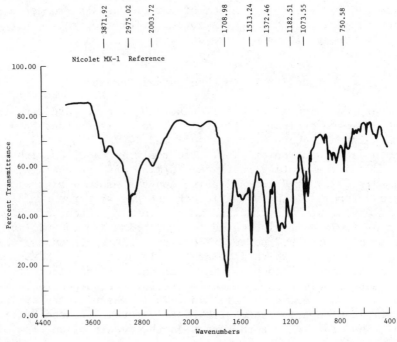

Figure 8.   IR of Synthesized LARC 160 with Curithane 103.

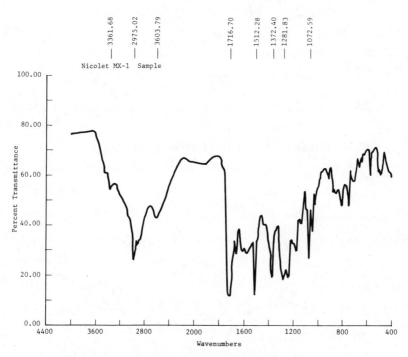

Figure 9. IR of Synthesized PMR-15.

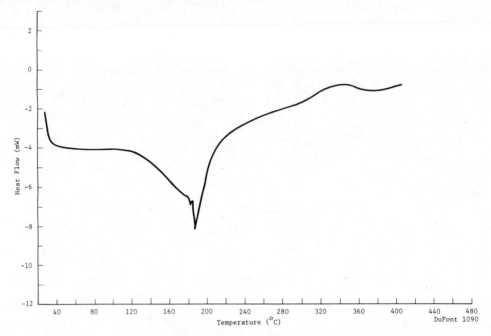

Figure 10.   DSC of PMR-15 Unstaged.

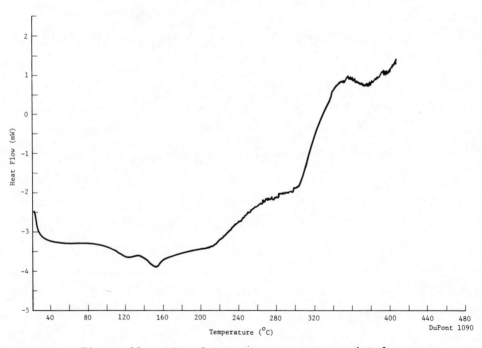

Figure 11.   DSC of PMR-15 Staged 0.5 Hr/120°C .

338

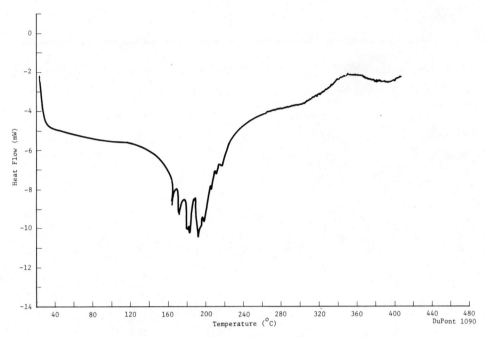

Figure 12.   DSC of LARC 160/22 Unstaged.

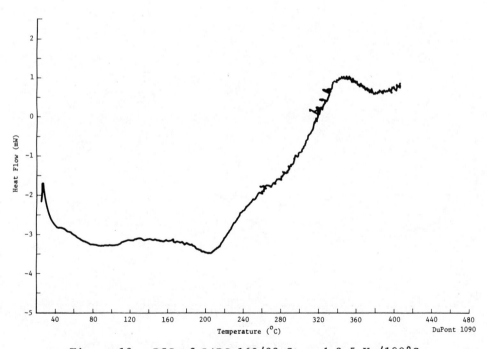

Figure 13.   DSC of LARC 160/22 Staged 0.5 Hr/120°C.

339

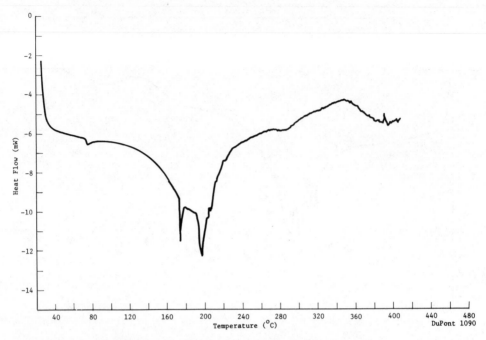

Figure 14. DSC of LARC 160/103 Unstaged .

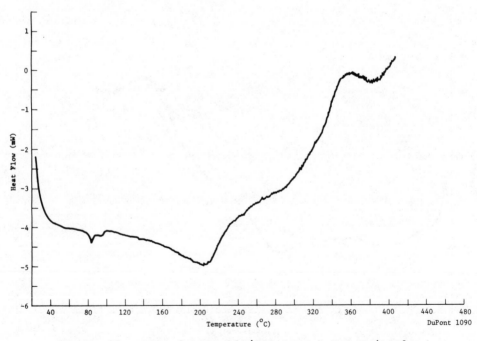

Figure 15. DSC of LARC 160/103 Staged 0.5 Hr/120°C .

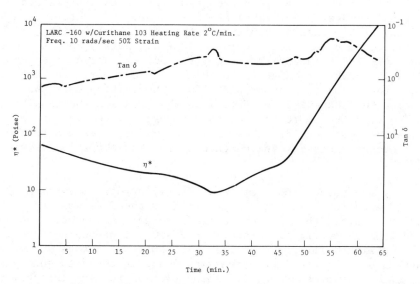

Figure 16. Effects of Volatile Content on Viscosity, original sample.

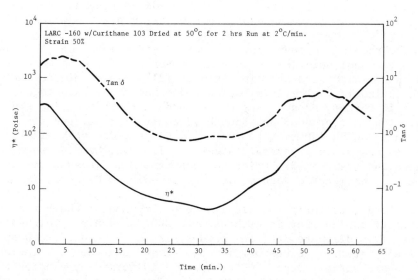

Figure 17. Effects of Volatiles on Viscosity, sample dried for 2 hrs. at 50°C.

341

within the sample cavity as the residual ethanol and condensation by-products boil off.

A similar curve of $\eta*$ versus time for an as-synthesized sample of LARC-160 with Jeffamine AP-22 is shown in Figure 18. A RT viscosity of 66 poise is similar to that of LARC 160 with Curithane 103, but the minimum viscosity at 84°C (13.6 poise) does not get as low as LARC 160 with Curithane 103 (9.2 poise). Also, the Tan δ imidization peak occurs at 124°C which is 7°C lower than that for the Curithane 103 version of LARC 160. Thus, it appears that substituting Curithane 103 for Jeffamine AP-22 causes the resulting LARC 160 to soften more and perhaps to react slower during the imidization reaction. Furthermore, the imidized resin appeared to have a higher viscosity in the case of the Curithane 103 version of LARC 160.

When as-synthesized samples of PMR-15 or either version of LARC 160 were run, excessive scattering of the data was encountered. In order to avoid the problem, a pressure cell transducer was obtained from Rheometrics. The cell had 500 psi pressure capability and was designed for use with resin systems that outgas during cure. Unfortunately, little improvement was found in the data generated while using the pressure cell. Inspection of the samples showed bubbling still occurred within the samples and the resin was forced out from between the parallel plates. An extensive effort was made to overcome the problem, including modification of the plates, the fixtures and the test geometry. When these efforts proved unsuccessful, work with the pressure cell was discontinued. Further studies were based on the use of staged or imidized samples of resin.

The rheological behavior of preimidized samples of LARC 160 and PMR-15 (ethanol solution) were next compared. The data were gathered on samples which were oven imidized at 121°C for 60 minutes prior to characterization. The imidization cycle had three advantages. First, it corresponded to the cycle used in the production shop. Second, it gave reproducible, smooth viscosity curves, and finally, additional staging (up to 2 hours at 121°C) provided only minor differences in the viscosity data. After imidization, samples were pulverized and desiccated. Prior to measurement, samples were further dried in a desiccated vacuum oven (RT/380-510 mm Hg) overnight.

In order to form a sample in the environmental chamber of the rheometer, the chamber is preheated to 200°C and a pre-weighed amount of imidized resin (fine powder) is introduced between opened parallel plates. The chamber is then closed and when the indicated temperature again reaches 200°C a sample is formed by closing the plates to a predetermined gap (0.6mm). The chamber is again opened and excess resin is removed from the edges of the plate. The chamber is reclosed and heating is resumed when the chamber reaches 200°C.

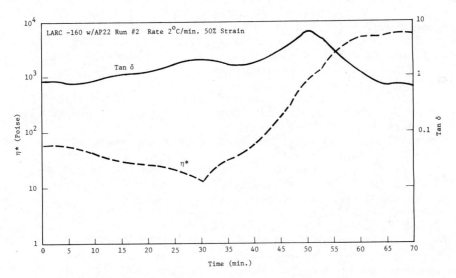

Figure 18. Complex Viscosity versus Time.

Complex viscosity as a function of temperature is shown in Figure 19 for LARC 160 (AP-22 and Curithane 103 versions) and PMR-15 (ethanol solution) at heating rates of 2 and 4°C/min. There appear to be distinct differences between all of the resins. The greatest difference was between the two types of LARC 160. Because of the greater amount of 4,4'-MDA in Curithane 103, one might logically expect this version of LARC 160 to have a higher minimum viscosity than its AP-22 counterpart. However, the data show the opposite. Furthermore, if the AP-22 version of LARC 160 is compared with earlier data obtained from preimidized LARC 160 prepreg the viscosity profiles of the two resins are significantly different. The resin from the prepreg reaches a minimum viscosity in the neighborhood of 400 poise at 292°C as compared to around 7,000 poise at 250°C for freshly synthesized resin. Although the heating rates were the same, i.e., 4°C/min., the prepreg resin was scanned and started at 239°C while the current material resin was scanned from 200°C. The additional time at higher temperatures tends to explain the observed differences and also presents another problem. The viscosity measured in a characterization run might not reflect the true viscosity during a process. It clearly points out the need for parallel heat histories of samples being compared.

Note that an improved viscosity characterization procedure would be desirable to eliminate the initial peak in complex viscosity during the first 10 to 20 minutes of a run. The resin is not very fluid at 200°C. The lower temperature was used to minimize any reactions during sample formation and hence give better reproducibility. This higher initial viscosity requires a low shear stress and low shear rate because of the transducer limitations. However, the deflection of the transducer shaft must exceed 0.001 of the radius in order to give reliable results. The shear strain of 3 used to generate the data in Figure 19 was later found to be slightly less than required by calculation ($\gamma = 5$). The data are thus questionable. The effect of shear strain was examined in subsequent tests and the data appear in Figure 20.

For the AP-22 version of LARC 160, shear strains of 3, 5, 10, and 50 were examined. A starting temperature of 230°C was required to get the initial viscosity low enough to accept the higher shear strains. The data shows a scatter band but no systematic variations as a function of shear strain. The only large difference occurs in the case of $\gamma = 3$. Similar data for the Curithane 103 version of LARC 160 at shear strains of 3, 10, and 50 are also shown in Figure 20. In this case, the largest difference occurs in the case of $\gamma = 50$. It would appear that a shear strain of 10 might be the best choice. Despite the scatter, it is clear that in all cases the Curithane 103 version of LARC 160 has a lower minimum viscosity in the process critical range as compared to AP-22.

344

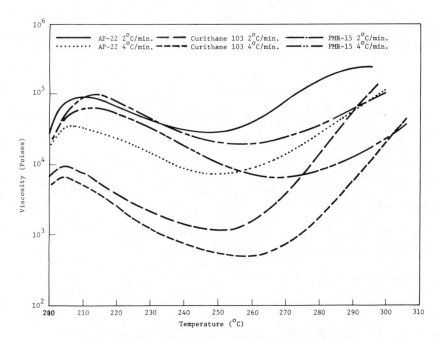

Figure 19.   Complex Viscosity versus Temperature.

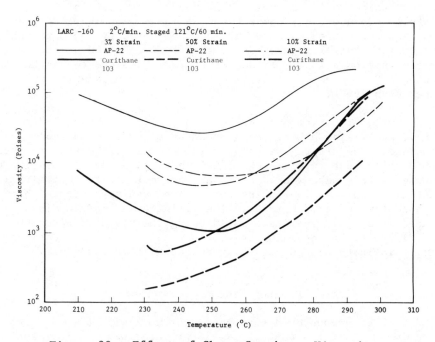

Figure 20.   Effect of Shear Strain on Viscosity.

Note also that the curves for PMR-15 and AP-22 based LARC 160 (Figure 19) resemble each other far more than they resemble that of the Curithane 103 based LARC 160.

Laminate Fabrication

Laminates were fabricated using both versions of LARC 160 as a matrix material.

Standard commercial Fiberite prepreg was used in the AP-22 based laminate. Here LARC 160 is used to impregnate woven T-300 (8 harness satin) epoxy finish graphite cloth. The following cure schedule was used:

1. Heat at 2°C/minute to a temperature of 121°C.
2. Hold 90 min. at 121°C, then heat at 2°C/min.
3. Pressurize to 1.2 MPa at 260°C.
4. Continue heating at 2°C/min. to 330°C.
5. Hold 90 min. at 330°C, 1.2 MPa, then cool under pressure to below 100°C

A well consolidated laminate was made by this procedure.

From the rheological data generated on the LARC resin it was apparent that the AP-22 cycle would need to be modified to successfully produce a Curithane based laminate. If the $\eta*$ curves are compared, the Curithane LARC 160 must be heated to 290°C to achieve a viscosity equivalent to that of the AP-22 LARC 160 at 260°C (pressurization point). It was, therefore, decided to use the AP-22 LARC 160 cure schedule listed above, but to pressurize at 290°C rather than 260°C.

A sample of Curithane LARC 160 was applied to woven T-300 (8 HS), epoxy finish cloth reinforcement by a "dip and squee-gee" technique. The prepreg was air dried at ambient temperature overnight, followed by 4 hours at 93°C in a circulating air oven. The purpose of this bake was to eliminate solvent and thereby minimize plasticization and flow during cure. The resulting prepreg had a resin content of 46.8 percent. The laminate was laid up in a notched metal mold and press cured using the standard LARC 160 (AP-22) cycle, but applying pressure at 290°C. An apparently high quality laminate was successfully molded by this procedure.

Prepreg Viscosity Studies

A problem encountered throughout the work with the addition polyimide systems was that of sample preparations. Almost any technique for preparing a neat resin sample from prepreg can significantly change the nature of the resin. Solvent techniques can pose serious problems in removing residual polar solvents that interact strongly with the polymer. Small amounts of such solvents, if not removed, act as plasticizers and reduce Tg of the resin. Heating to remove solvent will stage or alter the resin. Even physically scraping or grinding off resin flash can result in a sample contaminated with fiber or debris.

Another serious question must also be addressed. Laminates are fabricated from prepreg, not resin. How does the rheological response of the resin in the presence of the reinforcement differ from that of the neat resin? The presence of the fiber with its large surface area and the presence of active chemical species on the fiber surface can certainly affect the resin, both physically and chemically.

It was therefore decided to attempt to develop techniques for determining the rheological characteristics of reinforced prepreg samples on the Rheometrics mechanical spectrometer. Techniques were successfully developed, using well characterized graphite/epoxy materials and are reported elsewhere[6]. LARC 160/woven T300 prepreg samples were investigated, but efforts proved unsuccessful. Outgassing problems could not be overcome (as with neat resin samples) and the imidized systems were too high in viscosity to study with the torque transducers available to us at the time. However, samples of another commercial addition polyimide, U.S. Polymeric V378A, were available. Rheometrics curves were successfully run on V378A resin reinforced with Celion 3000 and T300 fibers. Although rheological curve of the neat resin from each batch showed little difference, a significant difference in the prepreg viscosity curves were observed as can be seen in Figure 21. The Celion reinforced prepreg has a significantly lower minimum apparent viscosity, and therefore higher flow. This is consistent with the processing characteristics for these prepregs observed in the LMSC shop and elsewhere in the industry.

Solid State Rheometrics Studies

In order to run solid state rheological studies on the two LARC 160 resins, it was necessary to prepare neat resin moldings. This was not an insignificant task, as a staging cycle was required that allowed the powdered resin to be cured under pressure. Simple casting and curing or pressing to stops resulted in high void moldings with incomplete consolidation. After several attempts, we developed cycles for both resin systems. The Curithane based LARC 160, however, required significantly more staging than the AP-22 version to achieve the low flow characteristics required to produce satisfactory moldings. The staging cycles*, shown in Table II were used:

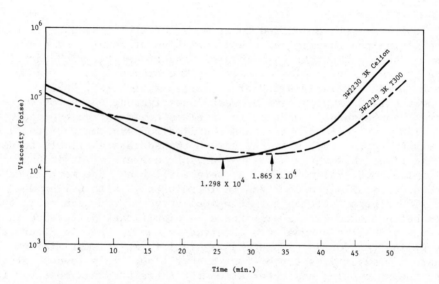

Figure 21.   V378A Prepreg Viscosity versus Time.

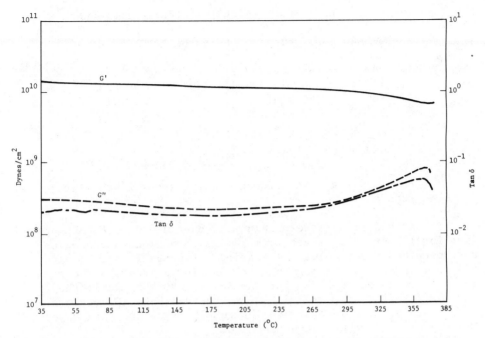

Figure 22. Solid State Rheometrics Data, LARC 160 w/AP-22
(Neat Resin Casting).

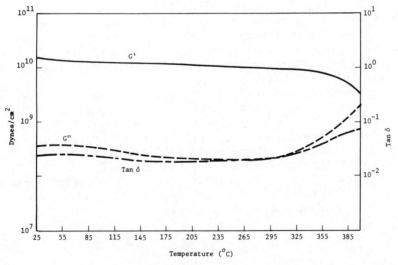

Figure 23. Solid State Rheometrics Data, LARC 160 w/Curithane
103 (Neat Resin Casting).

Table II.   LARC 160 Staging for Molding Compounds.

AP-22 LARC 160 Cycle	Curithane LARC 160 Cycle

Step:

	AP-22 LARC 160 Cycle			Curithane LARC 160 Cycle	
1.	140°C — 90 min.		1.	140°C — 90 min.	
2.	140°C — 30 min.		2.	140°C — 30 min.	
3.	180°C — 60 min.		3.	180°C — 60 min.	
4.	210°C — 30 min.		4.	210°C — 60 min.	
5.	250°C — 60 min.		5.	220°C — 60 min.	
	4.5 Hours Total		6.	250°C — 75 min.	
				6.25 Hours Total	

*Resin was ground between staging steps to facilitate removal.

The staged powders were then molded at 320°C, 0.83 MPa pressure for 1 hour. Satisfactory moldings were produced, samples were machined for rheological testing. Figures 22 and 23 give the solid state curves for the two LARC 160 polymers. The peak in Tan δ occurred at 367°C for the AP-22 LARC 160, while the curve went off-scale (at 397°C) for the Curithane 103 version. These data are consistent with TMA results that indicate a possibly higher Tg for Curithane LARC 160.

CONCLUSIONS AND RECOMMENDATIONS

Several important results were obtained and conclusions reached during the course of the work:

o   Imidized Curithane 103 based LARC 160 has significantly lower viscosity and therefore higher flow than AP-22 based LARC 160 during processing.
o   Rheological studies can be used to develop processing cycles for Curithane 103 based LARC 160 resin and prepreg.
o   Relatively small differences are found between the two uncured LARC 160 resins by conventional chemical characterization techniques.
o   Rheometrics rheological data were generated on V378A prepreg systems. The type of woven graphite was found to significantly affect the rheological characteristics of the prepreg and was consistent with their observed processing behavior.
o   Fully cured Curithane based LARC 160 may have a Tg that is higher than that of AP-22 based LARC 160.
o   Additional work is needed to develop efficacious pressure cell instrumentation to make rheological measurements on systems that outgas during cure.

ACKNOWLEDGEMENTS

Support of this work by NASA-Langley Research Center is gratefully acknowledged.

REFERENCES

1. C. A. May, Editor,"Resins for Aerospace", ACS Symposium Series No. 132, Washington, D.C., 1980.
2. T. L. St. Clair and R. A. Jewell, National SAMPE Tech. Conf. Series, 8, 82 (1976); Sci. Adv. Mat'l. & Proc. Eng. Ser., 23, 520 (1978).
3. J. D. Leahy, "Development and Demonstration of Manufacturing Processes for Fabricating Graphite/LARC 160 Polyimide Structural Elements", Contact NAS1-15371, 5th Quarterly Report, June through September, 1979. Contact: R. Baucom, NASA Langley Research Center, Hampton, VA.
4. R. Young and F. Sykes, "Anaysis of Aromatic Polyamine Mixtures for Formulation of LARC-160 Resin", 12th Natl. SAMPE Tech. Conf. Series, 12, 602 (1980).
5. J. D. Leahy, "Development and Demonstration of Manufacturing Processes for Fabricating Graphite/LARC 160 Polyimide Structural Elements", Contact NAS1-15371, 9th Quarterly Report, June through September 1980. Contact: R. Baucom, NASA Langley Research Center, Hampton, VA.
6. M. G. Maximovich and R. M. Galeos, "Rheological Characterization of Advanced Composite Prepreg Material", 28th National SAMPE Symposium, Anaheim, Ca., April, 1983.

ETCHING OF PARTIALLY CURED POLYIMIDE

C. E. Diener and J. R. Susko

IBM Corporation
General Technology Division
Endicott, New York 13760

The fabrication of patterns in partially cured polyimide films may be accomplished by wet etching with potassium hydroxide. In practice, problems encountered in obtaining uniform etching result from thickness variations in coatings and temperature variations in typical process ovens. The relationship between coating thickness, thermal exposure, degree of cyclization, and etchability was studied. The degree of cyclization was determined by Fourier Transform Infrared Spectroscopy and etch rates were measured by monitoring coating thickness. It was found that the level of imidization was directly influenced by coating thickness as well as thermal exposure. For films exposed to the same thermal conditions, the etch rate decreased as the thickness increased beyond 15 µm. However, when films of various thicknesses were brought to the same level of imidization, a constant etch rate was obtained. A significant improvement in the etch rate can be achieved by reducing the level of imidization, increasing the etchant temperature, or enhancing the solubility of the carboxylate salt of the polyamic acid produced during etching.

INTRODUCTION

The fabrication of patterns in polyimide films is often required for a variety of purposes. Generally, a film or layer

of polyamic acid is deposited on a substrate and partially cured
by exposure to controlled thermal environments. The application
and development of a photoresist or other suitable medium provides
a patterning mask for selective removal of the partially cured
polyimide which is accomplished with a variety of etchants, e.g.,
sodium and potassium hydroxide or tetralkyl ammonium hydroxide.
In practice, problems are often encountered in obtaining uniform
etched patterns, especially those involving small dimensions in
thick films (> 10 μm). These problems are known to result, at
least in part, because of thickness differentials within the
polyamic acid layer and variations in temperatures within the
ovens used.

The objective of the study was to further develop the base
data for understanding the etch process by determining the
relationship between coating thickness, thermal exposure, degree
of cyclization and etchability. In addition, the effects of some
process modifications were studied.

EXPERIMENTAL

Films of polyamic acid (a pyromellitic dianhydride/
4,4'-oxydianiline system in n-methyl pyrrolidone) were sprayed on
substrates at appropriate thicknesses to give approximately 5, 10,
and 15 μm final cure thicknesses. Immediately following coating,
the test specimens were subjected to thermal conditions to effect
a partial cure (11 minutes at 100° C, infrared oven; 13 minutes
at 130° C, convection oven). Some test specimens were given an
additional thermal treatment - 150° C for 15 minutes.

Preceding initial etching of the samples, the degree of
imidization was determined and coating thicknesses were measured.
The degree of imidization was determined by FTIR spectroscopy,
using coatings on silicon wafers. The carbonyl band at 1776 cm^{-1}
was monitored and ratioed with a reference band at 1012 cm^{-1}.
The ratios were normalized with that of a completely cured film
to give an imidization value. Coating thicknesses were measured
by light-section microscopy.

Initial etching was done with a spray etcher. Triplicate
samples of each cure condition were etched for 20, 30, 40, and
55 sec. duration with potassium hydroxide as etchant maintained
at 0.23 M concentration and 30° C.

The specimens were subjected to a second or interim cure
for 10 minutes at 200° C after which they were etched a second
time with potassium hydroxide at 30° C for 4 minutes. Following
each etch cycle, the test specimens were examined and, where
applicable, the thickness of the polyimide determined.

354

In subsequent experiments, polyimide films (of the same PMDA/ODA system) of varying thicknesses were spin coated on silicon wafers. Partial imidization was obtained by exposing them to oven temperatures of 130, 140, and 150° C. Initial thickness and degree of imidization measurements were made by the methods described above. Etch rates were determined by subjecting the test specimens to potassium hydroxide etchant in timed increments while monitoring film thickness.

RESULTS/DISCUSSION

Our initial interest was directed toward identifying the effect that thickness produced on the degree of cure, and subsequently the etching of the partially cured polyamic acid. It was found that for test specimens, given the same thermal exposure, the degree of cure varied as the thickness of the coating (see Figure 1-Curve A), an observation reported earlier.[1] Both the sprayed and spin-coated samples (Figure 2) exhibited this tendency. Experiments to date have failed to show conclusively a reason for the higher degree of conversion for thicker films.

The data showed significant scatter for the sprayed samples in the 20+ μm range -- both in thickness and degree of imidization.

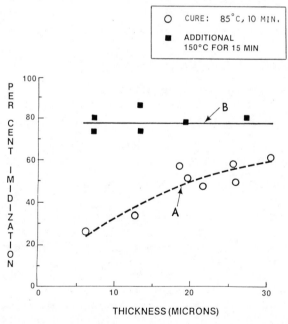

Figure 1. Per cent imidization values obtained by FTIR spectroscopy copy for sprayed coatings of various thicknesses.

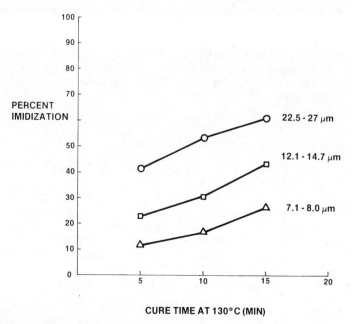

Figure 2. Percent imidization versus cure time at 130° C for spin-coated films of various thicknesses.

These differentials probably result from fluctuations in the spray equipment and variations in temperature which are known to exist within typical process ovens.

For test specimens given an additional cure (150° C for 15 minutes), there is really no influence of thickness on the degree of imidization (Figure 1-Curve B).

The etch data for the sprayed samples are graphically represented in Figure 3. A constant etch rate of 0.4-0.5 μm/sec is indicated for polyimide layers up to 15 μm thick; whereas above this thickness, a decrease in the etch rate is indicated (~0.1 μm/sec). The significance of this curve is that at pre-etch thicknesses of > 15 μm, an added unit or two of thickness extends the required etch time to a point that the actual allotted etching time is insufficient. For example, if the expected coating thickness was 20 μm or less, 70 seconds would be sufficient for etch. However, if an unanticipated coating thickness of 22 μm occurred, the selected etch time would be insufficient. Satisfactory removal of remaining polyimide would then be dependent on a second etch cycle.

356

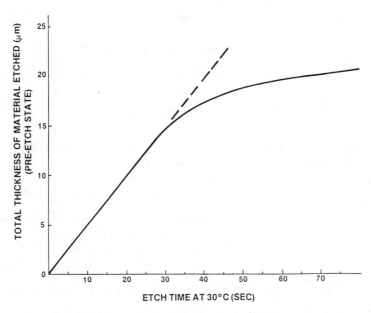

Figure 3. Graphical representation of thickness of material
etched vs etch time for spray-coated samples.

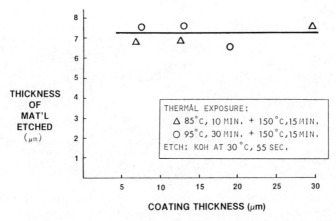

Figure 4. Etch data for spray-coated samples given additional
thermal exposure -- 150° C for 15 minutes.

The etch data for the samples given the additional cure (150° C for 15 minutes) is shown in Figure 4. This information, i.e., a constant quantity of material etched for polyimide films at the same cure state, independent of thickness, was the basis for later studies.

The function of a second etch is to remove residual material which has been converted from the polyamic acid to the potassium carboxylate salt during the initial etch. This material does not cyclize during the subsequent exposures to interim or final cure temperatures. For this reason the material becomes a contaminant with respect to the final desired polyimide pattern.

The effects of the initial and second etch are shown in the bar graph (Figure 5). Each individual bar (see arrow) represents the data obtained from one sample. The thickness of material removed during the initial etch (▨) tends to vary according to etch time. The thickness of material removed during the second etch (☐) is constant and is independent of the level of imidization or the duration of the initial etch to which the samples had previously been exposed. Since samples that did not completely etch during the initial etch were not cyclized in the subsequent interim cure, conversion to the carboxylate salt of the polyamic acid must have occurred (Figure 6-I). It is seen that this

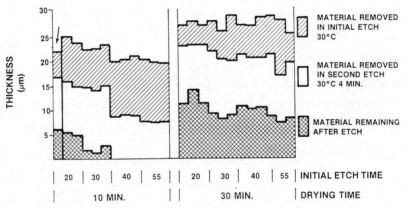

Figure 5. The effects of initial (▨) and second (☐) etch on polyimide films. Sample thickness: ~ 25 μm. Etchant was KOH at 30° C.

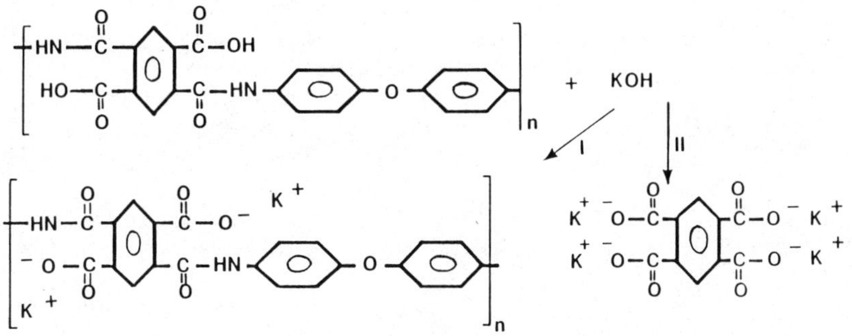

Figure 6. Mechanism of etching polyimide (partially cured polyamic acid) with potassium hydroxide as given by Dine-Hart[2], et al.

conversion occurs through the bulk of the coating with as little as 20 seconds initial etch, i.e., as rapidly as the etchant diffuses into the material. The actual removal of the carboxylate salt moiety is apparently limited by its solubility in the etchant.

These findings suggest that the etch rate could be enhanced by raising the temperature of the etchant or improving its solubilizing action on the salt by including a more efficient solvent. Co-solvents such as butyl carbitol, cellosolve acetate, diglyme and tetrahydrofuran have been suggested by polyimide manufacturers. The improved etch rate gained through a solubility increase at higher temperatures must be coupled with the fact that o-carboxyamido bonds are cleaved at temperatures above 40° C (Figure 6-II), so a temperature increase may perform a dual function.

Limited experiments mentioned previously had shown that when polyimide coatings were brought to the same level of imidization, a constant etch rate was observed, independent of the thickness of the coating. Later experiments were designed to measure the etch rate of polyimide films that had been prepared so as to obtain the following conditions:

1) constant imidization levels for different thickness films

and

2) constant thickness films for different imidization levels

Following the preparation of samples and determination of imidization level and thickness as described above, the specimens were "bracketed" according to the imidization level and thickness. Where possible, a sample set consisted of at least one coating of each of three thicknesses (approximately 7, 14, and 25 μm at pre-etch state) at the same level of imidization.

Etch rates were determined with potassium hydroxide etchant at 30° and 45° C. In contrast to the previous tests, samples were etched in 10 or 20 second increments and washed, dried, and measured between tests.

These tests confirmed the earlier findings with respect to an increase in the level of imidization as film thickness increases -- given the same thermal exposure. The effect of cure temperature on the degree of imidization for films of the same thickness is shown in Figure 7. As expected, imidization increases as the cure temperature increases.

In the thickness range studied (7-25 μm), polyimide coatings brought to the same level of imidization were characterized by a constant etch rate, independent of thickness. This was generally observed at all levels of imidization examined (30-70%) and for both etchant temperatures used (30°, 45° C). A typical set of curves is shown in Figure 8.

The effect of etchant temperature on one set of films is shown in Figure 9. Etch curves for films of various thicknesses cured to 40 per cent and etched at 30° and 45° are plotted together. The vertical lines represent the range of values at each time inverval for the 7, 15, and 22 μm thick films tested. A substantial reduction in the time required for etching results from the fifteen degree increase in temperature.

For polyimide films of the same thickness, an increase in the etch rate was observed as the degree of imidization decreased. Etch curves for films of the same thickness (25 μm) at three levels of imidization (40, 60, and 70%) are shown in Figure 10.

We have also consistently observed that a change in the etch rate occurs once the surface layers (2-6 μm) are removed and etching proceeds into the bulk material (see Figure 9). This

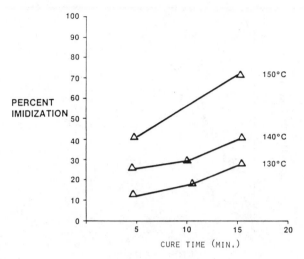

Figure 7. Per cent imidization versus cure time at 130, 140, 150° C for films of 7.4-7.8 μm thickness.

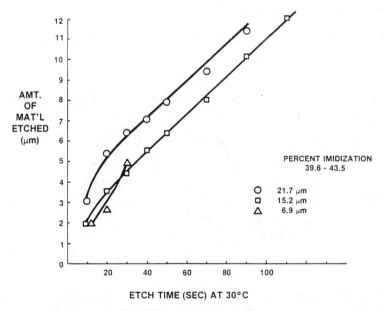

Figure 8. Etch rate curve for various thickness coatings at constant imidization levels (39.6-43.5%). Etchant temperature was 30° C.

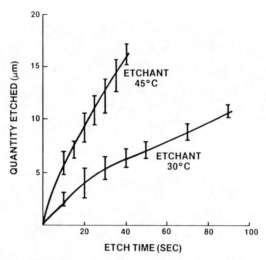

Figure 9. Etch curves for coatings cured to 40% imidization and etchant temperatures of 30 and 45° C. (The vertical lines represent the range of values for 22, 15 and 7 μm thick films.)

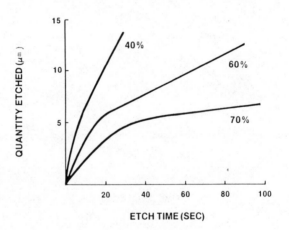

Figure 10. Etch curves for 25 μm thick films at various imidization levels. Etchant temperature was 45° C.

362

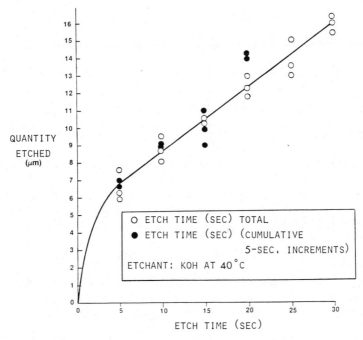

Figure 11. Etch rate for material etched in one operation compared to the same material etched in 5 second increments.

trend was more pronounced with 30° C etchant but was still evident at 45° C. Early in our studies it was thought that this change in etch rate resulted from the "gel" barrier formed by the conversion of the polyamic acid to the potassium carboxylate moiety. Accordingly, the "gel" barrier would block the diffusion of etchant into the material; however, this was not the case. A comparison of etch rates for a material which was etched in one operation and a material in which etching was done in 5 second increments is shown in Figure 11. The purpose of etching in 5 second increments was to minimize the barrier formed by the "gel" by washing it away after every increment. In this way, a "fresh" surface was available for etching. The data show that the etch rates are comparable whether or not the "gel" etch product is washed away. It was concluded that the change in etch rate from surface layers to the bulk material is real and not dependent on the amount of gel formed as etching progresses.

Hoagland and Fox[3], in their study of the hydrolysis of polyaspartimide, observed a high pseudo first order rate of hydrolysis which decreased as hydrolysis proceeded. They attributed this change in rate to the replusion of the negatively charged hydroxide ion by the electrostatic charge generated by the release of carboxylate groups. A mechanism such as this may be operative in polyimide as etching proceeds from the surface layers into the bulk.

SUMMARY

A significant relationship exists between the thickness, degree of imidization and the etchability of partially cured polyimide films. Generally thicker coatings imidize to a higher degree than thinner coatings given the same thermal exposure, and as imidization increases, etchability decreases. In the thickness range studied, polyimide films brought to the same level of cyclization were characterized by a constant etch rate, independent of thickness. It was also consistently observed that the surface layers (2-6 μm) etched at more rapid rate than the remainder of the film.

Since the thickness directly affects the degree of imidization at a given temperature, the uniformity of the applied film, variations in process ovens, and etch bath composition and temperature are critical factors in polyimide processing.

ACKNOWLEDGEMENTS

- D. Bolan, C. Goodwin, and M. Flandera for coordination of spraying, cure, and etching of sprayed samples.

- R. Ginsburg for FTIR spectroscopic determination of degree of imidization.

- R. Winslow for thickness and weight measurements and etch data.

REFERENCES

1. R. Ginsburg and J. R. Susko, these proceedings, Vol. 1, pp. 237-247.
2. R. A. Dine-Hart, D. B. V. Parker, and W. W. Wright, Br. Polym. J. $\underline{3}$, 222 (1971).
3. P.D. Hoagland and S.W. Fox, Experientia, $\underline{29}$, 962 (1973).

LASER MICROMACHINING OF POLYIMIDE MATERIALS

P. A. Moskowitz, D. R. Vigliotti, and R. J. von Gutfeld

IBM Thomas J. Watson Research Center
Yorktown Heights, New York   10598

A laser-enhanced chemical process has been developed for the removal of material from polyimide to produce vias, grooves, and cavities on the order of tens of μm in size. The process takes advantage of the highly absorbant nature of polyimide to blue light. The focused light of an argon laser is used to heat the polyimide locally in the presence of a caustic bath such as NaOH. The heating speeds the chemical reaction of the polyimide with the NaOH solution to form soluble compounds which are then dissolved. The effects of varying concentration of etchant have been investigated; removal rates have been measured for NaOH and other solutions; and etching of copper/polyimide laminates has been accomplished.

INTRODUCTION

Polyimide materials are used in ever increasing quantity for electronics applications because of their temperature stability, flexibility, and chemical inertness. Because of this chemical inertness, modification processes demanding the subtraction of material generally require reactive ion etching, mechanical machining, or etching in hot caustic solution.

We have developed a laser-enhanced chemical process for locally removing material from Kapton® films or bulk pieces of Vespel® to produce mil-sized vias, grooves and cavities. This process has been previously demonstrated to be effective for the laser induced etching of metals and ceramics.[1,2] By contrast with some other techniques, the gradual nature of our process leaves the material without visible damage.

It has been demonstrated that basic solutions (KOH, NaOH etc.) will react with polyimide to produce soluble compounds.[3,4] In the case of concentrated KOH (18N) at room temperature, the process proceeds at a rapid rate, yielding noticeable deterioration of the polyimide in a few minutes. However, for concentrated NaOH or weaker concentrations of KOH, the reaction proceeds more slowly.

In our experiments we immerse the material in NaOH and focus the beam of an argon laser on the surface to be processed. The heating induced by the laser speeds the hydrolysis and dissolution of the polyimide. Turbulence in the fluid caused by local temperature differentials brings fresh fluid in contact with the surface. Volume removal rates on the order of 10^4 μm^3/s have been measured for a 100 mW incident laser beam. Removal rates increase with laser power. However, increased power results in the decomposition of the polyimide accompanied by the release of gaseous products, probably carbon dioxide and hydrogen. The deposition of carbon compounds tends to inhibit further etching.

APPARATUS

The etching apparatus shown in Figure 1 uses an argon laser. The coherent light from this laser is continuous with principal lines in the blue-green at 488 nm and 516 nm. A system of lenses is used to focus the beam to a diameter of 10 to 20 μm on the surface of the sample to be etched. The sample is contained within a bath of the etchant solution, which in turn is mounted on a computer controlled x-y-z stage to facilitate focusing and the movement of the sample. An electrically controlled oscillating mirror allows rapid horizontal scanning of the laser for the production of grooves.

The laser light is strongly absorbed by polyimide. Measurements on Du Pont Kapton have shown a 60 percent absorption for a 15 μm thickness of the film. The results are shown in Figure 2. Relatively low laser powers, 100 to 200 mW, have been used in these experiments to produce a gradual etching of the polyimide. Examples are shown in Figure 3. Previous experimenters have demonstrated the possibility of making vias in polyimide in air using a 1.06 μm wavelength pulsed Nd-YAG laser.[5] In this process the pulse was absorbed by an underlying copper layer and the heat was used to burn-away the polyimide. We have found that we could use the same laser to burn vias in polyimide without the copper substrate beneath the polyimide. An SEM of one such via is shown in Figure 4.

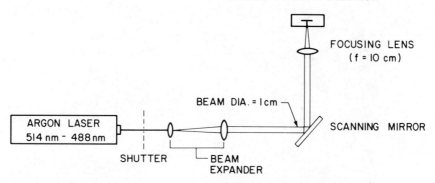

Figure 1. Top view of laser micromachining apparatus. The sample is mounted on a computer-controlled x-y-z stage (not shown).

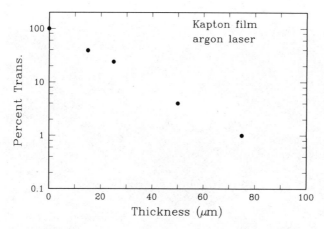

Figure 2. Measurement of transmission vs. film thickness for Kapton using an argon laser with principal lines at 514 nm and 488 nm.

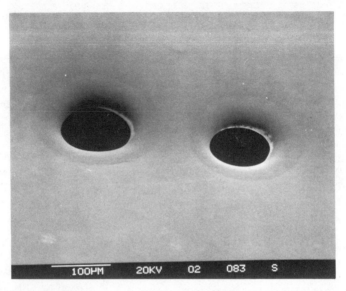

Figure 3a. SEM taken at a 45 degree angle of a pair of vias made in 75 μm thick Kapton. The vias were made with a 150 mW argon laser beam in 25 N NaOH.

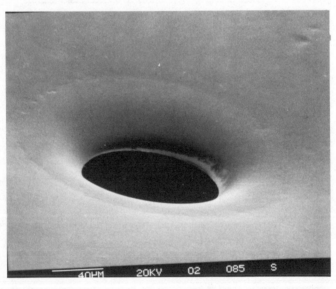

Figure 3b. Magnification of a via made under conditions similar to those shown in Figure 3a.

PROCEDURE

NaOH rather than KOH was chosen for most of the experiments because it was observed to have almost no effect on the polyimide film at room temperature, while KOH was sufficiently active at higher concentrations to cause rapid deterioration of the film. Thus, the laser beam could be used to etch polyimide film in a selective manner without the need for further protection of the film or a masking process. The data were gathered using 130 μm (.005 in) thick Kapton film. Using a 100 mW laser intensity, we etched blind vias in the film, and measurements were made of volume removal and linear depth as a function of time for concentrations of NaOH ranging from 1 N to 25 N solutions.

As can be seen in the data presented for 5 N NaOH in Figure 5, initial volume removal is linear but tends to slow somewhat with time. The depth of the blind vias as a function of time also displays this feature. Linear rates obtained from this data show an asymptotic time relationship. We believe this may be attributed to the loss of contact with fresh solution as the etching proceeds deeper into the material and perhaps also to an accumulation of non-soluble etching products. Agitation or stirring of the solution might improve the etching rates.

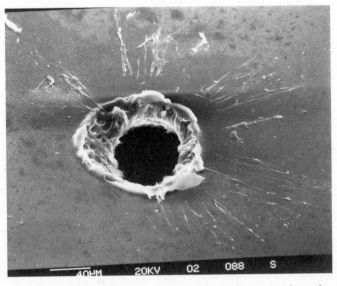

Figure 4. Via produced in 25 μm thick Kapton in air using a Nd-YAG laser, 20 ns pulse, 50 mJ energy, 1.06 μm wavelength.

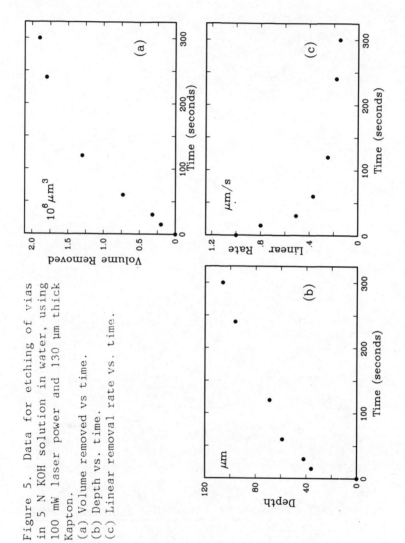

Figure 5. Data for etching of vias
in 5 N KOH solution in water, using
100 mW laser power and 130 μm thick
Kapton.
(a) Volume removed vs time.
(b) Depth vs. time.
(c) Linear removal rate vs. time.

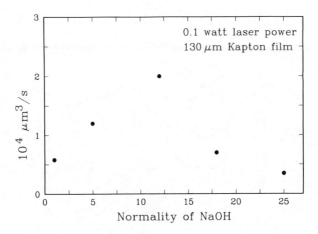

Figure 6. Volume removal rate vs. Normality of NaOH solution in
water.

The best removal rates were recorded for 5 N to 18 N NaOH with
slower rates for weaker and stronger solutions. This occurs as a
result of competition between two processes. More concentrated
solutions are more effective in breaking down the polyimide, while
less concentrated solutions are more effective in dissolving the
reacted material producing a trade-off for the complete process.
Volume removal rates for NaOH, plotted in Figure 6, peaked at about
2.0×10^4 $\mu m^3/s$, with corresponding linear rates about 1.0 $\mu m/s$. In
general, the less concentrated solutions showed much less in-plane
etching than the more concentrated solutions. For example,
comparison of the vias in Figure 7 etched at 12 N and 25 N shows the
first to be vertical, although with somewhat ragged edges, while the
25 N vias are more gradual with smooth shoulders. Thus, it is
possible to control the shape of vias by the proper choice of etchant
concentration.

KOH in Ethanol/Water Solution

Recent work on the properties of caustic etchants in
alcohol/water solutions has been carried out at Du Pont.[4] A 1.0 N
solution of KOH in 80 percent ethanol and 20 percent water at 70 C is
recommended for the etching of Kapton film.[6] The indicated etch rate
for this process is then about 10 $\mu m/min$. We have tried this
solution with our laser process and have measured a volume removal
rate of about 4.0×10^4 $\mu m^3/s$ with initial linear etch rates of over
3.0 $\mu m/s$ or about 200 $\mu m/min$. This is double our best rate with NaOH
and twenty times the rate measured by Du Pont. The results of this
experiment are plotted in Figure 8.

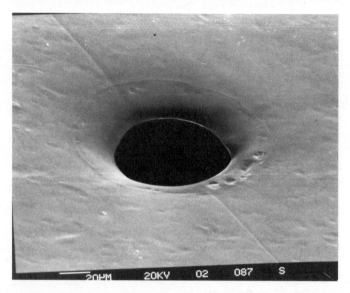

Figure 7a. Via etched through 25 μm thick Kapton in 25 N NaOH. The via is tapered with smooth shoulders. The line extending along the diameter of the via is an artifact of the SEM.

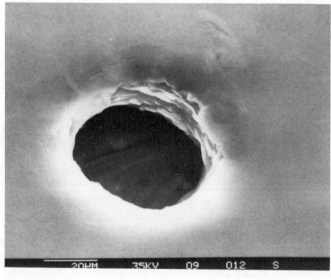

Figure 7b. Via etched through 25 μm thick Kapton in 12 N NaOH. The via is straight through the material.

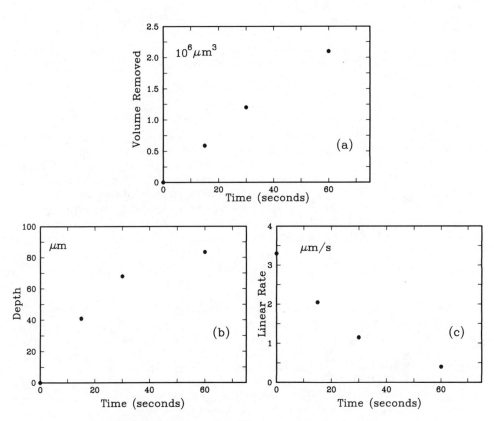

Figure 8. Data for etching of vias in 1.0 N KOH solution in 4.1 ethanol:water, using 100 mW laser power and 130 μm thick Kapton.
(a) Volume removed vs time.
(b) Depth vs. time.
(c) Linear removal rate vs. time.

Other Etchants

The properties of ammonium hydroxide as an etchant were investigated. An 8 N solution was found to etch the polyimide film at a rate approximately three times slower than the corresponding concentration of NaOH. This solution is easily decomposed by the heat generated in the etching process, liberating quantities of ammonia gas. Thus, ammonium hydroxide appears not to be a good candidate for this process.

Tetraethylammonium hydroxide, about 1.5 N solution, was also tried. No apparent etching of the polyimide took place.

VESPEL

Vespel is a bulk material which has the same chemical composition as Kapton. Experiments with Vespel showed that it responds to the laser etching process in much the same way as does the Kapton film. No detailed quantitative study of the parameters of the process was undertaken. However, by using the scanning mirror shown in Figure 1, we were able to etch grooves in Vespel. For example, one such groove 3 mm in length was etched to a depth of 300 μm, with a 175 μm width at the top, 55 μm width at the bottom.

COPPER/POLYIMIDE LAMINATE

One of the major applications of polyimide is for the fabrication of laminated copper/polyimide structures for use as electrical connectors or I/O cables. The polyimide serves well for this application because of its thermal stability, flexibility, and low dielectric constant of about 3.5. We have previously used a 1.4 meter long, eighty line, five layer copper/polyimide laminate for an I/O cable to carry high-speed electrical signals between room temperature electronics and superconducting circuitry operating at 4.2 K.[7] This cable remained flexible even while immersed in liquid helium.

For the present experiments a 15 μm thick Kapton sheet with an evaporated layer of approximately 15 μm thick copper was obtained. By using a focused beam of 100 mW and an etching solution of 12 N NaOH, vias were etched through the polyimide to bare copper in about 20 seconds. The effect of the copper was to draw heat from the area of the focused beam. As the etched region approached the copper substrate the process became less efficient, producing tapered vias considerably smaller at the bottom than at the top. This is in sharp contrast to the straight-through vias produced under similar conditions in non-laminated polyimide sheets. The copper was not affected by its exposure to the caustic bath and laser beam.

To confirm that the absorption of the laser light by the polyimide, and not by the copper, was the dominant process, the experiment was repeated with a 647 nm krypton laser. The film is virtually transparent at this wavelength. The laser provided a power of 140 mW. No etching was observed after several minutes of exposure to the focused krypton laser beam.

DISCUSSION

We have demonstrated the efficacy of a laser-enhanced etching process for the micromachining of polyimide materials. The process has been used to produce vias, blind vias, and grooves on the order of tens of μm in size. The process is maskless, requiring only the translation of the work piece or the movement of the optics to produce structures over the surface of the polyimide. It is possible to control the shape of vias through the concentration of the etching solution. Both smoothly tapered or straight through vias can be obtained. The process has also been shown to be applicable to copper/polyimide laminates where the production of vias in such laminates may be an important application.

Compared to reactive ion etching or ion milling, the apparatus needed is relatively inexpensive. Lasers and their associated optics are generally an order of magnitude less expensive than vacuum systems. The serial nature of the process is at a disadvantage when compared with vacuum processing systems where many pieces are processed simultaneously. However, the speed of the laser enhanced etching, measured in μm/s as opposed to ion etching speeds of .01 to 0.1 μm/min makes up for this in part. Liquid processes are also easier to contain than gaseous ones.

The laser process is an order of magnitude faster than recommended liquid etching methods. As a result, in instances where the number of vias needed is not excessive, the maskless laser process may be superior. The low power of 100 mW used compared to an easily obtainable 10 to 20 watts from an argon laser leaves open the possibility of multibeam operation where the high power laser beam is split into many lower power beams. The laser-enhanced process may, however, find its best application as a supplementary repair facility when used in conjunction with another batch manufacturing method.

REFERENCES

1. R. J. von Gutfeld, E. E. Tynan, and L. Romankiw, Electrochemical Society Extended Abstracts, Vol 79-2, Abstract No. 472, 1979.
2. R. J. von Gutfeld and R. Hodgson, Appl. Phys. Lett., 40, 352 (1982).
3. N. A. Adrova, M. I. Bessonov, L. A. Laius, and A. P. Rudakov, "Polyimides A New Class of Thermally Stable Polymers", Technomic Pub., Stamford, Conn., 1970.
4. J. A. Kreuz and C. M. Hawkins, Paper IPC-TP-411, Inst. for Interconnecting and Pack. Elec. Cir. 25th Ann. Mtg., Boston, Mass., April 1982.
5. P. B. Perry, S. K. Ray, and R. Hodgson, Thin Solid Films 80, 111 (1981).
6. "Faster Caustic Etching of Kapton® Polyimide Film", Du Pont Technical Information Data Sheet, March 10, 1982.
7. P. A. Moskowitz, R. W. Guernsey, and J. W. Stasiak, in "Advances in Cryogenic Engineering," R. W. Fast, Editor, Vol. 27, pp. 201-206, Plenum Press, New York, 1982.

®Trademark of E. I. Du Pont de Nemours & Co. Inc.

STRUCTURE - Tg RELATIONSHIP IN POLYIMIDES

M. Fryd

Finishes and Fabricated Products Department
E. I. du Nemours and Company, Inc.
Philadelphia, Pa. 19146

The Tg of polyimides is lowered by a decrease in the rigidity of the chain through the introduction of flexibilizing linkages, or by a reduction of interchain interactions. The interchain interactions in polyimides are charge transfer complexes whose strength is dependent on the electron affinity of the dianhydride and the ionization potential of the diamine. Charge transfer complex formation can be minimized, most effectively, by the introduction of "separator" groups in the dianhydride which reduce its electron affinity, or by the use of meta diamines which distort the polyimide chain and interfere with chain packing.

INTRODUCTION

Polyimides, because of their high thermal resistance and excellent mechanical and electrical properties, have attracted a great deal of interest from polymer scientists over the last thirty years. Much of that work has been directed at uncovering relationships between the structure and glass transition temperature (Tg) of this important family of polymers. These studies aimed at gaining insights which would identify approaches to lowering the Tg, thus making polyimides more tractable and processable, without giving up any or at least much of their other outstanding characteristics.

With a few exceptions it has been difficult to arrive at broad generalities unmarred by logical inconsistencies. Part of the problem is due to the complexity of the factors which affect the Tg of polyimides. Another part resides in the difficulty of accurately and reproducibly measuring the Tg of these polymers because it is significantly affected by a number of extraneous variables such as: molecular weight, nature of end groups, extent and method of imidization, processing atmosphere and temperature profile, presence of absorbed moisture or retained solvent, and finally but not least importantly, the method of measurement used by the investigator. However a reexamination of the literature in the light of known data about imide chemistry suggests a hypothesis which clarifies the structure-Tg relationship of polyimides.

Abbreviations for the dianhydrides and diamines discussed in the paper are presented in Table I.

DISCUSSION

Tg is the temperature below which the various groups and atoms in the polymer chain have insufficient rotational energy to overcome the forces which restrict torsional oscillations and hinder their transition to free rotation. The introduction of flexible linkages into a macromolecule increases the number of degrees of freedom available and tends to lower the Tg of a polymer. Because of the generality of this effect there is a commonly accepted correlation between the flexibility of a polymer and its Tg. Therefore most of the efforts at reducing the Tg of polyimides have focused on reducing their rigidity by the introduction of flexiblizing linkages, such as oxygen, sulfone, or perfluoroalkylene[1,2,3,4] segments, in either the diamine or dianhydride moieties.

All these changes in structure led to the expected and desired result-lowering of Tg. However they produced some data (Tables II and III) which cannot be explained if we assume that chain flexibility is the main determinant of a polyimide's Tg.

Table I. Diamine and Dianhydride Abbreviations

Diamine or Dianhydride	Abbreviations
	PMDA
	ODPA
	BTDA
	HFPDPA
	HDFODPA

(continued)

Table I. (Continued)

Diamine or Dianhydride	Abbreviations
	ODA
	DABP
	MDA
	PDA
	BABB
	BABDM

Table II. Effect of Placement of Flexibilizing Segment on Glass Transition Temperature.

Polymer	Dianhydride	Diamine	Tg (°C)	Reference
1	PMDA	p,p-ODA	399	5
2	ODPA	p-PDA	342	5
3	PMDA	p,p-DABP	412	6
4	BTDA	p-PDA	333	6
5	PMDA	m,m-DABP	321	7
6	BTDA	m-PDA	300	6

Table III. Effect of Isomerism versus the Introduction of Flexible Linkages.

Polymer	Dianhydride	Diamine	Tg (°C)	Reference
7	BTDA	p,p-DABP	290	6
8	BTDA	m,m-DABP	264	6
9	BTDA	p,p-BABB	286	6
10	BTDA	p,p-MDA	290	6
11	BTDA	m,m-MDA	234	7
12	BTDA	p,p-BABDM	256	6
13	HFPDPA	p,p-ODA	222	4
14	HFPDPA	m,m-ODA	178	4
15	HDFODPA	p,p-ODA	186	4

Perusal of Table II shows that the introduction of flexible linkages in the dianhydride portion of a polyimide has a greater impact on lowering Tg than the presence of the same flexibilizing segment in the diamine. Since in both cases we are introducing the same number of additional degrees of freedom, the difference in Tg cannot be due to differences in chain rigidity but must represent differences in intermolecular interaction.

An examination of the literature suggests the likely nature of that interaction. Monomeric aromatic imides and anhydrides are known to be strong electron acceptors, and to form charge transfer complexes (CTC) with electron donors such as amines [8,9]. The strength of these complexes is determined by the electron affinity of the dianhydride and the ionization potential of the diamine.

Polyimides can be looked at as polymeric chains with alternating donor and acceptor elements which can interact with each other to form interchain charge transfer complexes. The suggestion that polyimides form CTC's was first made by Dine-Hart and Wright[10] who compared the fusibility, solubility and color of model aromatic imides, and by Gordina et al.[11], who examined the relationship between the color and chemical structure of polyimides. The actual presence of CTC's in polyimides was eventually demonstrated by

Kotov et al.[12] who found absorption bands in the U.V. spectra of a series of polyimides similar to the bands observed for CTC's of monomeric imides. The wavelength and intensity of the bands were, as expected, dependent on the electron affinities of the dianhydrides and the ionization potential of the diamines which constituted the polyimides.

The presence of interchain CTC's in polyimides helps clarify the relationship between structure and Tg. While the rigidity of the chain is certainly an important factor for the individual isolated macromolecule it becomes less dominant in a condensed state (bulk) where the chains exist as aggregates held relatively parallel to each other by the CTC interactions. The stronger the CTC, the closer the chain packing and the greater the interchain interference with free rotation. Polyimides based on PMDA, whose imide has the highest electron affinity (1.55 eV), have a crystallographically determined[13] interchain distance of 4-5Å. Since the minimum interchain distance necessary for free rotation of phenylene groups in polyimides has been calculated[14] to be 7A, one can see the reason for their well known lack of tractability.

Following this line of reasoning, the greater impact on Tg by flexibilizing groups in the dianhydride can now be more readily understood. A "hinge" group in the diamine segment can either increase or decrease the strength of the CTC depending on whether it is electron donating or withdrawing. A "hinge" in the dianhydride, on the other hand, will always reduce the strength of the CTC relative to a pyromellitimide, irrespective of its electronic characteristics, because it reduces the electron affinity of the dianhydride by isolating the powerful electron withdrawing anhydride groups from each other.

A similar argument can be marshalled to explain the data in Table III which show that meta isomerism has a greater effect on Tg than the introduction of flexibilizing segments. As in the discussion above, the use of a meta diamine does not introduce additional degrees of freedom and thus cannot flexibilize the molecule. It does, however, distort the linearity of the polyimide chain and this interferes with chain packing and charge transfer formation. The resulting increase in interchain distance lowers the energy necessary for rotation and therefore the Tg.

REFERENCES

1. A. P. Rudakow, M. I. Bessonov, Sh. Tuichiev, M. M. Koton, F. S. Florinskii, B. M. Ginzburg, and S. Ya. Frenkel, Polymer Sci. USSR, 12, 720 (1970).
2. H. R. Lubowitz, U.S. Patent 3,699,075 (1972)
3. G. L. Brode, J. H. Kawakami, G. T. Kuratkowski, and A. W. Bedwin, J. Polym. Sci., Chem. Ed., 12, 575 (1974).

4. J. P. Critchley, and M. A. White, J. Polymer Sci., Part A-1, 10, 1809 (1972).
5. T. L. St. Clair, A. K. St. Clair, and E. N. Smith, in "Structure-Solubility Relationships in Polymers", F. W. Harris and R. S. Seymour, Editors, pp. 199-214, Academic Press, New York 1977.
6. V. L. Bell, L. Kilzer, E. M. Hett and G. M. Stokes, J. Appl. Polymer Sci., 26, 3805 (1981).
7. V. L. Bell, B. L. Stump, and H. Gager, J. Polymer Sci., Chem. Ed. 14, 2275 (1976).
8. M. Choudhury, J. Phys. Chem. 66, 353 (1960).
9. R. S. Davidson and A. Lewis, Tetrahedron Letters, 611 (1974).
10. R. A. Dine-Hart and W. W.Wright, Makromol, Chem., 143, 189 (1971).
11. T. A. Gordina, B. V. Kotov, O. V. Kolnivov and A. N. Pravednikow, Vysokomol Soed, B15, 378 (1973).
12. B. V. Kotov, T. A. Gordina, V. S. Voischchev O. V. Kolninov, and A. N. Pravednikow, Vysokomol. Soed. A19, 614 (1977).
13. L. G. Kazaryan, D. Ya. Tsvarkin, B. M. Ginzburg, Sh. Tuichiyev, L. Korzhavin, and S. Ya. Frenkel, Polymer Science USSR, 14, 1344 (1972).
14. P. Krasnov, A. Ye Stepanyan, Yu. I. Mitchenko, Yu. A. Tolkachev, and N. V. Lukasheva, Polymer Science USSR, 19, 1795 (1977).

SOFTENING AND MELTING TEMPERATURES OF AROMATIC POLYIMIDES

M. I. Bessonov and N. P. Kuznetsov

Institute of High Molecular Compounds

Academy of Sciences, Leningrad 199004, USSR

Softening temperatures T_s and melting temperatures T_m of a great number of aromatic polyimides and their chemical derivatives have been studied. The results obtained are generalized. The analysis is based on the chemical classification of polyimides taking into account the presence and location of "hinge" atoms and atom groups making polymer chain flexible. If there are such "hinges" in both components of a monomer unit then T_s and T_m are distinctly revealed. For these polyimides the dependence of T_s and T_m on the number of "hinges" in a monomer unit can be successfully explained by the well-known physical theories of transition temperatures of polymers. Other groups of polyimides have no distinct T_s and T_m and are usually considered as "infusible". Calculations and experiments have shown that it is because in these groups the values of T_s and T_m are close to or higher than the value of the polyimide thermal degradation temperature. The maximum values of T_s and T_m can be as high as $\simeq 1000$ and $\simeq 1300°C$, respectively. The influence of chemical structure on T_s and T_m for the whole class of aromatic polyimides can be understood if strong intermolecular interaction of benzimide rings and internal degrees of freedom in monomer units are taken into consideration. Simple models are used in this work for this purpose. The conclusions made here are thought to be valid both for aromatic polyimides and other related classes of polyheteroarylenes.

INTRODUCTION

Temperatures of the main physical transitions-softening temperature T_s and melting temperature T_m - are important characteristics of thermally stable polymers. In this paper the principal results of our investigations of the relationships between the values of T_s and T_m and the chemical structure for the class of aromatic polyimides are reported.

The values of T_s and T_m were mainly determined from the dependence of static and dynamic Young's moduli on temperature and from thermomechanical curves at a constant tensile stress of about 1 MPa. Polyimide films obtained by thermal cyclization of polyamic acids were used as samples.

While searching for the correlations of T_s and T_m with the chemical structure of polyimides we used the following classification[1,2]. In the general chemical formula of aromatic polyimides

$$- N \overset{\displaystyle CO \qquad CO}{\underset{\displaystyle CO \qquad CO}{\diagup \diagdown \diagup \diagdown}} Q \diagup \diagdown N - R -$$

aromatic radicals Q and R are of the following two main types:

"rigid" -   etc., and

"flexible" -   etc.

Inner rotation in rigid radicals is impossible or it does not influence the chain conformation. Flexible radicals always contain atoms or groups of atoms playing the role of "hinges" in the chain because inner rotation around them certainly changes conformation of chains and make them flexible. The hinges are usually such atoms and groups as $-O-$, $-S-$, $-SO_2-$, $-CO-$, etc. Measurements and calculations show that isolated polyimide molecules containing $-O-$, $-S$ and other hinges in the main chain are just as flexible as typical carbon chain polymers.[3-5]

In accordance with the presence and location of hinges in the repeating unit, all aromatic polyimides may be divided into four groups:

      Group A   in which no hinges are present in the repeating
unit,
      Group B   in which they are present only in the dianhydride
radical Q,
      Group C   in which they are present only in the diamine
radical R and
      Group D   in which they are present in both Q and R radicals.

    These groups of polyimides differ not only in this and other
chemical features but also in their physical properties. The
differences in softening and melting behavior are particularly
pronounced.

    Figure 1 shows typical dependence of Young's modulus on
temperature for samples from all four groups of polyimides. It
can be seen that polyimides of groups A and B have very smooth
temperature dependence of the modulus without any evidence of
softening or melting of polymer up to 800K.

    The temperature dependence of Young's modulus for polyimides
of group C exhibit inflexions at high temperatures showing physical transitions. However in most of the cases the transitions are
not very pronounced and moreover are masked by crosslinking and
degradation of polyimides at these high temperatures. Hence the
problem of the existence of transitions and the numerical values
of T_s and T_m for polyimides of group C has not been solved up to
the present.

    A characteristic example is the well-known poly-
pyromellitimide

$$-N\underset{CO}{\overset{CO}{\diagdown}}\!\!\!\!\bigcirc\!\!\!\!\underset{CO}{\overset{CO}{\diagup}}N-\!\!\bigcirc\!\!-O-\!\!\bigcirc\!\!-$$

(1C in Figure 1). It is used for Kapton films in the USA and for
PM films in the USSR[6,7]. Various values of T_s ranging from 600 to
800K are reported in the literature for this polyimide[8]. In the
light of this investigation the most probable value has to be
taken as T_s = 643K[9,10].

    In contrast to all the previous cases, softening and melting
of polyimides of group D are observed as distinctly and with the
same features as carbon chain polymers. However, after heating to
650-750K polyimides of group D also undergo crosslinking. This
leads to an irreversible increase of the modulus of elasticity in
the softened state (compare curves on heating and on cooling
for the amorphous 1-D polymer in Figure 1), to a hindrance
of crystallization of the melt on cooling (compare analogous
curves for the crystalline 2-D polymer in Figure 1), to a
noticeable displacement of T_s towards higher temperatures and
other effects.

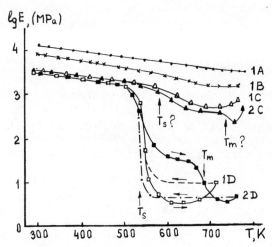

Figure 1. Temperature dependence of Young's modulus E for polyimides of various groups[10]:

1A		$æ \cong 45\%$
1B		$æ \cong 50\%$
1C		$æ \cong 20\%$
2C		$æ \cong 20\%$
1D		$æ \cong 0$
2D		$æ \cong 30\%$

$æ$ - degree of crystallinity of the samples at the beginning of experiments. The experiments both on heating and subsequent cooling of samples are shown for 1D and 2D.

The possibility and necessity of introducing this classification of polyimides becomes clear from a marked difference in the dependences of log E on T for polymers of groups C and D with the same chemical composition but different location of hinges in the repeating unit (compare polymers 1C and 1D and polymers 2C and 2D in Figure 1). This is also confirmed by the results of other authors who studied the thermomechanical properties of various polyimides based on pyromellitic ("rigid") and benzo-phenonetetracarboxylic and other "flexible" dianhydrides[8,11-15]. The results of our investigation of chemical derivatives of polyimides such as polybenzimidazoleimides, polyimidazopyrrolones, polybenzoxazoleimides, polyamideimides and polyesterimides show that the classification may be applied to a wide range of polyheteroarylenes[10,16-19].

RESULTS AND DISCUSSION

The foregoing considerations lead to many important questions. Is it possible to find quantitative relationships between the T_s and T_m values of polyimides and their chemical structure? Do polyimides have any specific features of softening and melting in comparison to carbon chain polymers? Why are these transitions not observed for polyimides of groups A and B? What is the physical meaning of the effect of the location of atomic hinges in a monomer unit on the values of T_s and T_m? In the following we will provide answers to these questions.

It is reasonable to begin with polyimides and their derivatives belonging to group D as they have clear T_s and T_m. Very often the chemical structure of polyimides is regulated by varying the number of some identical groups in both radicals of repeating unit. In this case they may be considered as regular copolymers of different composition. Thus, polyimides with the general formula

are the copolymers of imide

and paraoxyphenylene

Similarly, poly-esterimides with the structure

may be regarded as copolymers of esterimide

and paraoxyphenylene.

The dependence of softening temperature T_s of a copolymer on its composition is expressed by the Gordon-Taylor's equation[20]

$$T_s = \frac{(1-w_2) + Kw_2}{(1-w_2)T_{s2} + Kw_2 T_{s1}} \cdot T_{s1} \cdot T_{s2}$$

where T_{s1} and T_{s2} are the softening temperatures of the homo-polymers, w_2 is the weight fraction of the second comonomer and K is a constant. This equation is derived on the basis of the famous free-volume concept.

Figure 2 shows the dependence $T_s = f(w2)$ for polyimides and polyesterimides with the above chemical structure. In both cases $\ell = 0$, 1 or 2 and m=0, 1, 2, 3 or 4 in various combinations. The experimental points fall near solid curves corresponding to the Gordon-Taylor's equation. The precise obeying of this equation and the numerical values of the coefficients K (1.6 for poly-imides and 2.0 for polyesterimides) are the same as for carbon chain polymers. Hence, it may be assumed that the Gordon-Taylor's equation is applicable to polyimides and their derivatives to the same extent and with the same restrictions as it is to carbon chain polymers and copolymers.

According to Zhurkov[22], the softening of polymers upon heat-ing is due to the breaking of intermolecular bonds. The larger the number of groups in polymer chains capable of forming polar, hydrogen and other strong intermolecular bonds the higher is the softening temperature and vice versa. The chains of polyimides and polyesterimides considered above consist of strongly inter-acting units containing benzimide rings and weakly interacting units containing oxyphenylene groups. The softening temperature T_s of these polymers may be determined by the Zhurkov equation

$$T_s = T_{so} - \frac{2RT_{so}^2}{U} \cdot n$$

where T_{so} is the softening temperature of the reference polymer containing in chains only strongly interacting units, U is the

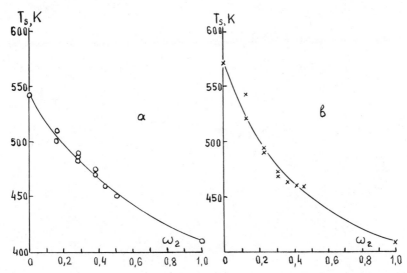

Figure 2. Softening temperatures T_s of a) polyimides and b) polyesterimides vs weight fraction w_2 of oxyphenylene[10]. $T_{S2} = 410K$[21].

energy of their interaction, n is the mole fraction of weakly interacting units. Figure 3 shows that for both polyimides and polyesterimides there is a linear relationship between T_s and n. It is possible to determine from the slopes of the straight lines in Figure 3 that U ≃ 35 kJ/mol for polyimides and U ≃ 30 kJ/mol for polyesterimides. These figures reasonably characterize the interaction between polyatomic benzimide groups capable of forming strong polar and complex bonds.

Hence, the main concepts used in polymer physics for quantitative analysis of softening temperatures – the free-volume concept and that of local intermolecular interactions – are quite applicable to polyimides, their derivatives and, evidently, to polyheteroarylenes in general. The available data do not show which concept is to be preferred. However, this problem has not been completely solved even for "ordinary" polymers either.

The equations considered above may be used only for well characterized homologous series of polymers. On the other hand there are many empirical equations for T_s without a rigorous physical confirmation but applicable to many polymers. We used the Askadsky and Slonymsky equation[23]

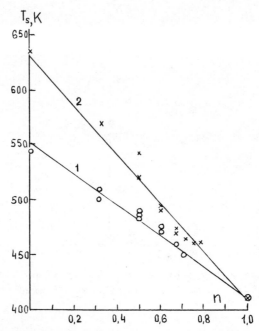

Figure 3. Softening temperatures T_S of 1) polyimides and 2) poly-esterimides vs molar fraction n of oxyphenylene groups[10,28].

$$\log T_s = \frac{\Sigma n_i K_i}{N_A \Sigma \Delta V_i} + 1.435$$

where K_i is the contribution to T_s of the i-th atom or group of atoms, n_i is their number in the repeating unit, $\Sigma \Delta V_i$ is the van der Waals volume of the unit, and N_A is Avogadro's number. The values of K_i and ΔV_i are tabulated[23,24]. We used this equation to calculate T_s for a wide range of polyimides and their deriva-tives, belonging to all four groups.

Figure 4 shows calculated and experimental values of T_s for about forty of polyimides, polyesterimides and polypyrrones of group D. It can be seen that the discrepancy between the calculated and experimental values does not exceed 20-40K for overall varia-tions of T_s from 400 to 700K. The points for polymers of group C, for which T_s is not shown quite distinctly experimentally, are in the same range.

This allows one to calculate, with good accuracy, values of T_s for polyimides of groups A and B for which no experimental data are available. Table I shows that the calculated values of T_s are $\sim$ 1000K for group A and $\sim$ 750K for group B.

Table I. The Transition Temperatures of Polyimides[25].

Group a. No.	Polyimide	τ_{10}^* K	T_s, K calc. exper.	T_m, K calc. exper.
A	1	800	975 / no	1300 / no
	2	790	850 / no	1130 / 1100**
B	3	790	750 / no	970 / no
	4	800	685 / no	915 / 900**
C	5	790	690 / 650	865 / 870**
	6	730	590 / 600	805 / 800 / 820**
D	7	770	550 / 543	725 / 700
	8	750	500 / 483	630 / 660, / 660**

*   Temperature at which 10% wt. loss occurs.
** By rapid heating and quenching.

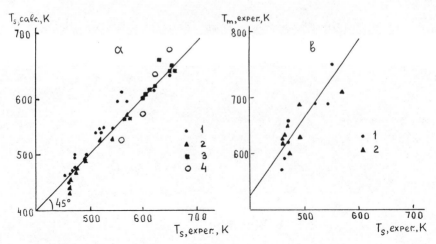

Figure 4. Comparison of a) calculated and experimental T_s and b) experimental T_s and T_m for heteroaromatic polymers of group D: 1 - polyimides, 2 - polyesterimides, 3 - polypyrrones, and of group C: 4 - polypyromellitimides.

Comparison of experimental values of T_s and T_m for poly-imides and polyesterimides of group D in Figure 4 shows that they may be related to each other by the well known Beaman equation

$$T_m = \alpha T_s$$

The coefficient α, calculated by the least-squares method, is 1.3 and the coefficient of linear correlation between T_m and T_s is 0.95. Assuming that this relationship is valid in all cases it is possible to evaluate T_m for polyimides of any group. Table I shows that the calculated values of T_m for polyimides of groups A, B and C are very high ~ 1300, ~ 1000 and $\sim 850K$, respectively. The problem arises whether these values are quite real.

Note that only for polyimides of group D are the values of T_m less than the temperatures τ_{10} corresponding to a 10% weight loss during thermogravimetric experiments (Table I)[26]. In all the other cases T_m is higher or even much higher than τ_{10}. Hence, it is natural to suggest that the infusibility of polyimides of groups A, B and C is due to the fact that during thermomechanical tests the samples undergo thermal degradation before the melting point has been attained. To check whether these polyimides can melt, the experiment should be carried out in such a way as to reduce the effect of degradation to a minimum.

As has been done for the determination of T_m of polyvinyl alcohol[27] the quenching phenomenon was used here. Films of

394

polyimides of groups A, B and C are always partially crystalline. Hence, if they are heated for a short time at successively increasing temperatures T_q and rapidly cooled, it might be expected that after the attainment of $T_q = T_m$ the amorphous structure would be retained. This can be established by monitoring any property sensitive to crystallinity. Since a time of 0.01-0.1sec is sufficient to heat a polymer film 10-20μm thick, degradation may be strongly suppressed.

An experiment carried out with a reference polymer of group D (8 in Table I) for which the melting temperature was calculated as well as reliably determined experimentally by several methods showed that the transition through T_m may be clearly identified by the proposed method. It can be seen in Figure 5 (curve 8) that the Young's modulus measured at room temperature each time after quenching the sample from a given temperature T_q to 273K sharply decreases at $T_q = T_m = 660K$. The same effect is observed for all the polyimides investigated belonging to groups A, B and C (curves 2 and 4-6 in Figure 5). In all the cases the sharp decrease in Young's modulus corresponds to T_q close to T_m calculated from the empirical equation. The measurements of density, calculations of the degree of crystallinity and X-ray diffraction patterns showed that after quenching from $T_q \simeq T_m$ polyimide films became amorphous.

Hence, it may be concluded that all aromatic polyimides including those usually considered to be non-softening and infusible are capable of physical transitions. The transition temperatures can be reliably evaluated from the usual empirical equations and determined experimentally if the interferences due to thermal degradation are avoided.

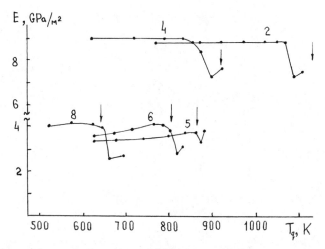

Figure 5. Young's modulus E at 293K vs quenching temperature T_q for polyimide films[25]. Numbers correspond to polymers in Table I. Arrows show calculated T_m.

It remains to elucidate the physical reason for the drastic change in the transition temperatures when the atomic hinges are rearranged in the repeating unit or are removed from the chain.

In the first attempt[28] we proceeded from the fact that transitions are impossible without high mobility of macromolecules. In the bulk, the mobility is determined by the flexibility of the macromolecules themselves and by the strength of intermolecular interactions and their distribution along the chain. We considered that the strongest are dipole-dipole interactions between the   C=O polar groups in imide rings of neighboring chains. Then the mobility of interacting polyimide molecules containing different groups can be described by the simple models as shown in Figure 6.

It is clear that in model A molecular mobility is extremely limited both because atomic hinges are absent from the chains and the chains are fixed by a dense "comb" of intermolecular bonds. In models B and C the chain shape can vary as a result of rotation about the hinges. However, only correlated movement of neighboring chains is possible because the position of rigid chain units between hinges is strictly fixed by pairs of strong interchain bonds. The larger the distance ℓ between bonds in these pairs, the greater the hindrance to chain mobility. In model D interchain bonds do not form pairs; they are always separated by one or several intrachain hinges. Hence single units of neighboring chains may move independently. The value of

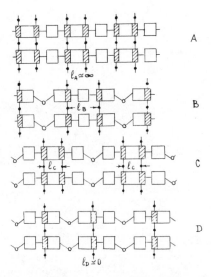

Figure 6. Schematic models for interacting pairs of polyimide molecules of various chemical structures[28].

reciprocal ℓ may be the measure of mobility of the models. Then according to Figure 6 we have

$$\frac{1}{\ell_A} \ll \frac{1}{\ell_B} < \frac{1}{\ell_C} \ll \frac{1}{\ell_D}$$

in accordance with the real values of T_s and T_m for polyimides of different groups.

Although this model is very primitive, it probably correctly reflects the overall character of the effect of intermolecular interactions on transition temperatures at passing from one group of polyimides to another.

In order to understand the effect of the hinges and their rearrangement in chain on transition temperatures more quantitatively we used[25] the well-known equation

$$T_m = \Delta H_m / \Delta S_m$$

relating the melting temperature to the enthalpy ΔH_m and entropy ΔS_m of melting. The enthalpy ΔH_m depends mainly on the energy of intermolecular interactions. Since the molecules of aromatic polyimides consist of the same rings, the value of ΔH_m should be approximately constant for the entire class. As a result T_m should be determined only by the value of entropy ΔS_m.

For carbon chain polymers ΔS_m is usually identified with the entropy S_{dis} of disorientation of the chain backbone on passing from a crystal to a melt. The value of S_{dis} for an individual chain may be calculated. The evaluations of T_m for carbon chain polymers based on these calculations give reasonable results[29]. However for polyimides the assumption that

$$\Delta S_m = S_{dis}$$

is not fulfilled. For polymers of group A the calculated values of S_{dis} are negative which is physically impossible and those for polymers of groups B, C, D are too low and do not agree with the experimental values of T_m and ΔH_m[25]. These discrepancies show that additional positive components should be present in the entropy of melting of polyimides.

Molecules of polyimides (and other polyheteroarylenes) have a complex structure. They contain planar rings exhibiting their own rotational and vibrational degrees of freedom. When such macromolecules are packed in a crystalline lattice, the following elements should be successively ordered: the chain backbone, the rings within each segment and ordered segments with respect to each other. During melting, the chain backbone and the rings should be disordered in the reverse sequence. Correspondingly, the entropy of melting should contain two additional terms

$$S_m = S_{dis} + S_1 + S_2$$

namely S_1 - the entropy of disorientation of internally ordered segments with respect to the ordered chain backbone, and S_2 - the entropy of disorientation of rings within individual segments. On these assumptions total entropies of melting were calculated for polyimide molecules of all groups[25]. For all the cases including group A, the values of ΔS_m were found to be finite, positive and inversely proportional to the experimental values of T_m. The heat of melting ΔH_m calculated from these values was found to be close to the experimental values.

These results show that the influence of chemical structure on T_m of polyimides on passing from one group to the other may be explained in terms of the change in the entropy of macromolecules, and the main contribution to entropy is provided in this case by inner degrees of freedom of the repeating unit.

SUMMARY

The analysis carried out here allows us to assert that the vigorous influence of the monomer chemical structure on main transition temperatures for the class of aromatic polyimides is related to two basic physical reasons: change in the contribution of internal degrees of freedom to thermodynamic functions, and change of the system of intermolecular interactions at passing from one classification group of polyimides to another. Both factors are interrelated and should be taken into account in examining any physical properties of polyimides and many other polyheteroarylenes.

REFERENCES

1. A. P. Rudakov, N. A. Adrova, M. I. Bessonov and M. M. Koton, Dokladi Academii Nauk SSSR, 172, No. 4, 899 (1967).
2. N. A. Adrova, M. I. Bessonov, L. A. Laius and A. P. Rudakov "Polyimides - A New Class of Thermostable Polymers", Technomic Publ. Co., Stamford, CT, USA, 1970.
3. M. I. Bessonov, Vysokomol. Soedin., 8A, No. 1, 206 (1967).
4. S. S.-A. Pavlova, G. I. Timofeeva and I. A. Ronova, J. Polym. Sci., Polym. Phys. Ed., 18, No. 6, 1175 (1980).
5. V. A. Zubkov, T. M. Birstein and I. S. Milevskaja, Vysokomol. Soedin., 17A, No. 9, 1955 (1975).
6. "Kapton" Polyimide Films, Du Pont Bulletins, H-1D; E-33798; F-1C; E-33337; H-2; A-78996.

7. Elektroizoliatsionnye Polyimidynye Plenki PM-1, PM-1E, PM-2, PM-4, PM-14/50, PM/212, Prospect VDNCh SSSR, (1976).
8. C. E. Sroog, J. Polym. Sci., Macromolec. Rev., 11, 161 (1976).
9. I. I. Perepechko and A. V. Prokazov, Vysokomol. Soedin., 20A, No. 2, 243 (1977).
10. N. P. Kuznetsov, Thesis, Institute of High Molecular Compounds, Academy of Sciences of the USSR, Leningrad (1979).
11. R. E. Conlehan and T. L. Pickering, ACS Polymer Preprints, 12, No. 1, 305 (1971).
12. J. K. Gillham, K. D. Hallock and S. I. Stadnicki, J. Appl. Polym. Sci., 16, No. 10, 2595 (1972).
13. J. K. Gillham and H. G. Gillham, Polym. Eng. Sci., 13, No. 6, 447 (1973).
14. B. I Azarov, G. M. Tseitlin, V. V. Korshak and B. V. Vorobiev, Plastic. Massy, No. 7, 37 (1971).
15. V. L. Bell, J. Polym. Sci., Polym. Chem. Ed., 14, No. 1, 225 (1976).
16. N. A. Adrova, M. I. Bessonov, M. M. Koton, A. Mirzaev, A. P. Rudakov, L. K. Prokhorova and N. P. Kuznetsov, Vysokomol. Soedin., 13B, 764 (1971); 14B, 819 (1972); 16B, 788 (1974).
17. A. P. Rudakov, F. S. Florinsky, M. I. Bessonov, M. M. Koton, N. P. Kuznetsov, L. A. Laius and V. E. Smirnova, Vysokomol. Soedin., 14A, No. 1, 169 (1972).
18. N. P. Kuznetsov, M. I. Bessonov, T. M. Kiseleva and M. M. Koton, Vysokomol. Soedin., 14A, No. 9, 2034 (1972).
19. N. A. Adrova, V. N. Bagal, A. M. Dubnova, I. L. Kvitko, M. M. Koton, N. P. Kuznetsov and F. S. Florinsky, Vysokomol. Soedin., 15B, No. 7, 509 (1973).
20. I. M. Ward, "Mechanical Properties of Solid Polymers", Wiley and Sons Ltd., New York (1968).
21. H. Lee, D. Stoffey and K. Neville, "New Linear Polymers", McGraw Hill Co., New York (1968).
22. S. N. Zhurkov, Doklady Academii Nauk SSSR, 47, No. 7, 493 (1945).
23. A. A. Askadsky and G. L. Slonimsky, Vysokomol. Soedin., 13A, No. 8, 1917 (1971).
24. G.L. Slonimsky, A. A. Askadsky and A. I. Kitaigorodsky, Vysokomol. Soedin., 12A, No. 3, 494 (1970).
25. M. I. Bessonov, N. P. Kuznetsov and M. M. Koton, Vysokomol. Soedin., 20A, No. 2, 347 (1978).
26. M. M. Koton and Ju. N. Sazanov, Vysokomol. Soedin., 15A, No. 7, 1654 (1973); 17A, No. 7, 1469 (1975).
27. M. I. Bessonov and A. P. Rudakov, Soviet Physics, Solid State, 6, No. 5, 1333 (1964).
28. A. P. Rudakov, M. I. Bessonov, Sh. Tuichiev, M. M. Koton, F. S. Florinsky and B. M. Ginzburg, Vysokomol. Soedin., 12A, No. 3, 641 (1970).
29. M. V. Volkenshtein, "Configurational Statistics of Polymer Chains", Interscience, New York (1963).

THERMAL STUDIES OF COMPOSITIONAL VARIATIONS OF SOME NOVEL SILICONE POLYIMIDES

B. Chowdhury

M&T Chemicals Inc.
Rahway, New Jersey   07065

Thermal behavior of polymers has long been recognized as a useful guide in designing materials for specialized applications and for predicting their end-use properties. As a general rule, polymer structural units can be purposely varied in synthesis to produce specific compositions with the desired properties. This paper is concerned with the application of thermal methods for unravelling the composition-property relationship for a newer class of polymers called polysilicone imides (PSI), resulting from modification of aromatic polyimides by specially equilibrated silicone blocks. These polymers are novel from a synthetic point of view and unique in their applicability in a wide range of service temperature, high thermal stability, significant mechanical strength and good adhesive and electrical properties. Although thermoplastic in nature, they are able to cross-link mechanistically through a silicone block and exhibit thermoset properties. Thus, PSI materials have opened up new possibilities in polymer applications not known for typical polyimides. Thermal studies show these properties to be directly related to the compositional variations of both the aromatic and the silicone moieties, which are discussed in some detail here.

Polymers, in general, exhibit a wide variety of thermal behavior as a result of structural differences. Such differences may be small, yet significant in end-use applications. Thermal methods are ideal for studying polymer compositional variations because of their ability to discern effects due to small changes in chemical structure.

A newer group of polymers called polysilicone imides (PSI), resulting from polycondensation reactions of aromatic dianhydrides with organic diamines and aminofunctional capped di- and/or polysiloxanes,

$$(X+Y) \quad O\!\!\bigwedge\!\!\stackrel{C}{\underset{C}{R}}\!\!\bigwedge\!\!O \;+\; X{\cdot}H_2N{-}R'{-}NH_2 \;+\; Y{\cdot}H_2NR''{-}\underset{CH_3}{\overset{CH_3}{Si}}{-}O{-}\underset{CH_3}{\overset{CH_3}{Si}}{-}R''{-}NH_2$$

Solvent →

$$\left[\; H_2N\!\!\stackrel{C}{\underset{C}{R}}\!\!\begin{matrix}C{-}NHR'\\ C{-}OH\end{matrix}\;\right]_X \;+\; \left[\; \begin{matrix}H\;O\\ N{-}C\end{matrix}\!\!\stackrel{}{\underset{}{R}}\!\!\begin{matrix}C{-}NH{-}R''{-}\underset{CH_3}{\overset{CH_3}{Si}}{-}O{-}\underset{CH_3}{\overset{CH_3}{Si}}{-}R''\\ C{-}OH\end{matrix}\;\right]_Y$$

↓ imidization

$$\left[\; N\!\!\stackrel{C}{\underset{C}{R}}\!\!NR'\;\right]_X \;+\; \left[\; N\!\!\stackrel{C}{\underset{C}{R}}\!\!NR''{-}\underset{CH_3}{\overset{CH_3}{Si}}{-}O{-}\underset{CH_3}{\overset{CH_3}{Si}}{-}R''\;\right]_Y$$

are finding a wide variety of applications because of their highly desirable properties. In general, silicone polyimides possess high thermal, hydrolytic and oxidative stability and exhibit good adhesion and electrical properties. Dependence of these properties on the structural aspects of polyimides has been discussed by several authors.[1-4] Most recently, excellent thermal stability based on structural considerations has been reported for polyimides prepared from novel dianhydrides containing a bis-pyrrole structure.[5]

The purpose of this paper is to report information obtained from thermal studies of some novel polyimides modified by silicone blocks.

EXPERIMENTAL

The DuPont 990 thermal analyzer and the various modules were used for differential scanning calorimetry (DSC), thermomechanical analysis (TMA), dynamic mechanical analysis (DMA), differential thermal analysis (DTA) and thermogravimetry (TGA). Standard techniques were used for obtaining and interpreting results.

A typical preparation procedure for a standard PSI polymer is as follows:

An equilibrate* (80.0g) of the disiloxane and a methyl tetramer having 2.34% (0.0586 mole) of $-NH_2$ (calculated) group is combined with 19.25g (0.059 mole) of BTDA. The mixture is diluted with 900g of monochlorobenzene. A catalytic quantity (0.3g) of toluene sulfonic acid is added and the entire reaction is placed on reflux such that the water formed during cyclization is removed azeotropically with the chlorobenzene. Water evolution ceases after about 2 hours, but refluxing is continued for an additional 3 hours, then cooled. The polymer is separated by methanol precipitation.

RESULTS AND DISCUSSION

DSC

The effect of molecular segmental variations on the glass transition temperature (Tg) and also on the mode of transition was studied by this technique. Taking the simplest case of a polyimide synthesized from an aromatic carbonyl dianhydride and a silicone-containing equilibrated diamine, Tg occurs between 103-110°C for polymers containing short segments of siloxane units in the oligomer. Expansion of these units to about 100 lowers Tg somewhat, down to around 80°C, but it rises again at a moderate pace for compositions up to 500 units. This pattern is shown in Figure 1 which demonstrates the flexibility silicone polyimides possess in controlling the service temperature of the polymer.

---

*Equilibration results when reshuffling of silicon-oxygen bonds in at least two types of siloxane monomers are reacted using an acid or a basic catalyst, so that after a given period of time, the composition remains constant.

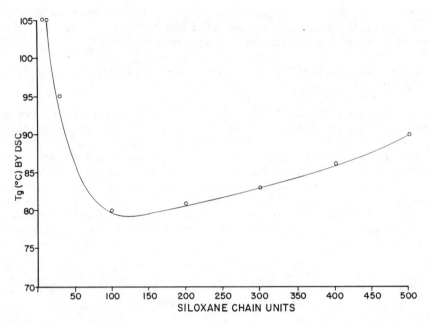

Figure 1. Hyperbolic relationship of Tg and siloxane chain units for PSI polyimides.

For all compositions, a sub-ambient Tg at about 10°C is thought to be due to the rotational aspect of the aromatic moiety when the sample is scanned at a slow rate from below zero degree centigrade. Increase of aromatic content increases Tg, possibly due to specific interactions causing skewing of the molecule from the planarity. An increase in Tg is also observed when the carbonyl group is re-placed with an ether or a thioether linkage in the dianhydride moiety. These substitutions are known to produce a tight-knit structure due to the zigzag pattern of these molecules. By these modifications, a Tg as high as 275°C has been obtained. The role of the siloxane block in these compositions can be seen when it is considered that a polyimide synthesized from a bis-dicarboxyphenoxy diphenylsulfide dianhydride (BDSDA) and an aminophenoxybenzene (APB) and containing no silicone, produces a Tg at about 153°C.[6] Some of these findings are summarized in Table I.

404

Table I. Tg Values for Changes in Polyimide Composition.

Polyimide   Tg (°C)

A.

$$\left[\begin{array}{c} \overset{\displaystyle O}{\underset{\displaystyle O}{\overset{\|}{\underset{\|}{C}}}} \quad \overset{\displaystyle O}{\underset{\displaystyle O}{\overset{\|}{\underset{\|}{C}}}} \\ -N \diagdown_{C} \diagup^{R} \diagdown_{C} \diagup NR'- \end{array}\right]\!\!-\!(\text{equilibrate})_{n\%}$$

                           103–110

B.   " $-(\text{equilibrate})_{n+x\%}$ 50–80
(depending on
the value of x)

C.   BDSDA $-(\text{equilibrate})_{n-y\%}$ 275

D.   BDSDA–APB   153

Controlling the molar ratio of an aminofunctional disiloxane with
various cyclic siloxanes gives rise to compositions of varying
degrees upon equilibration. These variations are referred to as
n, x, y, etc. The R group is not identified for proprietary
reasons.

At the other extreme, for very large siloxane block materials,
no above-ambient Tg is discernible by standard DSC technique right
up to decomposition.

An interesting point to note is that for materials in cate-
gory A in Table I, a first-order endothermic transition appears at
glass transition temperature on first heating, which is then re-
placed by the usual step-change at glass transition on second heating. Figure 2 shows typical DSC scans.

Endotherms of this nature are likely to be due to volume relaxation caused by short-range ordering of the amorphous molecules.[7]
Also, transitions over a broad range of temperature are indicative
of blends rather than pure high polymers consisting of uniform repeating units. Although PSI polymers are not pure blends, because
of the use of equilibrated siloxane amines which are polydispersed,
the behavior tends to be that of blends.

When the above transition widths are plotted against the percentage of silicone diamine equivalent weight required to produce
a given equilibrate, the transition broadening, considered to be a
function of molecular motions of microscopic origin and related to
the degree of compositional heterogeneity of the blend,[8] is found
to vary in a linear fashion. This is because of the polydispersity

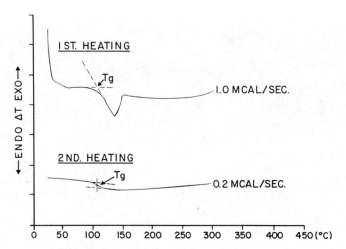

Figure 2. DSC scans showing transition due to Tg for standard PSI polyimide.

of the resulting polymers rather than their blend character in the case of PSI compositions. This plot is shown in Figure 3. It is evident that the dianhydride and the diamines are highly compatible for PSI polymers.

TMA

Glass transition occurring as a state change due to dimensional changes with temperature is found to be more easily discernible for silicone polyimides by TMA than by the calorimetric method of detection of changes in heat flow.

For polysilicone imides, in general, containing an intermediate number of siloxane segments than implied for categories A and B in Table I, as explained before, two distinct Tg's have been obtained by TMA, as shown in Figure 4. A lower Tg at 45-50°C and an upper Tg of 115-130°C are found, depending on the drying cycle used for preparing the sample. Such samples were run as thin films, solvent-cast on aluminum substrate. These two Tg's can be attributed to the silicone segment and the aromatic segment, respectively, but the lower Tg is rarely discernible along with the upper Tg on the

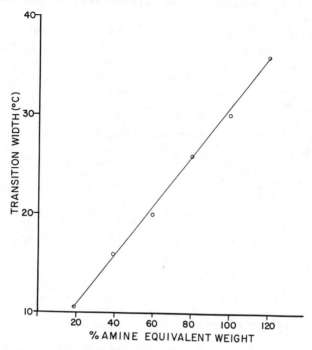

Figure 3. Linear relationship of Tg transition width and amine
equivalent weight for PSI compositions.

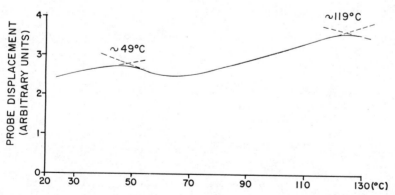

Figure 4. Tg values by TMA for solvent-cast polyimide on aluminum
substrate.

same thermogram by DSC. The shift in temperature, particularly for the upper Tg, from the DSC value is due to a degree of cure involving the drying process in the preparation of the film.

The changes in Tg values found by TMA for variations in the stoichiometric ratio of silicone diamine in the equilibrium composition for three different formulations marked A, B and C are shown in Figure 5 without further disclosure of composition. This plot confirms that optimum Tg is obtained for the 100% of theoretical diamine.

It is noteworthy that the same 153°C value has been obtained for Tg for a polyimide composed of a bis-dicarboxyphenoxy diphenyl-sulfide dianhydride and an aminophenoxybenzene by the TMA pene-trometer, as obtained by DSC. This value was obtained independently

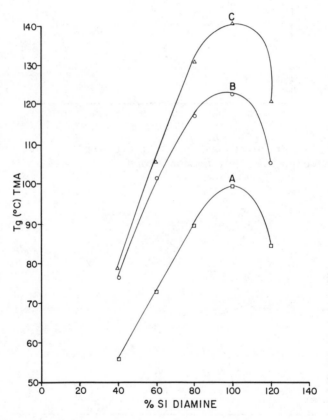

Figure 5. Changes in Tg values for compositional variations in PSI polyimides.

by the author and by NASA[9] using TMA and confirmed in a private communication. The absence of silicone is again evident here and the lower Tg is not obtained. Thus, silicone plays an important part in the service temperature of the polyimide.

The coefficient of linear expansion over a given temperature range for PSI materials can be tailored by compositional variations to match the expansion coefficient of the material of construction for a particular application. Comparative values of expansion co-efficients for different polyimides are shown in Table II.

DMA

The utility of silicone block copolymer polyimides in end-use applications is best studied by this method. The incorporation of silicone blocks in the polyimide system actually enhances thermal and dimensional stability of the material in use conditions. The modulus value near 5×10^{10} dynes/cm^2 in the glass transition region for a representative PSI material is somewhat greater than the mod-ulus at room temperature (generally 4×10^{10} dynes/cm^2). This is an unusual property and is thought to be due to the ability of silicone blocks to cross-link mechanistically, as mentioned before, under increasing temperature up to the softening point; i.e., well beyond the viscoelastic region. The cross-linking is evidenced by an increase in Tg value and the polymer becoming more solvent re-sistant by successive heating. Such cross-linking is not available to a typical polyimide.

A typical DMA scan is shown in Figure 6. A distinct transi-tion occurs at about -85°C (β-transition) which is in the true glassy region of the polymer. The strong intensity of this transi-tion and the modulus value of about 6×10^{10} dynes/cm^2 at this tran-sition indicates the polymer's ability to relieve strain through motion of the amorphous region side chains or branches from the main backbone. PSI polymers therefore possess good mechanical strength at low temperatures. This attribute however does not emanate from the silicone diamine equilibrate itself, which is found to have a low modulus at glass transition. Table III shows comparative values obtained by DMA for silicone diamine equilibrate having meta- and para-position linkages and a representative PSI polymer.

The Tg values at α-transition, measured by DMA damping, are found to be characteristically higher than those measured by DSC. This is due to the time dependency of these loss peaks causing them to move to higher temperatures with increased frequency.[10]

409

Table II. Expansion Coefficients for Changes in Polyimide Composition.

Polyimide	Expan. Coeff. (Temp. Range) /°C	Comments
BDSDA-APB (from reference 6)	5.43×10^{-5} (100–140°C) Tg 153°C	Measured below Tg, since probe penetration into sample occurs above Tg
DuPont Kapton® (all aromatic polyimide)	6.41×10^{-5} (100–140°C)	Measured below Tg for comparison
Polyamic Acid (containing components which were silicone diamines)	0.65×10^{-5} (100–160°C) 1.26×10^{-5} (220–280°C) Tg 189°C	High aromatic content, low free volume
BDSDA (equilibrate)	46.24×10^{-5} (80–140°C) Tg 150°C	High silicone content
[structure]	46.67×10^{-5} (75–100°C) Tg 105°C	Polysiloxane units have high free volume
Epoxy Matrix/100% NH_2 eq. wt.	7.54×10^{-5} (100–140°C) Tg 160°C	Behaves more like an organic diamine

Note: R group is not specified for proprietary reasons.

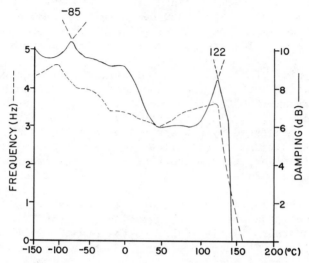

SIZE: 0.02 t X 0.55 W X 1.2 L CM
PROG: 5°C / MIN.
AMP.: 0.20 MM
CLAMP: HORIZONTAL

Figure 6.   DMA characteristics of a typical PSI polyimide.

Table III.   DMA Values for Silicones and Silicone Polyimides.

Polymer	Tg at α-Trans. (°C)	Peak at β-Trans. (°C)	Modulus at Tg (dynes/cm^2)	Modulus at Room Temp. (dynes/cm^2)
meta-siloxane amine	+95	−47	8.98×10^8	2.48×10^{10}
para-siloxane amine	+88	−57	2.50×10^8	1.81×10^{10}
Typical PSI Polymer as A. in Table I	+122	−85	4.81×10^{10}	3.81×10^{10}
Unspecified modification of above	+130	−75	5.22×10^{10}	1.52×10^{10}

411

High thermal stability also means low thermal activity for silicone polyimides. Lack of significant thermal transitions below decomposition, other than perhaps glass transition, under proper conditions, makes DTA not a very useful technique for studying PSI materials. However, since these are prepared via poly(amic acid) intermediates, DTA can provide useful information for the imidization process at the solvent reflux temperature, particularly in conjunction with TGA studies of the imidized polymer.

For a typical fully imidized polymer, significant weight loss occurs only after 400°C by thermogravimetry. A secondary weight loss occurs around 500°C, and decomposition is completed typically between 600 and 650°C. This is shown in Figure 7. Two major weight losses separated by a plateau are characteristic of silicone polyimide compositions. This allows for use of PSI materials, in certain applications, at temperatures higher than the initial decomposition temperature.

The DTA scan in Figure 8 is not only characterized by the onset of decomposition past 400°C, agreeing with the TGA finding, but also by an endotherm likely to be due to loss of small amounts of residual cyclics. The double-peaked endotherm and its general shape also indicate fractionation of these volatiles.

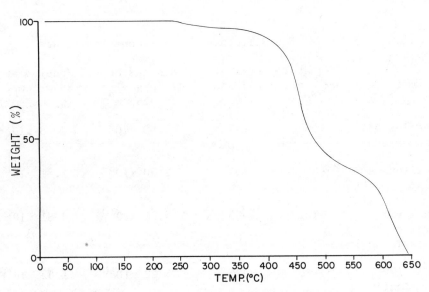

Figure 7.   TGA curve of a PSI polyimide.

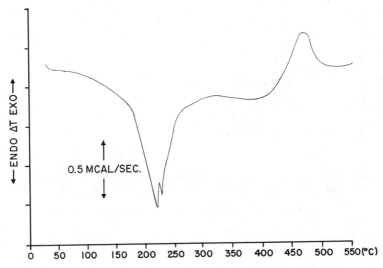

Figure 8. DTA curve of an imidized polymer.

Thermal imidization of the polymer in its amic acid form in a suitable solvent, viz. N-methylpyrrolidone, shows a complex behavior by DTA. The multiplicity of the broad endotherm between 190 and 225°C in Figure 9 is accompanied by intermittent gas evolution and is due to aberrant loss of small amounts of solvent, which is tenaciously retained over this temperature, plus loss of some residual volatile silicones mentioned above.

Although imidization is known to start at an earlier temperature, this endotherm is thought to represent its completion. This is because of the use of continuous scan rather than stationary reflux, as in synthesis. The endotherm centered at 425°C coincides with the procedural decomposition temperature for the imidized polymer obtained by TGA. Further endothermic activity centered at about 465°C, going into exothermic decomposition around 500°C, parallels the weight loss detected by TGA at this temperature.

Typical thermal inflections by DTA and TGA for standard PSI compositions are given in Table IV.

413

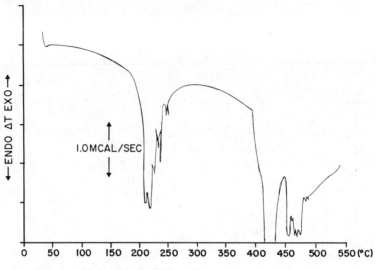

Figure 9.   DTA curve of a PSI polyamic acid.

Table IV.   TGA and DTA Characteristics of PSI Polymers.

Event	TGA	DTA
Completion of imidization	<1% Wt. loss	Endotherm onset ∿190°C and completion at ∿225°C
Procedural decomposition temperature	∿425°C (extrapolated onset)	Endotherm centered at ∿425°C
Secondary weight loss	∿500°C (extrapolated onset)	1) Endotherm centered at ∿465°C
		2) Exotherm onset at ∿500°C

A weight loss of <1% between 190 and 225°C by TGA is too small to account for total loss of water in the cyclization process.   On isothermal hold at 190°C and at 225°C for 30 minutes in air, N_2 or vacuum, no measurable weight loss is found by TGA, indicating that water loss actually occurs before 190°C.

CONCLUSION

It is clearly seen from the above results that thermal studies are extremely useful as a synthetic guide and for structure-property evaluation for segmental polymers. In the case of PSI polymers, the claim of thermal stability below 400°C and at still higher temperatures for solvent-based coating applications for in-situ imidization and cure is well justified.

REFERENCES

1. A. H. Frazer, "High Temperature Resistance Polymers," in Polymer Reviews, Vol. 17, Ch. 7, p. 315, John Wiley & Sons, New York, 1968.
2. C. E. Sroog, Macromolecular Reviews, 11, 161 (1976).
3. R. D. Deanin, "Polymer Structure, Properties and Applications," Ch. 8, p. 457, Cahners Publishing, Boston, 1972.
4. T. L. St. Clair, A. K. St. Clair, and E. N. Smith, in "Structure-Solubility Relationships in Polymers," F. W. Harris, Editor, pp. 199–214, Academic Press, New York, 1977.
5. R. W. Stackman, ACS Polymer Preprints, 22 (1), 58, March 1981.
6. Material obtained from NASA Langley Research Center, VA.
7. G. Wilkes, J. Appl. Phys., 49 (10), 5032 (1978).
8. B. Y. Min and E. M. Pearce, Proceedings of the 11th NATAS Conference, New Orleans, 1981, J. P. Schelz, Editor, Johnson & Johnson Products, New Brunswick, NJ, 2, 559 (1981).
9. T. L. St. Clair, NASA Langley Research Center, VA, (1982).
10. P. S. Gill and P. F. Levy, Proceedings of the 11th NATAS Conference, New Orleans, 1981, J. P. Schelz, Editor, Johnson & Johnson Products, New Brunswick, NJ, 2, 565 (1981).

POLYIMIDE CHARACTERIZATION STUDIES –

EFFECT OF PENDANT ALKYL GROUPS

Brian J. Jensen and Philip R. Young

NASA-Langley Research Center

Hampton, VA 23665

An investigation was conducted to determine the effect on selected polyimide properties when pendant alkyl groups were attached to the polymer backbone. A series of polymers were prepared using benzophenone tetracarboxylic acid dianhydride (BTDA) and seven different p-alkyl-m,p'-diaminobenzophenone monomers. The alkyl groups varied in length from C_1 (methyl) to C_9 (nonyl). The polyimide prepared from BTDA and m,p'-diaminobenzophenone was included as a control.

All polymers were characterized by various chromatographic, spectroscopic, thermal, and mechanical techniques. Increasing the length of the pendant alkyl group resulted in a systematic decrease in glass transition temperature (T_g) for vacuum-cured films. A 70°C decrease in T_g to 193°C was observed for the nonyl polymer compared to the T_g for the control. A corresponding systematic increase in T_g, indicative of crosslinking, was observed for air-cured films. Thermogravimetric analysis revealed a slight sacrifice in thermal stability with increasing alkyl length. No improvement in film toughness was observed.

417

INTRODUCTION

Aromatic polyimides have potential for extended use in various aerospace applications provided they can be reliably processed into useful articles. However, inherent problems with solubility, high glass transition temperature (T_g), and the evolution of volatiles during cure must be solved before this potential can be fully exploited. Several approaches have been taken to alleviate some of the difficulties associated with processability. Numerous polyimide structure property relationships have been reported[1-3]. Addition polyimides are being developed which are processed as oligomeric segments and then thermally cured through reactive end groups without the evolution of volatiles[4-6]. Stereoisomeric variations have been built into the polymer backbone and the effect on polymer properties determined[7]. Pendant phenyl groups were added to the polymer backbone in an effort to improve solubility[8]. Also, the effect of incorporating oxyethylene units in the polyimide backbone has been reported[9].

The objective of the present research was to investigate the effect on selected properties when alkyl groups of various lengths were attached pendant to the polyimide backbone. The eight variations studied are shown in Figure 1. The polyimide with no pendant group was included as a control. Although the data obtained

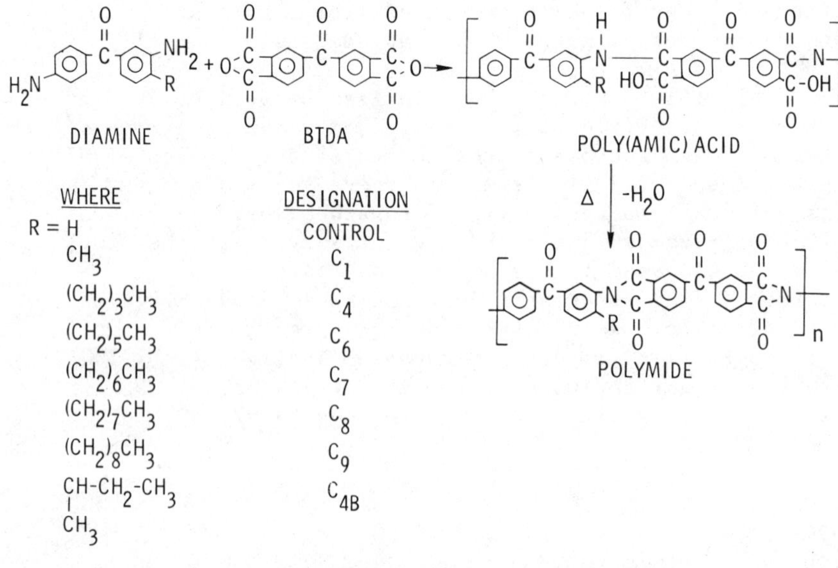

Figure 1. Polyimide synthetic scheme.

in this study were somewhat limited by the amount of sample available, fundamental information was gained on how particular variations in molecular structure affected such properties as T_g, thermal stability, and toughness.

EXPERIMENTAL

Characterization

Analytical chromatographic separations were obtained on a Waters Associates ALC/GPC-244 HPLC. Analyses were performed at 254 nm on either a Waters μPorasil or C_{18} - μBondapak column (3.9 mm id x 30 cm). Preparative separations were obtained on a Waters Prep LC/System 500 using a Prep PAK-500/Silica column and a refractive index detector. Gel permeation chromatography was performed on a 10^6, 10^5, 10^4, 10^3 Å μStyragel column bank. The mobile phase for all separations was prepared from chromatographic grade methylene chloride, methanol, 1-propanol, water, or N,N'-dimethylacetamide (DMAc). Poly(amic) acid inherent viscosities (η_{inh}) were measured at 0.5% concentration in DMAc at 35°C with a Cannon-Ubbelohde viscometer.

Thermal analyses were performed on a DuPont Model 990 Thermal Analyzer in combination with a standard differential scanning calorimetry (DSC) cell. Thermomechanical analyses (TMA) were performed on films in combination with a DuPont Model 940 Thermomechanical Analyzer. Torsional Braid Analyses (TBA) were performed using an in-house system[10] on glass braids impregnated with a 5% poly(amic) acid solution in DMAc and cured in vacuum at 100°C, 200°C, and 275°C for one hour each. Thermogravimetric analyses (TGA) were conducted on approximately 2 mg film samples with a Perkin-Elmer TGS-2 instrument in 15 cm^3/min flowing air.

Infrared spectra of monomers were obtained on KBr pellets with a Perkin-Elmer Model 297 Spectrophotometer. Spectra of polymeric films were recorded on a Nicolet 3600A FTIR System. Elemental analyses of monomers, shown in Table I, were performed by Galbraith Laboratories, Knoxville, Tennessee.

Mechanical properties were measured on an Instron Tensile Tester at a crosshead speed of 0.2 in/min and a gauge length of 3.0 in. Five one-inch wide and nominally one-mil thick film specimens were tested for each sample.

Monomers

The monomer 3,3',4,4'-benzophenone tetracarboxylic acid dianhydride (BTDA) was obtained from commercial sources and vacuum sublimed prior to use.

Table I. Elemental Analysis of Diamine Monomers.

$$H_2N-\bigcirc-\overset{\overset{O}{\|}}{C}-\bigcirc\overset{R}{\underset{NH_2}{}}$$

EMPIRICAL FORMULA	PENDANT GROUP	%C		%H		%N	
		CALC.	FOUND	CALC.	FOUND	CALC.	FOUND
$C_{13}H_{12}N_2O$	H	73.57	73.44	5.70	5.71	13.20	13.01
$C_{17}H_{20}N_2O$	C_4	76.09	75.92	7.51	7.61	10.44	10.39
$C_{17}H_{20}N_2O$	C_{4B}	76.09	76.24	7.51	7.64	10.44	10.60
$C_{19}H_{24}N_2O$	C_6	76.99	77.07	8.16	8.21	9.45	9.51
$C_{20}H_{26}N_2O$	C_7	77.38	77.54	8.44	8.22	9.02	9.00
$C_{21}H_{28}N_2O$	C_8	77.74	77.87	8.70	8.83	8.63	8.64
$C_{22}H_{30}N_2O$	C_9	78.05	78.24	8.95	9.00	8.28	8.29

Diamines were obtained from several sources. The m,p'-diaminobenzophenone (control) was synthesized by a previously reported method[7] and purified by recrystallization from benzene, toluene, and finally benzene. The p-methyl-m,p'-diaminobenzophenone monomer (C_1) was synthesized by the scheme shown in Figure 2. This synthetic scheme was typical for all diamines included in the study. The acid chloride I was reacted via a Freidel-Crafts acylation with toluene and the resulting compound II nitrated in mixed nitric and sulfuric acids. The dinitro compound III was reduced by catalytic hydrogenation at 50 psi with a 5% platinum on carbon catalyst yielding 35-40% of compound IV. The remaining product formed appeared to be a hydrate. Compound IV was purified by preparative liquid chromatography after unsuccessful recrystallization attempts. The p-sec-butyl-m,p'-diaminobenzophenone (C_{4B}) was prepared on a grant[11] with the Mississippi University for Women, Dr. J. Richard Pratt, Principal Investigator. It was purified by recrystallizing twice from benzene/cyclohexane. Five diamines were prepared on a grant[12] with the University of Southern Mississippi, Dr. Shelby F. Thames, Principal Investigator. Butyl (C_4) and nonyl (C_9) monomers were polymer grade as received, while the other three (C_6, C_7, C_8) were purified by preparative liquid chromatography.

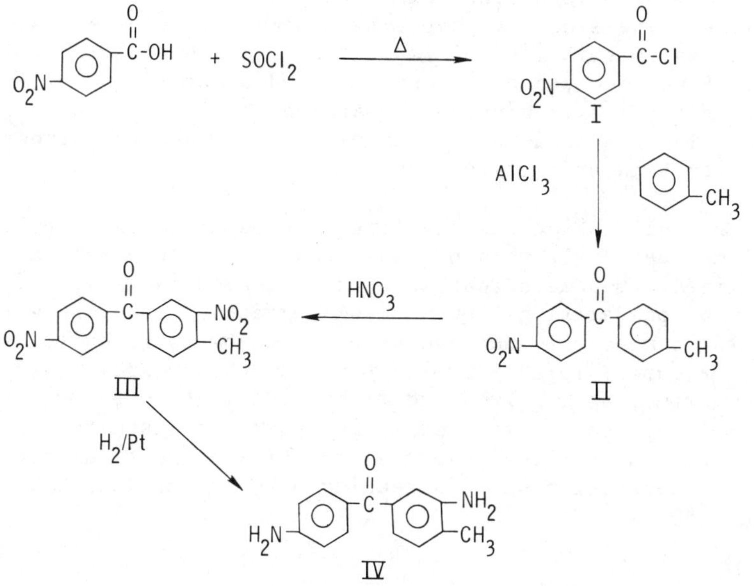

Figure 2. Synthetic scheme for C_1 diamine ($R=CH_3$).

Polymers and Film Preparation

Polymers were synthesized by solution polymerization in DMAc distilled over calcium hydride and stored under nitrogen. Polymerizations were run for 24 hr at 25°C and 15% solids content in a dry serum bottle equipped with a rubber stopper and a magnetic stirring bar. No attempts were made to increase viscosity other than by monomer purification. Solutions were stored at approximately -10°C until use.

Poly(amic) acid solutions were spread on glass plates using a doctor blade set at 15 mil. Films were dried overnight in a flowing-air dry box at 25°C. Each film was then cured in vacuum for one hour each at 100°C, 200°C, and 280°C. Three other films were cured in air for one hour each at 100°C, 200°C, and 300°C. Resultant films were approximately 1-mil thick.

RESULTS AND DISCUSSION

The primary concern in the polymer synthesis was whether
sufficiently high molecular weight polymer could be obtained to
cast films. Therefore, high purity monomers were essential to
eliminate the possibility of impurities limiting polymer molecular
weight. Since the pendant groups were adjacent to the amine
reactive site, steric hindrance was also considered a potential
problem. The C_{4B} (branched) diamine was included to further
investigate this area of concern.

Recrystallization was the first method employed to purify the
diamine monomers. If this was unsuccessful, vacuum sublimation
was utilized. Monomers that were still unacceptable were further
purified by preparative liquid chromatography. Monomers were subjected to both normal and reverse phase chromatography to test for
purity. Figure 3 is an example of normal phase chromatography on
the C_6 diamine, as received and after purification by preparative chromatography. Infrared spectra were consistent with the
corresponding structures and were essentially identical except for
increasing absorption with increasing alkyl group length from 2930
cm^{-1} to 2850 cm^{-1}.

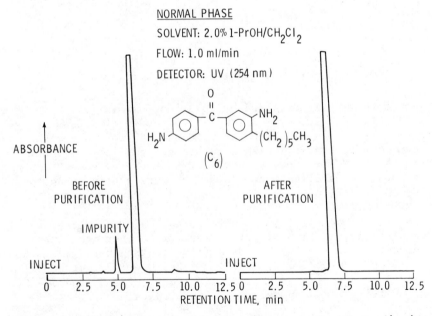

Figure 3. High pressure liquid chromatography of C_6 diamine
before and after purification.

After obtaining the purest monomers possible, polymerizations were run on approximately one gram of each monomer. Although the control diamine was the only monomer less than 99% pure, it produced the highest viscosity polymer. The C_{4B} diamine produced polymer with inherent viscosity (η_{inh}) of 0.19 dl/g and the C_1 diamine produced polymer of only 0.31 dl/g viscosity. The C_{4B} polymer was possibly limited by steric hindrance. This did not apply to the C_1 polymer. The latter polymerizaton was repeated with the same results. Neither of these two polymers could be included in much of the testing since they would not form flexible films . However, one polymerization of the C_9 diamine which resulted in polymer with η_{inh} = 0.28 dl/g did form a usable film. Figure 4 shows gel permeation chromatograms of the poly(amic) acids along with corresponding η_{inh}. As expected, a decrease in elution time with increasing η_{inh} is observed.

Characterization of the polymers in the cured imide form included infrared analysis on film samples which exhibited the same trend as the monomers, increased absorption in the 2930 cm^{-1} - 2850 cm^{-1} region as the pendant groups increased in length. These data also confirmed that imidization had taken place during cure by the appearance of characteristric imide absorptions at 1780 cm^{-1}, 1730 cm^{-1}, 1380 cm^{-1} and 720 cm^{-1}. No significant difference in the IR spectra for air-cured and vacuum-cured films of the same polymer was noted.

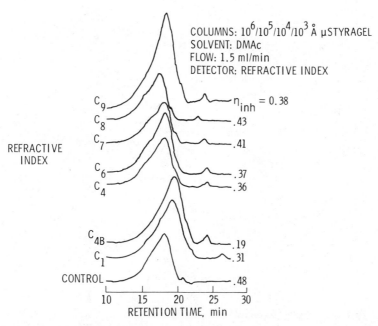

Figure 4. Gel permeation chromatography of poly(amic) acid solution.

One objective of this project was to study the effect the variation in diamine structure had on polymer T_g. Therefore, T_g was measured by several techniques. Torsional braid analyses data are summarized in Figure 5. The damping peak maximum and the corresponding loss of rigidity are indicative of polymer T_g. Vacuum-cured polyimides showed a decrease in T_g as the alkyl group increased in length. For the C_9 polyimide, a decrease of approximately 60°C from that of the control (T_g = 274°C) was noted. Further evidence for the decrease in T_g is shown in Figure 6 by DSC thermograms of vacuum-cured films. Figure 7 gives the same trends by thermomechanical analysis. The opposite trend in T_g was observed for air-cured samples. Although the effect of air cure was not as dramatic, a 20°C increase in T_g for the C_9 polyimide over the control was noted. These data are summarized in Figure 8. An increase in T_g is probably indicative of air induced crosslinking. Apparently, the longer pendant group polyimides crosslinked to a greater extent than the shorter pendant group polymers. These data suggest the possibility of varying the T_g of a polyimide over a wide temperature range by varying the length of an alkyl group pendant to the backbone and by choice of cure atmosphere.

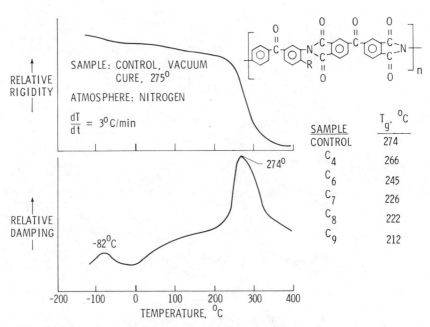

Figure 5. Polyimide T_g by torsional braid analysis.

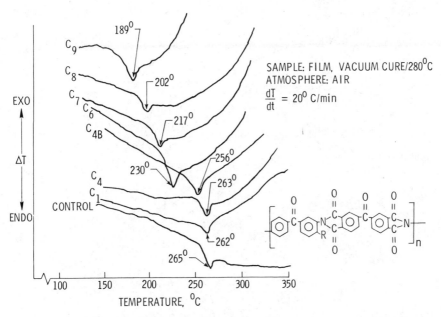

Figure 6. Differential scanning calorimetry of polyimide films.

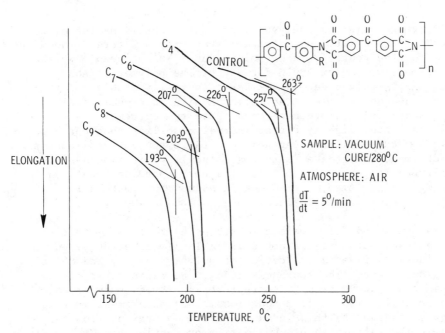

Figure 7. Thermomechanical analyses of vacuum-cured polyimide films.

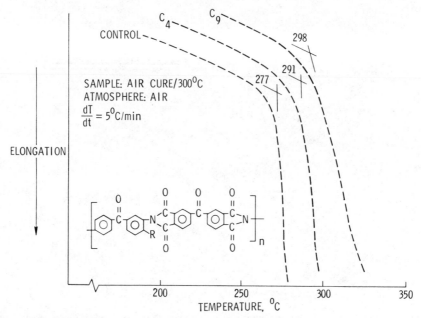

Figure 8. Thermomechanical analyses of air-cured polyimide films.

Thermogravimetric analyses were run to determine the effect
of structural variations on thermal stability. Figure 9 shows TGA
curves for the control and C_9 polyimides. No noticeable weight
loss is observed below 300°C for any of the polymers. Above
300°C, the C_9 polyimide was the least thermally stable and the
control the most stable. The polymers with intermediate alkyl
group length fell in order between these two extremes. Air-cured
films exhibited slightly better thermal stability than corresponding vacuum-cured films. Depending on the potential application,
the detrimental decrease in thermal stability with increasing
alkyl group length may be offset by a corresponding beneficial
lowering in T_g. Also, processing in vacuum but post curing in
air could result in an increase in T_g of about 100°C.

Vacuum-cured films were also subjected to tensile testing.
The resultant data are presented in Table II. Inherent viscosity
is included to give an indication of polymer quality and relative
molecular weight. The toughness index, the ratio of sample
stress-strain curve area to control stress-strain curve area,
gives an indication of film toughness. No improvement in toughness was observed by this index. One trend apparent from the
table is the relationship between film properties and η_{inh}. If
all polymers had the same approximate molecular weight as the control, an increase in toughness might have been observed.

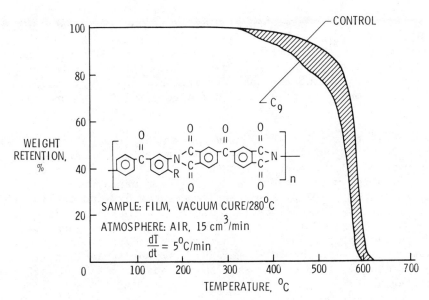

Figure 9. Thermogravimetric analyses of vacuum cured-polyimide films.

Table II. Mechanical Properties of Vacuum-Cured Polyimide Films[a].

SAMPLE	η_{inh},[b] dl/g	TENSILE[c] STRENGTH, MPa	ELONGATION, %[c]	TOUGHNESS[d] INDEX
CONTROL	.48	106.9	4.55	1.00
C_1	.31	(e)	(e)	(e)
C_4[f]	.36	71.6	3.39	0.44
C_{4B}	.19	(e)	(e)	(e)
C_6	.37	86.3	4.33	0.75
C_7	.41	85.8	4.96	0.93
C_8	.43	87.2	5.22	0.95
C_9	.38	82.9	4.90	0.82

(a) CURED AT 280°C
(b) 0.5% SOLIDS IN DMAc AT 35°C
(c) AVERAGE OF FIVE SPECIMENS
(d) RATIO OF AREA UNDER STRESS-STRAIN CURVES
(e) WOULD NOT FORM FILM
(f) POOR QUALITY FILM

CONCLUSIONS

An investigation was conducted to determine the effect on polyimide properties when alkyl groups are attached pendant to the polymer backbone. A control polymer with no pendant groups and seven different polymers with pendant groups ranging in length from one carbon to nine carbons were included in this study. The polymers were characterized in both the poly(amic) acid and polyimide stages.

The data obtained in this study prove the feasibility of varying T_g over a wide temperature range by attaching alkyl groups pendant to the polyimide backbone. A 70°C decrease was observed for the vacuum-cured C_9 polyimide compared to the control as measured by TMA. Also, an increase in T_g, indicative of crosslinking, was observed for air-cured films as the alkyl group increased in length. However, the incorporation of pendant alkyl groups in the polymer chain led to a slight sacrifice in thermal stability as measured by TGA. Mechanical property data indicated no improvement in film toughness with increasing pendant alkyl group length.

REFERENCES

1. C. E. Sroog, Encyclopedia of Polymer Science and Technology, 11, 247 (1969).
2. N. A. Adrova, M. I. Bessonov, L. A. Laius, and A. P. Rudakov (K. Gingold, and A. M. Schiller, transl.),"Polyimides, A New Class of Thermally Stable Polymers," Technomic, Stamford, 1970.
3. C. E. Sroog, Macromolecular Reviews, 11, 161 (1976).
4. E. A. Burns, R. J. Jones, R. W. Vaughn, and W. P. Kendrick, TRW-11926-6013-RO-00, NASA CR-72633, January 1970. (71N11657).
5. T. T. Serafini, P. Delvigs, and G. R. Lightsey, J. Appl. Polym. Sci., 16, 905 (1972).
6. T. L. St. Clair, and R. A. Jewell, SAMPE Tech. Conf. Proceedings, 8, 82 (1976).
7. V. L. Bell, Jr., B. L. Stump, and H. Gager, J. Polym. Sci., 14, 2275 (1976).
8. F. W. Harris, W. A. Feld, and L. H. Lanier, Appl. Polym. Symp., 26, 421 (1975).
9. W. A. Feld, B. Ramalingam, and F. W. Harris, J. Polym. Sci., 21, 319 (1983).
10. S. K. Dalal, G. L. Carl, A. T. Inge, and N. J. Johnston, Polmer Preprints, 15(1), 576 (1974).
11. NASA Grant NSG-1539, Mississippi University for Women, Columbus, MS 39701 (1980-present).
12. NASA Grant NSG-25-005-008, University of Southern Mississippi, Hattiesburg, MS 39401 (1971-76).

POLYIMIDE THERMAL ANALYSIS

M. Navarre

IBM France, Essonnes Plant
Department of Chemistry
B.P. 58, 91102 Corbeil Essonnes Cedex, France

Polyimides possess high thermal stability, good barrier properties and resistance to solvents. Because of these properties these polymers are becoming attractive in electronic devices. This work describes the study of the curing of a commercial polyamic acid after thermal treatment, by using differential scanning calorimetry (DSC), thermogravimetric analysis (TGA) and Infra-red spectroscopy to monitor the curing reaction leading to polyimide formation.

DSC and TGA allow measurements of the degree of cure of polyamic acid films in the temperature range of 100-250°C and the determination of the completion of imidization. Kinetic parameters have been determined and the reaction exhibits a first order kinetics in the range of 200-250°C.

Infra-red spectroscopy allows monitoring of the polyimide formation during cure. DSC and IR results are compared and the advantages and disadvantages of both methods are discussed; DSC is more precise but is also more time-consuming thean IR.

INTRODUCTION

The purpose of this study was to investigate the kinetics of the conversion of polyamic acids into polyimides as a result of thermal treatment.[1,2]

The polymer used was an aromatic polyamic acid resulting from the condensation of a diamine (4,4'- oxydianiline) and a dianhydride (pyromellitic dianhydride), dissolved in an organic solvent (N-methyl pyrrolidone).

Infra-red spectroscopy and differential scanning calorimetry techniques were used in order to follow the chemical changes at various temperatures and to determine the best conditions for having a fully cured polyamic acid (polyimide).

DSC is of particular value in following quantitatively the course of the reaction at temperatures in the range 200°C - 300°C.

A/ DIFFERENTIAL SCANNING CALORIMETRIC STUDY OF POLYAMIC ACID

1) Experimental Procedure :

The Dupont 1090 Thermal Analyser in conjunction with the 910 Differential Scanning Calorimeter (DSC) and the Thermogravimetric Analyser (TGA) were used in this work.

The polyamic acid analysed was a commercial product (Dupont RC 5878) dissolved in an appropriate solvent (N-methyl pyrrolidone).

For DSC analysis, the samples were contained in aluminum pans. The heating rates of 10° C/min were established under nitrogen by programming the temperature. The starting temperature was 30° C and the upper limit was 500° C.

The measurements were made on as-received samples and on cured samples. Curing was performed on a hot plate. Each sample was weighed in order to take care of solvent evaporation.

2) Results and Discussion :

Typical DSC and TGA thermograms of the initial polyamic acid are shown in Figure 1 and Figure 2, respectively. Endothermic peaks are observed up to 200° C. This is mainly due to the solvent evaporation (N-methyl pyrrolidone) and results in a large decrease of the sample weight.

430

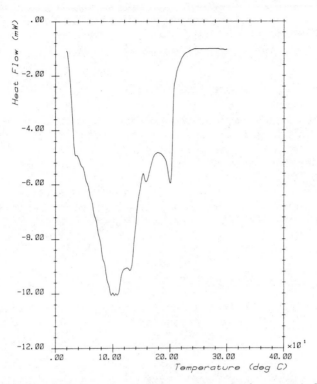

Figure 1. Differential calorimeter scans of the initial polyamic acid up to 300°C.

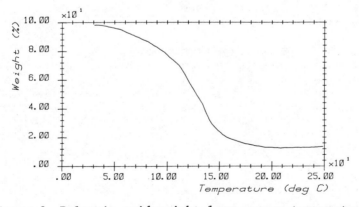

Figure 2. Polyamic acid weight loss versus temperature.

Above 500° C, polyamic acid degradation is observed (Figure 3).

On previously cured polyamic acid films, the product evolution with temperature is quite different and depends on the heating conditions which vary from 120° C up to 320° C at various times ($0 < \min < 100$).

In the case of an initial curing at 200° C an endothermic peak is observed in the range 200° C to 300° C (Figure 4) and can be attributed to the end of the polyamic acid transformation into polyimide. The peak intensity decreases with the time of curing. This event can be used as a means for characterizing the cure stage of the product or for determining the final cure conditions.

At 350° C, the peak does not exist irrespective of the time. So, the peak evolution, which is a characteristic of a chemical change in the polyimide, can be used, by quantitative measurement of the enthalpy of reaction, as a means of following cure advancement .

3) Polyimide Kinetic Study

For each sample, the enthalpy values can be calculated using the interactive DSC data analyser and are expressed by the peak area in the range 200° C to 300° C.

Figure 5 shows the plots of enthalpy (J/g) versus time after curings from 135° C up to 320° C.

From this curve, one can determine the time needed to obtain a fully cured polyimide at a given temperature.

It should be noted that long time curing near 135° C does not result in the total disappearance of the peak, corresponding to a complete reaction.

At 350° C the decrease of the peak is rapid. Only a few minutes are needed to produce the complete chemical change.

Figure 6 is the same as Figure 5 but reported to relative enthalpy variation :

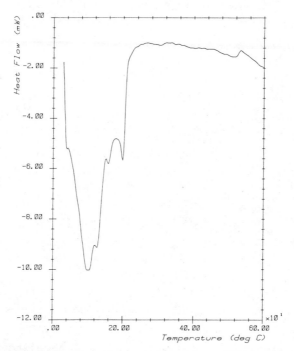

Figure 3. Differential calorimeter scans of the initial polyamic acid up to 600°C.

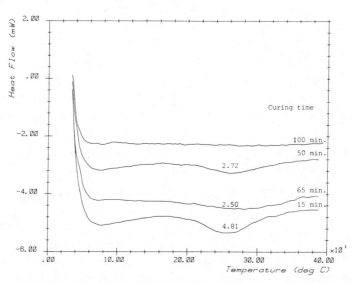

Figure 4. Increase in the intensity of the peak observed between 200°C and 300°C of a previously cured polyamic acid (200°C) with time of curing.

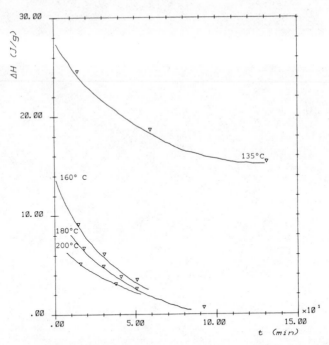

Figure 5.   Influence of curing time on the chemical changes in polyamic acid with temperature.

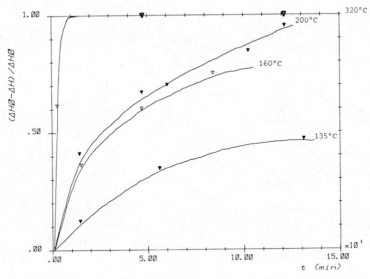

Figure 6.   Influence of the curing time on the chemical changes in polyamic acid.

434

$$\frac{\Delta Ho - \Delta H}{\Delta Ho} = f(t) \text{ at } T° C$$

$$\frac{\Delta Ho - \Delta H}{\Delta Ho} \rightarrow 1$$

if $\Delta H_o \rightarrow o$ (complete reaction)

with ΔH_o : peak area at $T°$ C with $t = o$

ΔH : peak area at $T°$ C (t variable)

The kinetics can also be followed by plotting, at a given time, the enthalpy (ΔH) as function of temperature ($T°$ C) (Figure 7).

As previously shown, whatever the time, the complete chemical reaction does not occur below 150° C.

A linear relationship exists in the log of the reaction rate (Log $\Delta H/ \Delta Ho$) and the cure time indicating a first order reaction (Figure 8).

Data presented in Table I represent values of the first order rate constants of the reactions at 135° C and 200° C determined by the slope of the straight lines and defined by :

$$\text{Log} \frac{\Delta H}{\Delta Ho} = -kt$$

Table I : Rate constants vs curing temperature .

$k\ (s^{-1})$	$T\ (°\ C)$
$0.11\ \ 10^{-3}$	135
$0.32\ \ 10^{-3}$	200

4) Conclusion :

Thermal analysis, using the DSC technique, provides useful

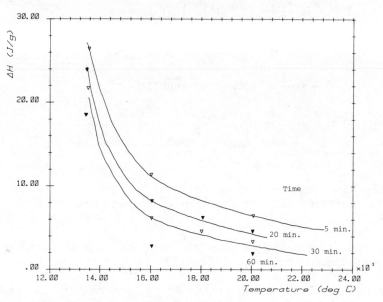

Figure 7. Influence of temperature on the chemical changes in polyamic acid.

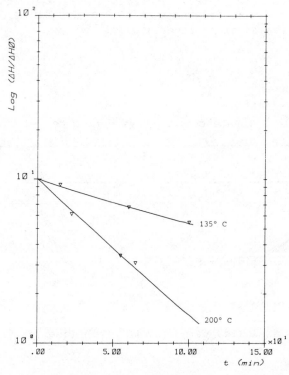

Figure 8. First order rate plots of polyamic acid at two selected temperatures: 135°C and 200°C.

information on polyamic acid chemical changes including solvent evaporation, chemical reaction leading to a fully cured polyimide and degradation.

The method proposed in this paper can easily be used to follow or determine the degree of cure or the imidization level. Although it takes long time to set it up, this technique is more precise than other techniques, such as Infra-red, for evaluating the end of the reaction providing a fully cured product.

B/ INFRARED STUDY OF CHANGES IN POLYAMIC ACID UNDER THERMAL TREATMENTS

1) Experimental Procedure

Two infra-red methods have been utilized to follow the formation of the product and to monitor the rate of formation of polyimide. (Both methods depend upon the apparatus used) :

The first one consists of studying the film of polyamic acid itself, after lifting it from substrates by mechanical means.

Two types of substrates were investigated : ceramic substrates such as those used on module manufacturing lines, and silicon wafers commonly used in semiconductor devices.

The instrument used was an Infra-red spectrometer model 580 from Perkin-Elmer. Samples were cured on the substrates (ceramics or silicon wafer) before being lifted and submitting for spectral measurements. Changes in transmittance were recorded between 4000 and 400 cm^{-1}.

The second method consists of performing the analysis on polyamic acid films deposited directly on silicon wafers and cured for spectral measurements. Data were recorded directly in terms of absorbance.

The advantage of the second method being the ability to work under the standard process conditions. The instrument used was an FT IR Spectrometer Model 7199 C from Nicolet equipped with a calculator.

This instrument offers a number of advantages due to a higher accuracy on wavelentgth and absorbance, a higher throughput, spectra storage, and spectra comparisons.

2) Results and Discussion

The progress of the imidization reaction was followed by

437

comparing spectra before and after curing and by measuring the decrease or the increase of the polyimide characteristic absorption bands[4] . Table II summarizes the main changes observed.

Typical spectra are shown in Figure 9 which represent FT-IR spectra of the polyamic acid after curing at 120° C 20 min (a), and 120° C 20 min plus 200° C 30 min, (b). Figure 10 corresponds to an FT-IR spectrum of polyamic acid after curing at 120° C 20 min, 200° C 20 min, 350° C 5 min.

To determine the imidization level, two methods were investigated.

The first one consists of using an external reference sample defined by its absorption bands at 1770 cm-1 and 720 cm-1 and corresponding to a fully cured polyimide.

IR absorption bands at 1770 cm-1 and 720 cm-1 are attributed to the imide carbonyl group CO-NH.

By following the evolution of the imide absorption bands at 1770 cm-1 and 720 cm-1 for a not fully cured polyamic acid, the imidization level or cure advancement can be determined. Figure 11 summarizes the data obtained.

The second method consists of measuring for each sample the absorbance observed at 1720 cm-1, due to the imide carbonyl band ; and the absorbance at 820 cm-1 due to C-H vibrations. By determining the increase in the ratios 1770 cm^{-1}/820 cm^{-1} or 720 cm^{-1}/ 820 cm^{-1}, for more or less cured samples, the imidization level can be evaluated. Figure 12 presents the data obtained.

Both methods exhibit two distinct stages of the imidization reaction. Initially the reaction rate is rapid, and then the rate decreases considerably as the reaction proceeds. Similar two stage imidization has been described in the literature[4].

4) Conclusion :

Both methods (IR and DSC) allow one to follow the rate of formation of polyimide, but large differences have been observed between them.

The method which requires the Infra-red absorption band at 820 cm-1, is less precise because of the weakness of this band.

Nevertheless, irrespective of the conditions used because of secondary effects such as competing reactions, interacting sol-

Table II.   Typical FT-IR Spectra of Polyamic Acids.

CHARACTERISTIC ABSORPTION BANDS	BEFORE CURING	AFTER CURING
$3200 - 3400$ cm^{-1} secondary amide	strong 3280 cm^{-1}	very weak ↘ if Δ ↗
$3200 - 2400$ cm^{-1} carboxylic acid	strong	"
1780 cm^{-1} C = 0 imide	no	yes ↗ if Δ ↗
1720 cm^{-1} C = 0 stretching vibration (coupled)	no	yes ↗ if Δ ↗
1669 cm^{-1}	strong	weak ↘ if Δ ↗
1640 cm^{-1} secondary amide	strong	very weak
1540 cm^{-1} secondary amide	strong	very weak
1370 cm^{-1} imide stretching vibration	strong	very weak
820 cm^{-1} parasubstitued C-H vibration	weak	weak
720 cm^{-1} C = 0 imide	no	weak

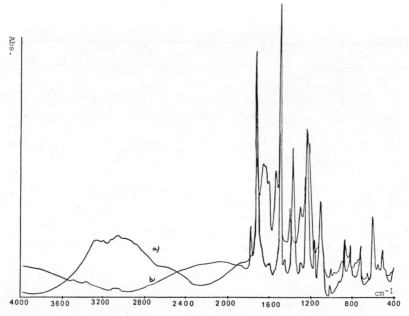

Figure 9. FT-IR spectra of the polyamic acid after curing at
(a) 120°C, 20 min; (b) 120°C, 20 min + 200°C, 20 min.

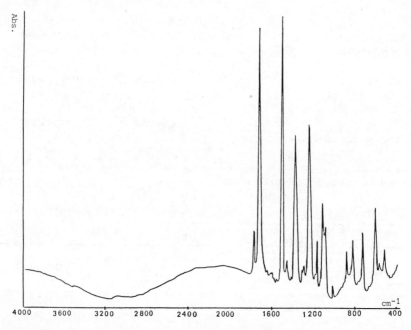

Figure 10. Infra-red spectrum of polyamic acid after curing at
120°C, 20 min; 200°C, 20 min; 350°C, 5 min.

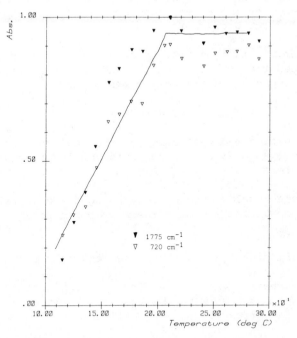

Figure 11. Increase in Infra-red absorption bands versus curing temperatures by using an external reference.

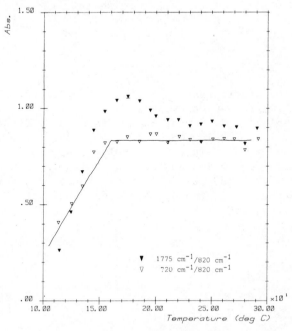

Figure 12. Increase in Infra-red absorption bands versus curing temperatures by using an internal reference.

vents, hydrolysis of products, it seems very difficult to determine precisely, by infra-red technique, the real temperature to obtain a fully cured polyamic acid.

However, it is important to note that a rather good correlation between the DSC measurements and infra-red results was obtained by using the external reference method.

CONCLUSION

The kinetics and energetics of chemical changes involved in converting polyamic acid into polyimide during thermal treatments have been determined.

From the results obtained, by using differential scanning calorimetry and infra-red spectroscopy we can conclude that for the polymer used :

. The imidization occurs in the temperature range of 100-250° C.

. The reaction exhibits a first order kinetics in the range 200-250° C which agrees with the results obtained in the literature on similar aromatic polyimide films[5].

· By using the infra-red technique, only semiquantitative data can be obtained, whereas differential scanning calorimetric measurements permit one to obtain more precise information on imidization.

REFERENCES

1. H. Lee. D. Stoffey and K. Neville, "New Linear Polymers", McGraw Hill Book Company, New York, 1967.
2. L. H. Tagle, J. F. Nevra, F. R. Diaz and R. S. Ramirez, J. Polym. Sci. Polym Chem. Ed., 13, 2827 (1975).
3. R. W. Lauver, J. Polym. Sci. Polym Chem. Ed., 17, 2529 (1979).
4. S. Hummel, "Infra-Red Analysis of Polymers, Resins and Additives", p. 183, Wiley - Interscience, New York, 1971.
5. J. A. Kreuz, A. L. Endrey, F. P. Gay and C. E. Sroog, J. Polym. Sci. A-1, 4, 2607 (1966).

SORPTION AND TRANSPORT OF PHYSICALLY AND CHEMICALLY INTERACTING

PENETRANTS IN KAPTON® POLYIMIDE

L. Iler, W. J. Koros, D. K. Yang and R. Yui

Department of Chemical Engineering
North Carolina State University
Raleigh, North Carolina 27650

Gravimetric sorption and desorption data are reported for water and anhydrous ammonia in Kapton® at 30°C. The data for water at low activity are described well by Fick's Law with a concentration dependent diffusion coefficient. At high vapor activity, evidence of small extents of chemical reaction of the water with imide groups was discovered. Even after exposure to a relative humidity of 81.4% for two weeks, however, only 0.31% of the imide structures were affected. Anhydrous ammonia, on the other hand, at rather low relative saturations (0.012 to 0.032) interacted strongly with some of the imide structures of Kapton®. Infrared and gravimetric sorption/desorption measurements were used to characterize the locus and extent of reaction. It was found that after exposure of a 0.5 mil sample to an ammonia relative saturation of 0.0316 for 16 days, roughly 17% of the imide structures were disturbed. Analysis of the coupled diffusion and chemical reaction suggests that only a fraction of the total number of imide groups (~20%) appear to become "activated" or "labilized" during the solid state curing step in which stresses may arise. Stress relief by the limited ammonolysis reaction presumably deactivates the remaining imide groups. The possiblity of using Kapton® as a combined protective barrier and scavenger coating or film is explored briefly.

443

INTRODUCTION

Aromatic polyether diimides are known for their excellent high temperature resistance to degradation. Our interest in these high glass transition polymers stems from an ongoing study of permselective membranes for various sampling and separation applications [1-3]. In the course of our research, we have encountered an extremely interesting range of physical and chemical interactions which several rather common penetrants exhibit with an important commercial polyimide, Kapton®, whose repeat unit is shown below.

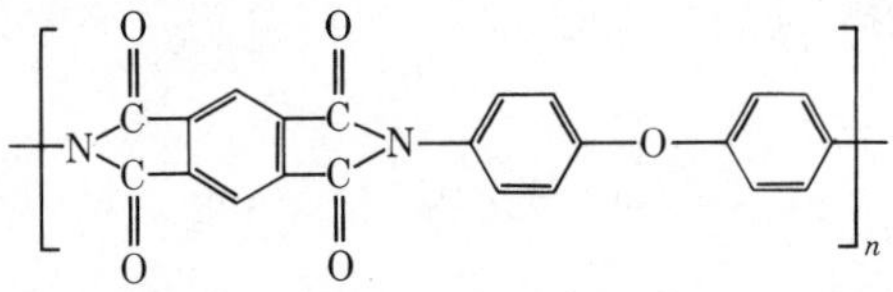

Imide linkages are known to be somewhat sensitive to attack by basic penetrants, and this fact is actually used to an advantage in the erosion of Kapton® to produce extraordinarily thin films (<0.0001 in) for electronics applications, by exposure to an alkaline etching medium. The present study will focus on much milder penetrant exposure conditions. In such cases, the small molecule acts primarily as a probe of the distribution of the physically and chemically heterogeneous environments which appear to be present in Kapton®[4].

EXPERIMENTAL

Materials

The Kapton® poly(ether-imide) used in the present study was kindly supplied by the E. I. DuPont Company, Circleville, Ohio. Films of thickness 2.0 ± 0.01 mil, 0.5 ± 0.01 mil, and 0.3 ± 0.01 mil were used.

The water used in this study was triple distilled and exposed to three freeze/thaw cycles to remove dissolved gases. The anhydrous ammonia was obtained from Air Products and Chemicals, Inc., Raleigh, North Carolina at a purity of 99.9% and was used as received.

Apparatus

The design and operation of the McBain quartz spring balance used in this study has been described previously[4]. With this system, a direct gravimetric measure of penetrant uptake within the polymer as a function of time can be determined by observing changes in the spring extension.

The sorption cell and supporting apparatus are shown in Figure 1. A quartz spring with a spring constant of 0.500 ± 0.006 mg/mm extension was used. The polymer sample was hung at the base of the spring, and a glass reference fiber was hung parallel to the spring to compensate for small shifts in the spring support position which might occur during the experiment. A precision microscope, capable of detecting spring deflections as small as 0.005 mm, was used to observe spring extension; therefore, deflections represented by masses as small as 2.5 µg were detectable.

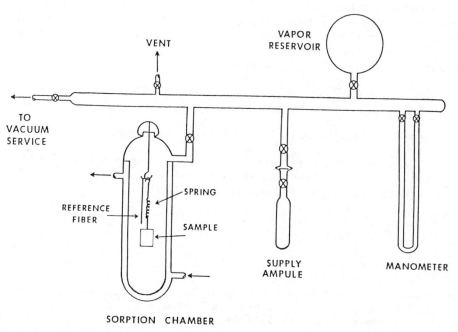

Figure 1. McBain balance apparatus used for gravimetric determination of vapor sorption data.

The sorption cell was maintained at a constant temperature by circulating a silicone fluid from a bath through a fluid jacket surrounding the cell. Brass screen was wrapped around portions of the cell and grounded to reduce static attraction between the cell wall and the polymer film. A large vapor reservoir was used to increase the total volume of the sorption system, thereby eliminating measurable fluctuations in pressure within the sorption cell.

Procedures

A Kapton® sample of known weight was hung from the quartz spring and degassed for 24 hours. A constant weight of the degassed sample was obtained within 10 hours. The system, excluding the sorption cell, was pressurized with penetrant to an empirically determined value that would give the desired sorption pressure when the stopcock to the sorption cell was opened. At time zero, the penetrant was admitted into the sorption cell, and the spring extension was measured as a function of time. The system pressure was monitored during the experiment and was permitted to vary no more than ±5 mm Hg. System pressure was adjusted by either introducing more penetrant or by raising or lowering the heating tape temperature in the exterior lines. To begin a desorption experiment, the valve connecting the sorption cell to the vacuum line was opened, and the contraction of the spring was observed as a function of time.

BACKGROUND AND THEORY

Simple Nonreactive Fickian Transport

The transport of relatively noninteracting penetrants such as N_2, CO_2 and SO_2 in glassy polymers may be described using Fick's Law (Equation (1)) for transport:

$$N = -D(C) \frac{dC}{dx} \qquad (1)$$

where N is the one-dimensional diffusional flux in the x-direction with D(C) equal to the local diffusion coefficient of the penetrant which can be a function of local concentration, C. The normalized sorption or desorption kinetics for a slab of thickness 2ℓ can be expressed as the infinite series, shown below for the case where D is a constant, independent of concentration:

446

$$\frac{M_t}{M_\infty} = 4 \left[\frac{Dt}{\ell^2} \right]^{1/2} \left\{ \frac{1}{\pi^{1/2}} + 2 \sum_{n=1}^{\infty} (-1)^n \text{ ierfc} \left[\frac{n\ell}{2(Dt)^{1/2}} \right] \right\} \quad (2)$$

where M_t and M_∞ are the mass sorbed (or desorbed) at time t and at "infinite" time, respectively[5]. The value of the constant diffusion coefficient in Equation (2) can be evaluated easily using the well-known "half-time" relation (Equation (3)) which is derived by evaluating the time ($t_{1/2}$) corresponding to $M_t/M_\infty = 0.5$ using Equation (2).

$$D = \frac{0.0492 \ \ell^2}{t_{1/2}} \quad (3)$$

For simple Fickian behavior described by Equation (2), the normalized sorption and desorption responses are linear functions of $t^{1/2}$ up to at least values of $M_t/M_\infty = 0.5$ to 0.6[5]. In situations for which the diffusion coefficient is a function of local concentration (C), Equation (2) can still be used with reasonable effectiveness if one realizes that the diffusion coefficient appearing in that equation represents an average or "effective" coefficient typical of the concentration range being considered.

The gravimetric sorption/desorption kinetic runs in Figures 2 and 3 for SO_2 in Kapton® at 35°C show a simple Fickian transport process typical of a case in which D(C) increases with local concentration. Specifically, if the data in Figure 2 are plotted in the normalized form as M_t/M_∞ (shown in Figure 3), then the response is a linear function of $t^{1/2}$ up to values of M_t/M_∞ ranging from 0.5 to 0.6 for both the sorption and desorption runs. Furthermore, the sorption data lie above the desorption data indicating that the diffusion coefficient increases with increasing concentration. The lines drawn through the normalized sorption and desorption curves in Figure 3 were calculated from the standard infinite series expression (Equation (2)) for simple Fickian uptake. Effective average values for the diffusion coefficients, corresponding to 6.75×10^{-11} cm^2/sec and 4.37×10^{-11} cm^2/sec for the sorption and desorption runs, respectively, were used to obtain the fit shown. The departure from the simple Fickian model's prediction at longer times for the desorption run is presumably a consequence of the considerable concentration dependence of the SO_2 diffusion coefficient. The diffusivity approaches its "zero concentration" or infinite dilution limiting value in the last stages of desorption[3].

447

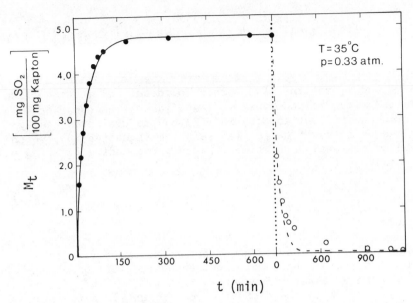

Figure 2. Sorption (●)/desorption (O) kinetic runs for a 0.5 mil Kapton® polyimide film at 35°C and at a SO_2 pressure of 0.33 atm.

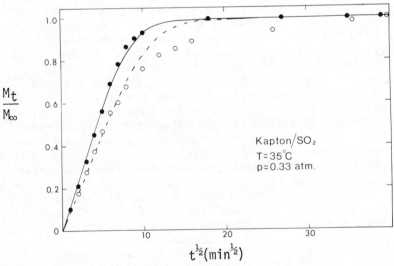

Figure 3. Normalized sorption (●)/desorption (O) kinetic runs for a 0.5 mil Kapton® film at 35°C and at a SO_2 pressure of 0.33 atm. Note that all of the SO_2 sorbed into the polymer film is removed during desorption.

As noted by Berens[6], the van der Waals volume, b, of a
penetrant serves as a useful measure of the size of the molecule.
The parameter, b, therefore, is a useful correlating tool for
"infinite dilution" diffusion coefficients of relatively
noninteracting penetrants in a given polymer, since it is more
difficult to move large molecules than small molecules through a
given environment. Strong interactions of the penetrant with the
polymer could reduce the infinite dilution diffusion coefficient
by imposing a large activation energy for execution of a typical
diffusional jump. A simple correlation for infinite dilution
diffusion coefficients, D_0, in terms of b is shown in Figure 4.
On the basis of this plot, we anticipated a diffusion coefficient
for ammonia in the range of 10^{-9} cm^2/sec. Such a diffusion
coefficient would yield an equilibration time of roughly 20
minutes for a 0.5 mil film. As will be discussed, radically
different behavior was observed with extremely protracted uptake.
Moreover, total desorption was not possible, leading to the
hypothesis of a chemical reaction. This hypothesis has been
verified by infrared spectroscopy work which will also be
described.

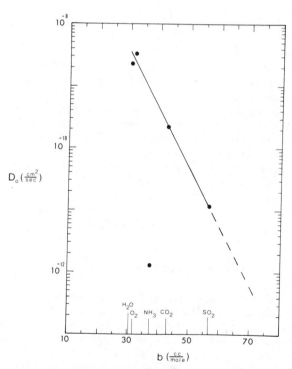

Figure 4. Correlation of diffusion coefficients at infinite
dilution for various penetrants in Kapton® at 30°C, using the van
der Waals volume, b, as the correlating parameter. Note that
both the H_2O and NH_3 points deviate from the correlation line for
the nonreactive gases.

449

Coupled Fickian Transport with Chemical Reaction

For cases in which a chemical reaction accompanies the diffusional invasion of the penetrant, the transient diffusion equation may be written as

$$\frac{\partial C}{\partial t} = \frac{\partial}{\partial x}\left[D(C)\frac{\partial C}{\partial x}\right] - r \tag{4}$$

where r is the local rate of consumption of the invading penetrant by the reaction when the local penetrant concentration is C. Although the assumption of first order kinetics simplifies the analysis of the response, the pseudo first order approximation may break down and second order kinetics must be considered, when a significant fraction of the functional groups in the polymer are being consumed by the reaction. For the case where psuedo first order kinetics provides a satisfactory description of the reaction phenomenona, the approach suggested by Danckwerts[7] can be used to model the combined diffusion and chemical reaction process. The mass uptake at any time, t, per unit cross sectional area is termed M_t and can be represented by Equation (5) in such a case:

$$M_t = \left[\frac{4C_o D}{\ell}\right]\left(\sum_{n=1}^{\infty}\frac{kt}{k+\alpha_n} + \sum_{n=1}^{\infty}\frac{\alpha_n}{(k+\alpha_n)^2}\{1-\exp[-t(k+\alpha_n)]\}\right) \tag{5}$$

where C_o is the equilibrium solubility of penetrant in the polymer in the absence of reaction, and k and D are the local first order reaction rate constant and diffusion coefficient of the penetrant, respectively. In cases where D is concentration dependent, an average value of the coefficient, averaged over the interval from zero to C_o, may be used approximately. The film thickness, ℓ, and D and k combine to yield the composite parameter $\alpha_n = \pi^2 D(2n+1)^2/4\ell^2$. Analysis of data for M_t vs t permits evaluation of D and k for the system[7].

RESULTS AND DISCUSSION

Water Vapor Sorption

Sorption and desorption kinetics for water vapor in 2.0 mil Kapton® at two different activities are shown in Figures 5 and 6. The low activity run is described well by the Fickian model, and the lines through the data points were calculated using average diffusion coefficient values of 2.86×10^{-9} cm^2/sec and 2.43×10^{-9} cm^2/sec for the sorption and desorption runs, respectively. The high activity run in Figure 6 indicates that only about 98% of

the originally sorbed water could be desorbed from the film even
after prolonged evacuation. The line through the sorption data in
Figure 6 was calculated using Equation (5) along with a value of D
= 3.609×10^{-9} cm^2/sec and k = 9.027×10^{-7} sec^{-1}. Since the
concentration in the film drops rapidly in the case of desorption,
relatively minor amounts of reaction should occur during penetrant
removal, so the simple nonreactive desorption model represented by
Equation (2) was used to calculate the line through the desorption
data. For the desorption data, a value of D = 2.70×10^{-9} cm^2/sec
was evaluated from the half-time formula (Equation (3)), and M_∞
was taken to be simply equal to the actual amount of penetrant
desorbed from the film. Desorption times extending for very long
periods (> 48 hours) failed to remove more of the sorbed water
than had been removed after the 400 minutes shown in Figure 6. The
effective values of the diffusion coefficients estimated from the
low and high activity runs appear to be of the same magnitude.
Therefore, while concentration dependence of the diffusion
coefficient is clear from the fact that the normalized
sorption/desorption responses do not superimpose in Figure 5, the
effect is not too large. Moreover, while the diffusion

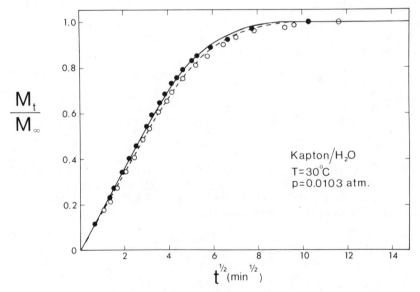

Figure 5. Low activity normalized sorption (●)/desorption (○)
kinetic runs for a 2 mil Kapton® film at 30°C and at a H$_2$0
pressure of 0.0103 atm. Note that all of the sorbed H$_2$0 is
removed during desorption.

coefficients evaluated from the low activity runs (Figure 5) are not rigorously characteristic of "zero concentration" conditions, it is clear that they do correspond to low sorbed concentration levels and their average value is reasonably consistent with the data in Figure 4 for relatively noninteracting penetrants. The fact that the diffusion coefficient for water lies somewhat below the correlation line may be due to the fact that hydrogen bonding interactions between water and imide groups tend to increase the activation energy for diffusion, thereby reducing the value of D.

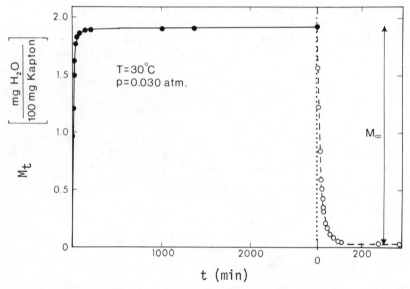

Figure 6. Sorption (●)/desorption (○) kinetic run for a 2 mil Kapton® film (high activity). Note that at desorption equilibrium (M_∞), 0.03 mg H_2O/100 mg of Kapton® remained in the film after protracted desorption under vacuum.

The small value of the reaction rate constant, k, determined from the Danckwerts analysis of the high activity run, explains why the reaction was undetectable in the low activity run. An increase in the temperature of measurement might increase the importance of the reaction significantly. These effects are currently under investigation.

Anhydrous Ammonia Sorption–0.5 mil Sample

The sorption and desorption runs shown in Figure 7 illustrates clearly that a considerably different response is observed for sorption of anhydrous ammonia in a 0.5 mil Kapton sample compared to water or SO_2. Since the van der Waals volume of ammonia is 37.1 cc/mole, one expects the effective diffusion coefficient for ammonia to be roughly 7×10^{-10} cm^2/sec on the basis of size alone. The sorption equilibration time in such a case would correspond roughly to 25 minutes (or 5 $min^{1/2}$). Clearly, the sorption process in Figure 7 is greatly protracted and is still occurring at 14,400 min (or 120 $min^{1/2}$). Furthermore, desorption for extended periods under high vacuum (approximately 12 days) is ineffective for the removal of roughly 50% of the originally sorbed penetrant.

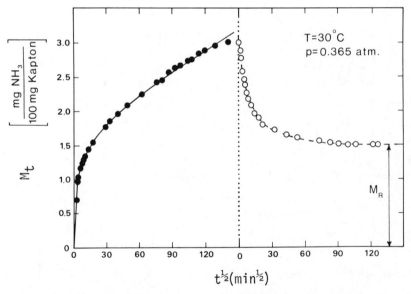

Figure 7. Sorption (●)/desorption (○) kinetic runs for a 0.5 mil Kapton® film at 30°C and at a NH_3 pressure of 0.365 atm. Note that at desorption equilibrium, 1.49 mg NH_3/100 mg of Kapton® (M_R) remained in the polymer after protracted desorption under vacuum.

As discussed earlier[4], the point of attack of ammonia appears to be the imide linkage, as illustrated below:

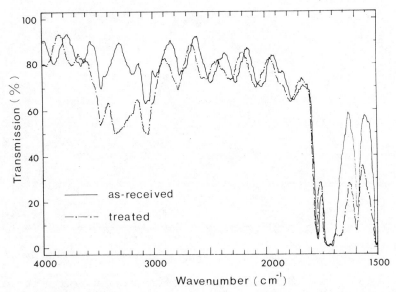

Based on the magnitude of the undesorbable component in Figure 7 (M_R), it appears that approximately 17% of the total imide groups in the sample have been attacked over the time scale of the experiment. A similar, but much less extensive reaction involving H_2O is presumably responsible for the small amount of undesorbable penetrant shown in Figure 6. In the case of water, only 0.31% of the imide groups were affected over the time scale of the experiment.

The infrared spectra shown in Figure 8 suggest that in the case of ammonia, the long term uptake leading to the undesorbable component is primarily due to random imide scission according to the above reaction. The presence of the increased absorbances in

Figure 8. Infrared spectra of the as-received and the NH_3 exposed Kapton® samples. Note the increased absorbance in both the 1630-1700 cm^{-1} and 3200-3500 cm^{-1} regions, consistent with the postulated reaction between NH_3 and the imide structures.

454

the range 1630-1700 cm^{-1} and 3200-3500 cm^{-1} are characteristic of primary and secondary amides, consistent with the above reaction. Similar infrared scans on the sample corresponding to Figure 6 revealed no significant difference from the as-received film, presumably since only 0.31% of the imide structures were affected by the water exposure.

A Danckwerts analysis of the sorption run in Figure 7 according to Equation (5) produced values for $D = 1.29x10^{-12}$ cm^2/sec and $k = 3.46x10^{-6}$ sec^{-1}. Again, one may assume that since the sorbed concentration drops rapidly during desorption, any continued reaction occurring during penetrant removal may be neglected to a first approximation. In this case, a value of $D = 2.35x10^{-11}$ cm^2/sec can be evaluated from the half-time formula for desorption (Equation (3)) using the actual amount of penetrant desorbed from the film for the value of M_∞.

Although the above two estimates of diffusion coefficients are not in very good agreement, it is clear that both estimates suggest that the mobility of NH_3 lies substantially below the level expected on the basis of simple size related factors. The extraordinarily low values of the NH_3 diffusion coefficient relative to water and the other "noninteracting" components in Figure 4 may arise from diffusional resistance caused by specific NH_3/polymer interactions. Such interactions may increase the effective activation energy for diffusion, thereby hampering diffusive motion.

In an attempt to resolve the reason for the large difference (a factor of 18) between the above estimates of the ammonia diffusion coefficient based on the Danckwerts analysis and the simple desorption analysis, an additional study of this system was undertaken using thin (0.3 mil) films for comparison with the 0.5 mil case.

Anhydrous Ammonia-0.3 mil Samples

As shown in Figure 9, replicate ammonia sorption runs on identical as received samples of 0.3 mil Kapton®films are highly reproducible. The desorption run for sample I in Figure 9 indicates clearly that considerable reaction has occurred even after 200 minutes for the 0.3 mil sample, since a substantial amount of ammonia (designated as $(M_R)_I$) is irreversibly retained in the film after prolonged evacuation times. By continuing the sorption experiment for 11,500 minutes with sample II and then desorbing, the irreversibly retained ammonia, $(M_R)_I$, is not increased proportionally to the reaction periods for sample I and II. It is clear, therefore, that the rate of reaction slows markedly as time progresses although only a small fraction (<14%)

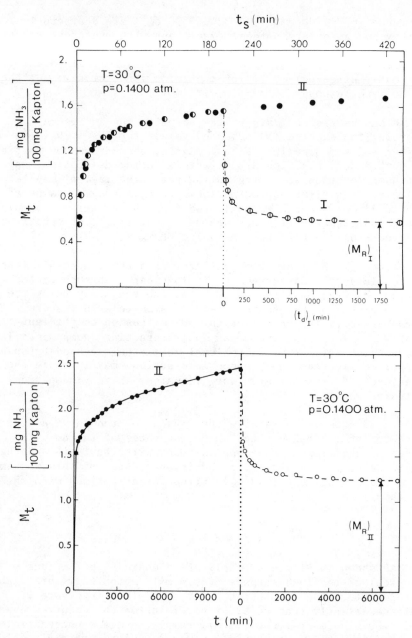

Figure 9. Replicate NH_3 sorption/desorption kinetic runs on identical as-received 0.3 mil Kapton® films (sample I - sorption (◑)/desorption (◐); sample II - sorption (●)/desorption (○)). Note that as sorption time (t_s) is increased from 200 min (I) to 11,500 min (II), the undesorbable NH_3 increases from 0.584 mg NH_3/100 mg of Kapton® (($M_R)_I$) to 1.234 mg NH_3/100 mg of Kapton® (($M_R)_{II}$), respectively.

of the originally present imide groups have been attacked even for
sample II.

If one performs a simple Danckwerts analysis of the short
time data for sample I & II ($\leq$200 min), values of $D = 2.9\times10^{-11}$
cm^2/sec and $k = 191\times10^{-6}$ sec^{-1} are obtained, in rather poor
agreement with the 0.5 mil Danckwerts' results at higher activity.
On the other hand, if one performs a Danckwerts analysis on the
long term run for sample II, values of $D = 0.48\times10^{-12}$ cm^2/sec and
$k = 3.2\times10^{-6}$ sec^{-1} are obtained which are in reasonable agreement
with the long term run at the higher activity for the 0.5 mil
sample. The k values are in very good agreement, and the somewhat
higher value of the diffusion coefficient at the higher activity
for the 0.5 mil sample may simply be a manifestation of the
increase in D(C) with concentration (or activity). As will be
discussed later, however, it is believed that the short term
values of D and k are more representative of physically meaningful
coefficients rather than the long term values derived for both the
0.5 and 0.3 mil samples.

From consideration of the "locked-in" concentrations shown in
the short time $((M_R)_I)$ and long time $((M_R)_{II})$ desorption plots for
the 0.3 mil film, one can show that approximately 6.6% of the
total imide groups have been attacked after 200 min, while 13.8%
of the total imide groups have been attacked after 11,500 minutes.
It appears that in both the 0.3 and 0.5 mil film the imide group
reactivity drops dramatically when close to 20% of the groups have
been reacted. Recall that 17% of the groups have reacted in the
0.5 mil film after close to 15,000 minutes (Figure 7), and the M_t
vs t plot had flattened out greatly at that point. The above
suggestion is reasonable in light of the fact that during the
curing process, involving closure of the imide rings, the solid
polyamic acid precursor film becomes a very rigid material whose
effective Tg rises to 400°C, which is well above the 150°C curing
temperatures. In such a case, the final curing process is likely
to introduce a considerable amount of unrelaxed stress in the
glassy polymer and thereby "labilize" a selected fraction of the
imide groups. Exposure of the polymer to an agent such as NH_3
then provides a means of stress relaxation via selective imide
scission. It has been shown earlier that reheating the sample to
120-150°C under vacuum causes reimidization as proven by IR
spectroscopy[4].

Kapton®, therefore, has an interesting ability to combine a
good barrier property with a limited, but nontrivial, capacity for
scavenging of undesirable basic pollutants or residuals which
might otherwise damage the component being covered by the film.
The change in properties such as modulus and Tg resulting from the
degradation of 5-20% of the imide groups is likely to be rather
modest since the main polymer chain stays intact. This topic is

currently under study and will be reported in a later paper.

The disagreement between the values of D estimated from desorption using Equation (1) and from the standard Danckwerts analysis for the long term sorption runs in the 0.3 and 0.5 mil samples may arise from a breakdown in the accuracy of the assumed pseudo first order kinetics. If the initial concentration of "labilized" groups, C_{Io}, amounts to only 20% of the total imide groups present, the pseudo first order expression in Equation (4) could begin to fail seriously when ammonolysis reaches 5% of the total groups (i.e., 25% of the active groups). The correct form for the reaction expression in such a case would be given by Equation (6):

$$r = k^* C_I C \qquad\qquad (6)$$

where k^* is a true intrinsic rate constant for reaction of a stress activated imide, C is the local concentration of ammonia, and C_I is the time and position dependent concentration of stress activated imide groups in the polymer. Interestingly, the instantaneous values of C_I in a reacting system does not necessarily equal the difference between the initial number of stress activated groups (C_{Io}) and the number of reacted groups, since stress relaxation accompanying the ammonolysis may aid in deactivating some of the groups that originally would be candidates for attack. Conversely, if the film is exposed to a macroscopic external stress, additional imides may be activated compared to the unstressed film. The situation is, therefore, potentially very complex and the challenge is to simplify it to the point where useful approximate calculations are possible, at least for an unstressed film. Note from Table I that in the limit as t→0, the D's estimated from the Danckwerts' analysis and those from the desorption half time analysis become similar (200 min. run), thereby suggesting that the diffusion coefficients estimated by the two methods ($3-4 \times 10^{-11}$ cm^2/sec) are physically reasonable and consistent. It appears that the effect noted earlier (with respect to an activation energy increase due to NH_3/polymer interaction) still causes the diffusion coefficient for ammonia to fall considerably below the corrresponding line of the noninteracting penetrants in Figure 4.

The above conclusions are reasonable since it is clear that by considering only short time cases (such as in Figure 9) using the Danckwerts analysis, the psuedo first order reaction assumptions become valid. Evaluation of $k^* C_I$ for a series of short run times, followed by extrapolation of the values back to t=0 will then permit estimation of $k^* C_{Io}$, characteristic of the inherent as-received film. The value of C_{Io} initially can be estimated independently by performing a series of long term measurements and noting the asymptotic limit to which the undesorbable component appears to be approaching. On the basis of the limited measurements performed here, it appears that this number corresponds to roughly 20% of the originally present imide links at 30°C. This corresponds to $C_{IO} = 1.5 \times 10^{-3}$ mole activated imides /cc. If one assumes that the value of $k = k^* C_I$ determined from the 200 minute Danckwerts' run is reasonably representative of the initial value, then the value of the intrinsic rate constant, k^*, can be calculated to be 0.13 cc/mole sec. Clearly, using these parameters (or more carefully determined ones as outlined above), the complete M_t vs t response can be derived by straightforward numerical methods for any film thickness or external vapor activity of interest. Similarly, by evaluation of the temperature dependence of D and k^* (and C_{IO} if necessary), the analysis can be extended to cover a wide range of temperatures as well. These numerical calculations are underway and will be described in a future paper.

Table I: Kinetic Parameters $D_{Danckwerts}$, k and $D_{Desorption}$ for H_2O and NH_3 Sorption in Kapton® Polyimide at 30°C.

Penetrant	Film Thickness (mils)	$t_{SORPTION}$ (min)	$D_{DANCKWERTS}$ (cm^2/sec)	k (sec^{-1})	$D_{DESORPTION}$ (cm^2/sec)
H_2O	2.0	2750	3.6×10^{-9}	9.03×10^{-7}	2.7×10^{-9}
NH_3	0.5	14,400	1.29×10^{-12}	3.46×10^{-6}	2.35×10^{-11}
	0.3	200	2.9×10^{-11}	191×10^{-6}	3.97×10^{-11}
		11,500	0.48×10^{-12}	3.2×10^{-6}	1.25×10^{-11}

CONCLUSIONS

Kapton® polyimide displays an interesting range of transport behaviors at 30°C ranging from simple Fickian diffusion to coupled diffusion with chemical reaction. The relatively noninteracting penetrants, SO_2, CO_2 and O_2 exhibit diffusion coefficients whose magnitudes can be correlated well on the basis of the relative molecular sizes of the penetrants. In the case of water, some evidence for physical (hydrogen bonding) and weak chemical interactions were observed. These interactions appeared to reduce the effective diffusion coefficient slightly for water in the film compared to the coefficient expected strictly on the basis of size alone. Polymer/penetrant interactions can, in principle, reduce the effective diffusion coefficient by increasing the average activation energy required for the penetrant to execute a diffusional jump. Based on the Eyring theory of rate processes, an increase in activation energy will tend to reduce the diffusion coefficient.

In the case of ammonia sorption into Kapton®, substantial chemical interactions are apparent. Infrared spectroscopy and gravimetric sorption/desorption measurements were effective for monitoring the ammonolysis reaction. The point of ammonia attack appears to be selected imide groups which comprise a relatively small fraction (~20%) of the total imide group population present in the initially cured film. It is suggested that a fraction of the imide groups become "activated" or "labilized" during the final solid state curing step in which unrelieved stresses may be introduced. The ammonolysis reaction serves to relieve these stresses and thereby deactivate the remaining imide groups substantially. It appears, therefore, that Kapton® has a useful ability to combine good barrier properties (low D's (Figure 1)) with an additional nontrivial capacity for scavenging undesirable basic pollutants or other residuals that might otherwise damage the component being protected. A complementary approach might be used to protect against the invasion of an acidic component (e.g., H_2S) by using a film with a stress-activated acid sensitive group.

ACKNOWLEDGEMENT

The authors gratefully acknowledge the financial support of this project by the Army Research Office under Grant Number DAA29-81K-0039.

REFERENCES

1. W. J. Koros, C. J. Patton, R. M. Felder and S. J. Fincher, J. Polym. Sci.: Phys. Ed., 18, 14 (1980).
2. W. J. Koros, J. Wang and R. M. Felder, J. Appl. Polym. Sci., 19, 2805 (1981).
3. R. M. Felder, C. J. Patton and W. J. Koros, J. Polym. Sci.: Phys. Ed., 19, 1895 (1981).
4. L. R. Iler, R. C. Laundon and W. J. Koros, J. Appl. Polym. Sci. 27, 1163 (1982).
5. J. Crank, "The Mathematics of Diffusion", Second Edition, Clarendon Press, Oxford, 1975.
6. A. R. Berens, J. Vinyl Technol., 1(1), 8 (1979).
7. P. V. Danckwerts, Trans. Faraday Soc., 47, 1014 (1951).

X-RAY SCATTERING MEASUREMENTS DEMONSTRATING

IN-PLANE ANISOTROPY IN KAPTON POLYIMIDE FILMS

R.F. Boehme and G.S. Cargill III

IBM T.J. Watson Research Center

Yorktown Heights, New York 10598

X-ray scattering measurements were made on Kapton®[1] films with thicknesses of 8, 25 and 50μm. These measurements indicate that the atomic scale structure is basically the same for the three Kapton films. However, the measurements reveal in-plane anisotropy of polymer chain alignment with preferred orientation parallel to an optical extinction direction. Anisotropy of x-ray scattering and therefore of polymer chain alignment within the polyimide film plane is greatest for the thinnest film, 8μm, and least for the thickest film, 50μm. The degree of anisotropy in tensile strength shows a similar dependence on film thickness. These appear to be the first x-ray scattering measurements demonstrating in-plane anisotropy for Kapton films. Furthermore, our measurements agree with previous measurements which have shown these polymer chains align in-plane rather than out-of-plane.

X-ray scattering measurements were made in a transmission geometry. Sharp peaks occur in the scattering patterns at $2\theta=2.5°$ and $2\theta=5.0°$ (Mo Kα radiation $\lambda=0.711$Å). A repeat period of d=16Å is derived by simply using the Bragg equation, $n\lambda=2d\sin(\theta)$. Since this is only slightly smaller than the monomer length, 18Å, we associate these peaks with the intrachain repeat distance. A broad peak occurs centered at $2\theta=9°$ corresponding to a repeat period, $d\approx4.8Å$, which can be associated with the monomer mean diameter of 5.6Å, and therefore with the interchain spacing between adjacent parallel polymer chains.

A variation in scattered intensity of the intrachain and interchain peaks occurs with rotation about the normal to the film plane, with maxima and minima occurring every 90°. Maxima in the intrachain peak scattering coincide with minima in the interchain peak scattering. It is this variation in scattering intensity which is associated with the in-plane anisotropy in the polymer chain alignment.

INTRODUCTION

Polyimide films are of particular technological interest because of their useful physical properties. These films are stable to elevated temperatures, have very good dielectric properties, are resistant to organic solvents, are strong, and have good dimensional stability.[2] The Kapton monomer unit of pyromellitic acid (PM), Figure 1, has a mean molecular diameter, determined from crystallographic data, of 5.6Å, and a spacing L of rigid units in the polymer chains of 18.0Å.[3] Few studies of atomic scale structure, or attempts to relate atomic structure and physical properties have been made. Ikeda[4] and Isoda, et al.,[5] observed the wide angle x-ray scattering for polyimide films to be isotropic within the film plane, and anisotropic out-of-plane. Ikeda[4] proposed that the observed mechanical properties for polyimide films result from a uniplanar structure of the film sample with "diffracted planes" parallel to the film surface, treating the films as isotropic within the film plane. The size and concentration of voids in polyimide films have been studied by Russell.[6] This work reports in-plane anisotropy of x-ray scattering in commercially available polyimide films and correlates this anisotropy with in-plane anisotropy of polyimide film mechanical properties.[7]

EXPERIMENTAL DETAILS

X-ray scattering measurements were made using a Rigaku small angle scattering goniometer and rotating anode x-ray generator with

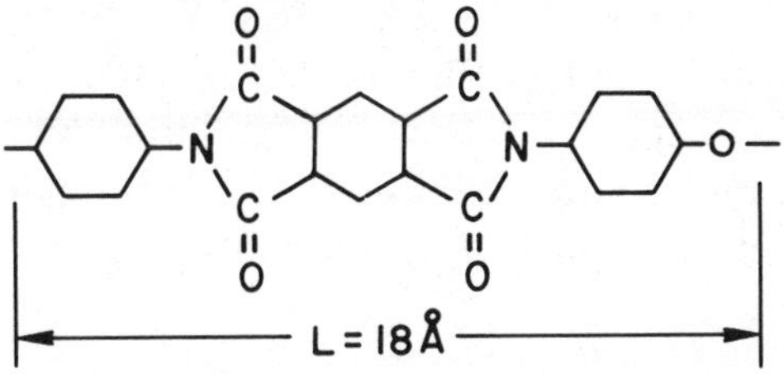

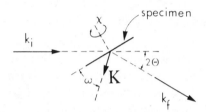

Figure 1. Pyromellitic acid monomer unit (PM) for Kapton.

$L = 18\text{Å}$

specimen

k_i

χ

2Θ

ω $\mathbf{K}$

k_f

Figure 2. Scattering geometry. $\mathbf{k_i}$ is the incident x-ray beam, $\mathbf{k_f}$ is the diffracted x-ray beam, $\mathbf{K}=\mathbf{k_f}-\mathbf{k_i}$ is the scattering vector, $K=4\pi\sin(\Theta)/\lambda$, 2Θ is the scattering angle into the detector, λ is the x-ray wavelength, ω axis rotates film specimen away from the scattering vector, χ axis rotates film specimen about its normal.

slits and line focus. Zirconium filtered molybdenum Kα radiation, $\lambda=0.71069\text{Å}$, or nickel filtered copper Kα radiation, $\lambda=1.54178\text{Å}$, were used with the generator at 50 kV and 60 mA. Scattered intensities were measured with a scintillation detector using pulse height analysis and counting techniques. The experiment was controlled by an IBM System/7 computer.

The scattering configuration is defined by three angles, Θ, ω, and χ, as shown in projection in Figure 2. $\mathbf{k_i}$ denotes the incident beam direction; $\mathbf{k_f}$, the direction in which the scattered intensity is measured. The magnitudes of these vectors are defined as $2\pi/\lambda$, where λ is the x-ray wavelength. $\mathbf{K}=\mathbf{k_f}-\mathbf{k_i}$ is the scattering vector, with magnitude $K=4\pi\sin(\Theta)/\lambda$, where 2Θ is the scattering angle. The angles ω and χ describe two independent rotations of the specimen. The ω-axis is normal to the scattering plane, defined by the incident and scattered beams, and in transmission geometry $\omega=0$ when the scattering vector, $\mathbf{K}$, is in the film plane. Except where noted, all experiments were carried out in the transmission geometry. The χ-axis is normal to the specimen surface, and $\chi=0$ may be arbitrarily chosen. Both the ω and χ axes pass through the intersection of the incident beam, the diffracted beam, and the specimen. If the specimen were macroscopically isotropic, the scattered intensity would be independent of the specimen orientation apart from simple absorption and irradiated volume factors.

Scattered intensity measurements were corrected for background (parasitic) scattering, for specimen thickness, and for angle-dependent absorption and irradiated volume factors. No correction was made for the different relative scattering strengths of molybdenum and copper radiations; therefore, intensities can not be compared in measurements using the different radiation sources.

Specimens for x-ray scattering measurements were prepared by cutting the Kapton films into 30mm x 50mm pieces and making stacks of 45 layers of 8μm film, 42 layers of 25μm film, and 20 layers of 50μm film, with care to maintain relative orientations of layers within each specimen. However, subsequent examination of these specimens revealed that one film of the 8μm specimen, one film of the 25μm specimen, and five films of the 50μm specimen had been reversed, back-to-front, in making the stacks. This complicates quantitative interpretation of the χ-dependent scattering by these specimens. A correction for these mis-orientations has been applied where necessary.

Optical examination of single layers of the Kapton films between crossed polarizers indicated that there were four film orientations, separated by $90°$ rotations, for which light normal to the film surface was extinguished. No variation in transmitted light intensity was seen when only one polarizer was used. This indicates that the Kapton films are birefringent, as previously reported by Argon and Bessonov.[3] This birefringence presumably results from preferential alignment of molecular units along a particular in-plane direction during the Kapton's manufacture or subsequent processing. A pattern of fine, bright or dark lines was also visible when the films were examined between crossed polarizers and were oriented to extinguish transmitted light. These lines are straight and parallel, are oriented along the roll direction in a roll of Kapton film, and are therefore referred to as "rolling marks". $\chi=0$ has been defined for these "rolling marks" being in the plane of diffraction. The four angles at which polarized light is extinguished are oriented with the "rolling marks" 60°, 150°, 240° and, 330° from parallel with the plane of polarization of one of the polarizers.

EXPERIMENTAL RESULTS

Transmission scattering measurements with $\omega=0°$, i.e. with $\mathbf{K}$ in the plane of the specimen, give patterns like those shown in Figure 3 for the 50μm specimen. Three components of this pattern are (1) the sharp (intrachain) peak at $K=0.4\text{Å}^{-1}$ with its much weaker second order at $K=0.8\text{Å}^{-1}$; (2) the broader (interchain) peaks extending from $K=1\text{Å}^{-1}$ to $K=2\text{Å}^{-1}$; and (3) the small angle scattering which decreases monotonically for $0.04\text{Å}^{-1}<K<0.2\text{Å}^{-1}$ for both of the thicker film specimens.

Intrachain Peaks

The sharpest features in the scattering patterns are the peaks at $K=0.4\text{Å}^{-1}$ and $K=0.8\text{Å}^{-1}$. Simply using the Bragg equation, $\lambda=2d\sin(\Theta)$, or equivalently $d=2\pi/K$, to associate a spatial repeat period d with the $K=0.4\text{Å}^{-1}$ peak yields $d\approx16\text{Å}$. Since this is only slightly smaller than the expected monomer length 18Å, these peaks are most reasonably attributed to the periodically repeating structure within the polymer chains, either individually or in coherent, crystalline bundles. These are therefore described as *intrachain* interference peaks. These results agree with those of Ikeda[4] and of Isoda, et al.,[5] except that in-plane anisotropy in the x-ray scattering was observed in our study, whereas in-plane x-ray scattering in the previous two studies was observed to be isotropic. The strength of these peaks is maximized for $\omega=0°$, i.e. for the diffraction vector $\mathbf{K}$ within the film plane, and this strength further depends on the angle χ, as shown in Figures 4 and 5. Note that the data shown for the 8μm and 50μm specimens were obtained with molybdenum Kα radiation, but that the data shown for the 25μm specimen were obtained with copper Kα radiation. The angular dependences shown in Figures 4 and 5 are consistent with the polymer chains being preferentially aligned at an angle of about 60° to the "rolling marks". The polymer chains are probably aligned along the in-plane optical axis. There is, however,

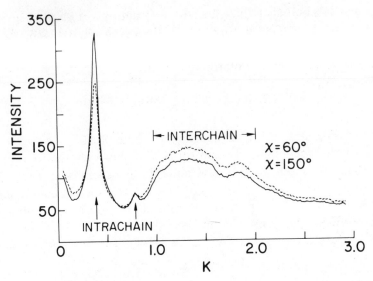

Figure 3. X-ray scattering within Kapton polyimide film plane for 50μm specimen. Peaks at K=0.4Å^{-1} and K=0.8Å^{-1} correspond to intrachain repeat distance along polymer chain of 16Å. Peaks from K=1Å^{-1} to K=2Å^{-1} correspond to interchain repeat distances of approximately 4.8Å. χ=60° corresponds to maximum intensity in the K=0.4Å^{-1} peak, and χ=150° to the minimum intensity. Mo Kα radiation, λ=0.711Å.

an ambiguity of 90° in our determination of the optical axis direction using simple crossed polarizers.

The full-width-at-half-maximum, ΔK, of the K=0.4Å^{-1} intrachain peak provides a measure of the average effective number of aligned repeat periods within chains. The data measured with copper Kα radiation are the data least affected by slit resolution smearing effects. Using the Scherrer equation with ΔK=0.06Å^{-1} for the K=0.4Å^{-1} peak, Figure 6, corresponds to about seven monomer repeat periods,

K/ΔK$\approx$7, or $\approx$112Å. This is in agreement with values reported by Isoda, et al.,[5] of 100Å to 125Å.

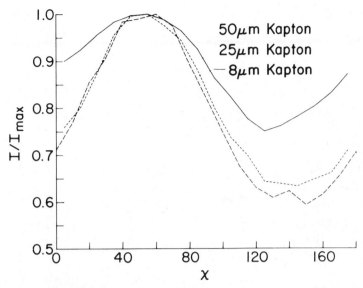

Figure 4. Normalized x-ray scattering intensity within Kapton polyimide film plane. Roll direction is at $\chi=0°$. High intensity
corresponds to preferred polymer chain alignment direction.
$K=0.4\mathring{A}^{-1}$ peak, $\omega=0°$. 8μm and 50μm films were measured
with Mo Kα radiation, $\lambda=0.711\mathring{A}$. 25μm films were measured
with Cu Kα radiation, $\lambda=1.54\mathring{A}$.

The strength of the χ-dependence of the intrachain peak intensity
provides a measure for the degree of polymer chain alignment along
the preferred in-plane direction.   This χ-dependence, shown in
Figure 4, has been characterized for each of the film thicknesses as
the ratio of the minimum and maximum intensities in χ-scans with
$K=0.4\mathring{A}^{-1}$ and $\omega=0°$. The values are $I_{min}/I_{max}=0.59$ for the 8μm
film, 0.63 for the 25μm film, and 0.75 for the 50μm film specimen.
However, correction for the five reversed layers of the 50μm film
specimen reduces this value of I_{min}/I_{max} to 0.68. Corrections for
the single reversed layers in the 8μm and 25μm specimens are much
smaller, yielding I_{min}/I_{max} of 0.58 and 0.62 respectively. These data
therefore suggest that the degree of in-plane alignment decreases
slightly with increasing film thickness.

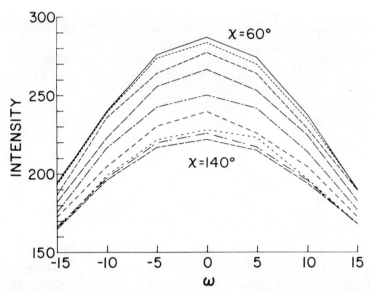

Figure 5. X-ray scattering within 8μm Kapton polyimide film plane, $\chi=60°$ to $\chi=140°$, and out-of-plane ,$-15°<\omega<15°$. Measurement shows polymer chains align within the film plane, and preferentially at an angle of approximately 60° to the roll direction. $K=0.4\text{Å}^{-1}$ peak, Mo Kα radiation, $\lambda=0.711\text{Å}$.

Interchain Peaks

The broader peak centered at $K=1.3\text{Å}^{-1}$ corresponds, with the simple Bragg equation, to a spatial repeat period $d\approx4.8\text{Å}$. Attributing this peak to interchain interference, i.e. to the spacing between adjacent parallel chains, is supported by two facts. First, this period is near the 5.6Å molecular diameter quoted by Argon and Bessonov[3] from crystallographic data. Second, the intensity of this peak for $\omega=0°$ is maximized at χ values for which the *intrachain* peak is minimized, Figures 3 and 6. Furthermore, the intensity of this peak is independent of ω in the range of measurement, $-15°$ to $+15°$.

In a "reflection geometry" scattering measurement, for which the scattering vector **K** is kept perpendicular to the specimen surface, at

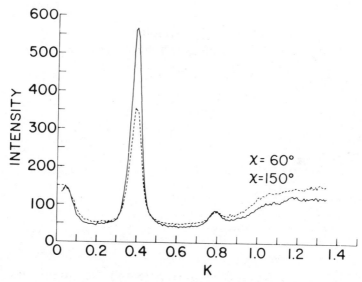

Figure 6.   X-ray scattering within Kapton polyimide film plane for
25μm specimen. Cu Kα radiation, λ=1.54Å.   Scherrer equation, L=K/ΔK, for the K=0.4Å^{-1} peak shows average coherent
domain in the Kapton polyimide chain direction to be approximately 100Å.

$\omega=0°$, only the interchain peak is seen; the intrachain peak is absent.
Although the above description is in terms of "an interchain peak,"
more careful examination of the scattered intensity in this 2Θ angular
range (see Figure 3) shows that the peak actually consists of a broad
maximum at K=1.3Å^{-1}, a weaker, overlapping maximum at
K=1.9Å^{-1}, and an unresolved shoulder at K=1.1Å^{-1}.   This structure
may result from there being several distinct chain-chain spacings
depending upon the mutual orientations of the particular chains.

Small Angle Scattering (SAS)

The present scattering measurements using Mo Kα radiation
extend down to K=0.04Å^{-1}.   In the angular region K<0.25Å^{-1},

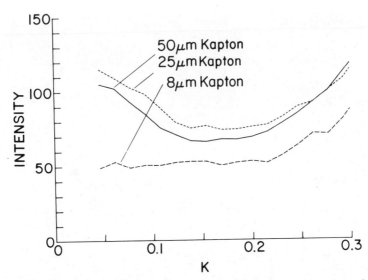

Figure 7. Small angle x-ray scattering (SAS) for $50\mu m$ and $25\mu m$ Kapton polyimide films indicates microstructural inhomogeneities. Mo Kα radiation, $\lambda=0.711\text{Å}$.

results for the $8\mu m$ specimen differ qualitatively from those for the two other, thicker specimens. As shown in Figure 7, the $8\mu m$ specimen has an almost flat intensity $I(K)\approx50$ in this region for **K** parallel to the chain alignment direction. The intensity level increases to $I(K)\approx65$ for **K** perpendicular to the chain alignment direction. For the $25\mu m$ specimen, the SAS intensity increases with decreasing K for $K<0.15\text{Å}^{-1}$, rising from $I(K=0.15\text{Å}^{-1})\approx70$ to $I(K=0.03\text{Å}^{-1})\approx120$. For the $50\mu m$ specimen, the SAS is very similar to that for the $25\mu m$ specimen, but with a weak (10%) dependence on χ. These measurements have not been corrected for slit length smearing.

The small angle x-ray scattering for the $50\mu m$ and $25\mu m$ specimens is essentially that described by Isoda, et al.,[5] as arising from a morphological two-phase system having periodic electron density functions and phase boundaries less than 10Å wide. The $8\mu m$ speci-

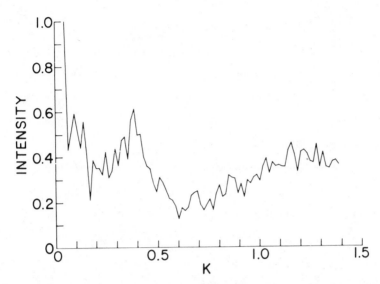

Figure 8.   X-ray scattering within single layer of polyimide film cast
on silicon substrate. General scattering pattern is the same as
for the three Kapton films. Mo Kα radiation, λ=0.711Å.

men looks more like that which they describe as arising from an
amorphous structure.

Russell[6] has observed small angle x-ray scattering in the region
$K<0.04$ Å^{-1} and attributes it to voids ranging in size from 50 to 150Å
with a volume fraction $<7x10^{-4}$

We attempted to make x-ray scattering measurements using a
single-layer specimen of 30μm thick polyimide film spun onto a
0.4mm thick silicon substrate (Figure 8). Scattered intensities, corrected for absorption in the silicon substrate, were a factor of about
50 weaker than for the multilayer Kapton specimens, i.e., about the
same as the ratio of the specimen thicknesses. Although the statistics
for the spun-on film data were poor, the 2Θ dependence of the scattered intensity was found to be similar to that for the Kapton specimens, with a sharp intrachain peak and a broader interchain peak. We

471

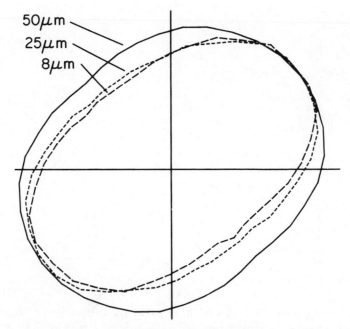

Figure 9. Polar plot of normalized x-ray scattering intensities within Kapton polyimide film plane. Roll direction is vertical. High intensity corresponds to preferred chain alignment direction. $K=0.4\overset{\circ}{A}^{-1}$, $\omega=0°$.

did not investigate the ω and χ dependence of the scattered intensity for the spun-on film specimen.

DISCUSSION AND CONCLUSIONS

Our x-ray scattering measurements demonstrate that the atomic scale structure for the three Kapton films studied is basically the same. These measurements confirmed the in-plane/out-of-plane anisotropy in polymer chain alignment previously reported by Ikeda,[4] and by Isoda, et al..[5] However, our measurements also showed that polymer chains were preferentially aligned parallel to a particular in-plane

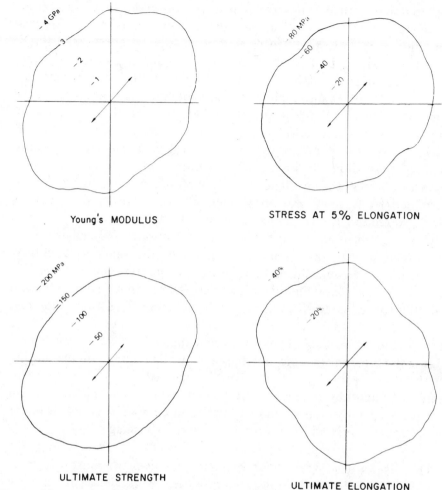

Young's MODULUS

STRESS AT 5% ELONGATION

ULTIMATE STRENGTH

ULTIMATE ELONGATION

Figure 10.   Polar plots of physical properties within polyimide film
plane.   Roll direction is vertical.   (From Reference 7 for 75μm
Kapton films.)

direction.   This was not apparent in the previously reported x-ray
scattering studies, although optical in-plane birefringence[3] and in-
plane anisotropy of mechanical properties[7] for Kapton films were well
documented.

Polar plots in Figures 9 and 10 show the in-plane anisotropy
observed by Blumentritt[7] for several mechanical properties of Kapton

films, together with the in-plane anisotropy observed in x-ray scattering measurements on our Kapton films. Directions of preferential chain alignment correspond to directions of highest normalized intensity in the polar plot. For all plots, the vertical axis is the roll direction. Larger degrees of in-plane anisotropy for the thinner films are apparent in Figure 9, although correction for the reversed layers in the 50μm specimen brings its anisotropy closer to that of the other two, thinner films. These corrected anisotropies are listed in Table I, together with anisotropy ratios for several mechanical properties, including those shown in Figure 10. Examination of Figures 9 and 10 demonstrates that the in-plane direction of polymer chain alignment, determined from x-ray scattering measurements, corresponds closely with the direction for maximum ultimate strength, maximum Young's modulus, and maximum stress for 5% elongation. This chain alignment direction also corresponds to a minimum in ultimate elongation, hygroscopic expansion, linear thermal expansion and long term dimensional shrinkage. Examination of Table I shows that several physical properties become more anisotropic with decreasing film thickness from 125μm to 75μm to 25μm (stress at 5% elongation, ultimate strength, and hygroscopic expansion) as expected following our observations of more anisotropic polymer chain alignments for decreasing film thickness from 50μm to 25μm to 8μm. However, anisotropy of ultimate elongation decreased with decreasing film thickness, and other properties showed no monatonic trend (Young's modulus, linear thermal expansion, and long term dimensional shrinkage).

The results described above indicate that in-plane anisotropy can have important effects on physical properties of polyimide films. We have shown that x-ray scattering measurements can be used to characterize this type of structural anisotropy in Kapton polyimide films. Such measurements should be useful in understanding the structural effects of manufacturing and processing variables and the mechanisms by which these variables affect other physical properties.

We began this work on polyimide films as a result of discussions with R. Schaefer and O.C. DeHodgins. We gratefully acknowledge their interest and helpful suggestions.

Table I. In-plane Anisotropy of X-ray Scattering and Physical Properties in Polyimide Films. Note: Anisotropy of physical properties (from Reference 7) is tabulated as value at 135° from machine direction divided by value at 45° from machine direction.

Film Thickness (μm)	8	25	50	75	125
$I_{min}(\chi=150°)/I_{max}(\chi=60°)$, $K=0.4\text{Å}^{-1}$, $\omega=0°$	0.58	0.62	0.68	-	-
Stress at 5% Elongation	-	0.72	-	0.76	0.8
Ultimate Strength	-	0.70	-	0.75	0.8
Ultimate Elongation	-	1.30	-	1.35	1.4
Coefficient of Hygroscopic Expansion- 23° C, 8 to 80% relative humidity	2.61	-	2.01	1.4
Young's Modulus	-	0.77	-	0.73	0.7
Coefficient of Linear Thermal Expansion, 25° to 50° C.	-	1.50	-	1.64	1.2
Long Term Dimensional Stability, 168 hours at 60° C, % shrinkage	-	1.24	-	1.18	1.1

REFERENCES

1. Kapton is a registered trademark of E.I. duPont de Nemours & Company.

2. N.A. Adrova, M.I. Bessonov, L.A. Laius, and A.P. Rudakov, "Polyimides, a New Class of Thermally Stable Polymers," Technomic Pub. Co., Stamford, CT, 1970.

3. A.S. Argon and M.I. Bessonov, "Plastic Deformation in Polyimides, with New Implications on the Theory of Plastic Deformation of Glassy Polymers," Phil. Mag., 35, 917 (1977).

4. R.M. Ikeda, "A Mechanical Effect of Orientation," Polymer Lett., 4, 353 (1966).

5. S. Isoda, H. Shimida, M. Kochi, and H. Kambe, "Molecular Aggregation of Solid Aromatic Polymers. I. Small-Angle X-Ray Scattering from Aromatic Polyimide Film," J. Polymer Sci.: Polymer Phys. Ed., 19, 1293 (1981).

6. T.P. Russell, "Concerning Voids in Polyimide," to be published.

7. B.F. Blumentritt, "Anisotropy and Dimensional Stability of Polyimide Films," Polym. Eng. Sci., 18, 1216 (1978).

THERMALLY STIMULATED DISCHARGE CURRENTS FROM CORONA CHARGED

POLYPYROMELLITIMIDE FILM

J. K. Quamara, P. K. C. Pillai* and B. L. Sharma

Department of Applied Physics
Regional Engineering College
Kurukshetra-132119, India

Thermally stimulated discharge current (TSDC) technique has been applied to investigate the polarization mechanism due to corona discharge in as-received and heat-treated Kapton® samples. The samples were charged on the upper free surface (non-metallized surface) by exposure to a negative corona discharge in air. Three peaks (termed as P_1, P_2 and P_3) were observed in the TSDC spectra. The peaks P_1 and P_2 appear at 46° and 60°C, respectively and are in the normal current direction opposite to that of charging current. The heat treatment causes a decrease in the peak current which may be due to a reduction in the number of traps due to high temperature aging. The peak P_3 appears at 249°C in as-received Kapton samples and is in the normal current direction. For heat-treated samples this peak appears around 225°C and is in the abnormal current direction. For iodine doped samples the peaks P_1 and P_2 appear at 38° and 50°C, respectively and are in the normal current direction. The peak current is nearly 10 times that in as-received samples which may be due to an increase in the carrier mobility because of"handing-on"of carriers as a result of diffusion of I_2 ions in the polymer matrix. The peaks P_1, P_2 have been attributed to shallow trap levels and peak P_3 to deep trap levels present in Kapton .

*Department of Physics, Indian Institute of Technology, New Delhi-110016, India.

INTRODUCTION

The corona discharge method for obtaining polarization in polymers is now widely being used[1,2,3]. Sawa et al.[4] have shown that the charging current during corona charging is much greater than that obtained by applying the same potential to a metal-polymer-metal system. The study of corona charged polymer electrets is also helpful from the electrophotographic point of view. Kapton (Poly-4,4' oxydiphenylene pyromellitimide) whose thermo and photo-electret properties have already been studied by the authors[5,6] and which shows good photoconduction[7,8] is potentially useful in the electrophotographic process. The study of the various traps present in Kapton polyimide is considered useful for an understanding of the conduction and dielectric properties of this polymer. Thermally stimulated discharge current technique, which is mainly used for the study of electret state in polymers, has been applied to study the corona charged Kapton electrets in the present communication.

EXPERIMENTAL

Kapton samples 2 cm x 2 cm in size were cut from a 3 mil thick film which was vacuum aluminized on one side. Heat treated samples were prepared by heating the as-received film at 150° and 300°C, respectively for 2 hours. Iodine doping was done by keeping the as-received film in an atmosphere saturated with iodine vapor for 20 hours at 100°C.

The Kapton samples were charged on their upper free surface (non-metallized surface) by exposure to a negative corona discharge for 5 seconds in air at atmospheric pressure and room temperature (Figure 1). The humidity was 60%. After charging, the sample was short circuited for 5 minutes to remove frictional and stray surface charges. For TSDC studies, the layout of the apparatus was the same as used by Von Seggern[3]. The samples were depolarized at a linear heating rate of $0.033K \ sec^{-1}$. The constant heating rate was achieved with the help of an automatic temperature controller. A Keithley Model 610C electrometer was used to measure the discharge current. The polarizing charge was calculated by measuring area under the TSDC curve. Activation energy values were calculated by the initial rise method of Garlick and Gibson[9].

RESULTS

Typical TSDC spectra for the as-received corona charged and heat-treated corona charged samples are shown in Figure 2. In all three cases there appear two peaks (termed P_1 and P_2) at temperature 46 ± 1°C and 60 ± 1°C, respectively. These peaks are in the normal current direction, opposite to that of the charging current. The magnitude of the peak current decreases from as-received sample

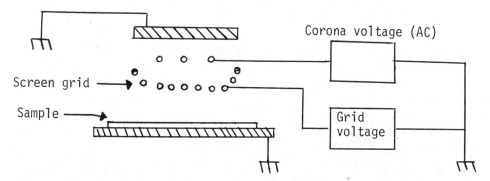

Corona voltage (AC)

Screen grid ⟶

Sample ⟶

Grid
voltage

Figure 1. Schematic diagram of the apparatus for corona charging
of Kapton film.

to heat-treated sample. In the region 100° to 180°C the discharge
currents are very small. A third peak (termed P_3) also appears in
the high temperature region. For as-received sample this P_3 peak
appears at 249°C and is in normal current direction. For heat-
treated samples this peak occurs at 225 ± 1°C and is in the abnormal current direction, i.e. in the same direction as that of the
charging current. A behavior similar to this has also been reported
by Mizutani et al.[10] in the TSDC study of corona charged high density polyethylene. The total filled volume trap density for as-
received and heat-treated corona charged samples comes out to be
nearly 6.1 x 10^{12} and 1.7 x 10^{12} traps/cm³, respectively.

Figure 3 shows the TSDC curve for the iodine doped corona
charged sample. Here also, the peaks P_1 and P_2 occur but at somewhat lower temperatures than those in the undoped sample. The
magnitude of the peak current is nearly ten times than that in the
undoped sample. In the abnormal current direction one more peak
appears at 120°C. Above 160°C, the current again reverses and
increases continuously with temperature. The total filled trap
density is estimated to be 11 x 10^{13} traps/cm³.

The activation energy for P_1 peak varies from 0.7 eV to 0.9 eV
and that for P_3 peak from 1.15 to 1.4 eV (from heat-treated to
as-received sample). The activation energy of the P_1 peak for the
iodine doped sample is 0.65 eV.

A few representative TSDC curves of thermo-electrets of Kapton
are illustrated in Figure 4. A single sharp peak, termed α, appears

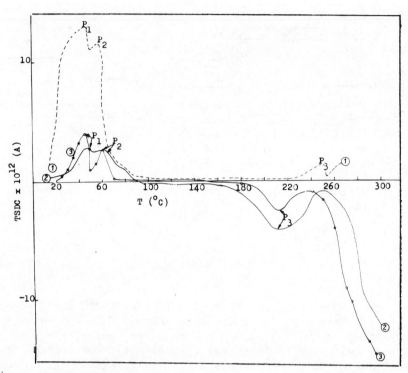

Figure 2. Thermal depolarization spectra for corona charged Kapton samples: (1) as-received, (2) heat-treated at 300°C, and (3) heat-treated at 150°C.

around 196°C in all cases. In addition to this peak one more peak, termed ρ-peak, appears at 249°C, for only high value of E_p (400 kV/cm).

DISCUSSION

The polarizing process in polymers due to corona discharge has been discussed by many workers. According to Carlsson[11] a high field established by corona in its proximity causes local air ionization. Corona ions impinging on the polymer surface exchange

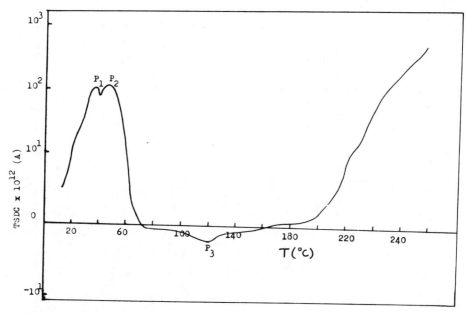

Figure 3. Thermal depolarization spectra for I_2-doped corona charged Kapton sample.

charge with surface atoms. Consequently the charges get trapped in the surface states.

According to Baum et al.[12], polymers under corona discharge may suffer chemical changes on the surface due to the formation of double bonds which, in turn, become trapping sites. The existence of shallow and deep trap levels in Kapton has already been discussed by the authors[6]. The appearance of the P_1 and P_2 peaks is due to these shallow trap levels. It has still not been possible to describe the nature of shallow traps. Several factors may be responsible for producing these traps, viz, chemical impurities, oxidation products, broken chains, and absorbed molecules[13]. Bauser found indications that the corona discharge itself can create new surface traps during charging[14]. The authors have suggested a possibility that the existence of shallow traps in Kapton may be due to the nitrogen atoms present in its structure[6]. The reduction in the peak current in heat-treated samples is related to the reduction in the volume trap density. Tanaka et al.[15] have also presumed that high temperature aging of Kapton would decrease the number of trap levels in it.

481

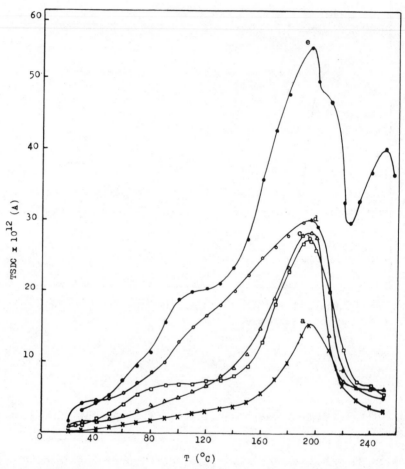

Figure 4. Thermal depolarization spectra for thermoelectrets of Kapton sample (T_p = 210°C) at different polarizing fields. a – 133.3 kV/cm; b – 200 kV/cm; c – 266.6 kV/cm; d – 333.3 kV/cm; e – 400.0 kV/cm.

Perlman and Unger[16] have suggested that the electrons are accommodated on unsaturated groups intrinsically available in polymers. Thus, the phenylene and ketonic (>C = O) groups present in the main chain of Kapton may be held responsible for providing deep trap levels. The trapping mechanism could be as follows:

(a) schematic chemical reaction of Kapton structure with electron capture:

$$>N - C \overset{\text{ring}}{\text{—}} C - 0- + 3e^- \rightarrow >N -C:\Theta \ldots \cdot C - 0-$$

(b)

$$\underset{/\ \backslash}{C} + e \rightarrow \underset{/\ \backslash}{\overset{0}{\underset{\|}{C}}{}^{\Theta}$$

During heating the electrons trapped at these localized states gain enough energy to jump into the conduction band and be trans-ported by the internal field to the shorted electrodes and give a current maximum (peak P_3). A peak (termed ρ-peak) in the TSDC spectra of thermoelectrets of Kapton was observed by the authors[5] at 249°C at high polarizing fields (Figure 4). The peak was assoc-iated with injection of charges from the electrodes to the sample surface. The existence of peak P_3 confirms the above results. The normal and abnormal TSDC behavior can be explained following Mizutani et. al.[10]. The distribution of injected homo space charge $(q_{s,t})$ and field strength $(E_{s,t})$ in the negative corona charged sample during TSDC measurement can be represented as in Figure 5. The density of $q_{s,t}$ drops with increasing s so that the zero field plane, S_0, is located near the corona charged surface. The field in the left region of S_0 drives carriers towards the right one. The high internal field in the left region of S_0 enhances the flow of carriers to the left electrode because of high drift velocity and field assisted detrapping resulting in normal TSDC (peaks P_1 and P_2). Since the rate of arrival of carrier at the left electrode increases with the rise of temperature, because of higher mobility and/or the higher detrapping rate, it may exceed the charge ex-change rate of electrode. The left electrode shows a partial block-

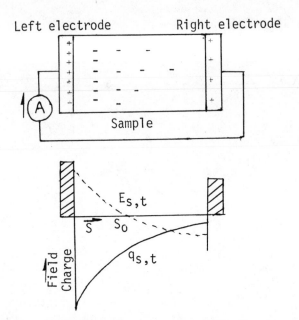

Figure 5. Schematic diagram for the distribution of injected space charge (electrons) and the internal field.

ing, which may result in the net flow of carriers to the right due to diffusion and internal field in the region to the right of S_0. This flow drives the zero field plane, S_0, to the right (non corona charged surface) and the abnormal TSD current appears (peak P_3). However, the occurrence of peak P_3 in normal current direction for as-received sample may be due to the high conduction current prevailing in this region of temperature dominating abnormal TSDC.

The iodine doping of Kapton does not change the behavior of TSDC spectra qualitatively. All the three peaks (P_1, P_2, P_3) appear but at lower temperatures and with a current magnitude nearly ten times than that in case of as-received samples. This may be explained as follows:

It is established[17] that iodine diffuses preferentially in the less dense volume units of polymers and is present either interstitially or is trapped between the chain configuration to act as a strongly electronegative acceptor and forms the charge transfer complex within the structure. As a result of the electrostatic interaction between the chain and iodine, the intermolecular interactions are reduced which results in increasing the carrier mobility and hence the conductivity. As a result of this a decrease in the relaxation time of trapped carriers, i.e. lowering of peak temperature would be observed. At the same time, the magnitude of TSD current is also increased.

CONCLUSIONS

Corona charged Kapton samples show three TSDC peaks. These peaks have been ascribed to the shallow and deep trap levels present in the material.

Iodine doping increases the conductivity of the material.

There is a good possibility that Kapton may be used in electrophotographic processes.

ACKNOWLEDGEMENT

Kapton samples were a gift from Du Pont (USA). The financial support was granted by the Ministry of Education, Government of India.

REFERENCES

1. R. A. Moreno and B. Gross, J. Appl. Phys., 47, 3397 (1976).
2. S. S. Bamji, K. J. Kao and M. M. Perlman, J. Electrostatics, 6, 373 (1979).
3. H. von Seggern, J. Appl. Phys., 50, 2817 (1979).
4. G. Sawa, D. C. Lee and M. Ieda, Jpn. J. Appl. Phys., 14, 643 (1975).
5. J. K. Quamara, P. K. C. Pillai and B. L. Sharma, Acta Polymerica, 33, 205 (1982).
6. J. K. Quamara, P. K. C. Pillai and B. L. Sharma, Acta Polymerica, 33, 501 (1982).
7. P. K. C. Pillai and B. L. Sharma, Polymer, 20, 1431 (1979).
8. R. H. Barlett, G. A. Fulk, R. S. Lee and R. C. Weingart, IEEE Trans. Nuclear Science, NS-22, 2273 (1975).
9. G. F. J. Garlick and A. F. Gibson, Proc. Phys. Soc., 60, 574 (1948).
10. T. Mizutani, S. Ito and M. Ieda, TIEE of Jpn., 51 (July/Aug. 1977).
11. D. J. Carlsson and D. M. Wyles, Can. J. Chem., 48, 2397 (1970)
12. E. A. Baum, T. J. Lewis and R. Toomer, J. Phys. D., 10, 487 (1977).
13. J. Fuhrmann, J. Electrostatics, 4, 109 (1977/78).
14. H. Bauser, Het Ingenieursblad, 44, 321 (1975).
15. T. Tanaka, S. Hirabayashi and K. Shibayma, J. Appl. Phys., 49, 784 (1978).
16. M. M. Perlman and S. Unger, "TSC Study of Traps in Electron-Irradiated Teflon and PE", Paper presented at the Intern. Conf. on Electrets, Charge Storage and Transport in Dielectrics, Fall Meeting Electrochemical Society, Miami Beach, 1972.
17. P. C. Mehedru, J. P. Agarwal, K. Jain and A. V. R. Warrier, Thin Solid Films, 78, 251 (1981).

THE PHOTOINDUCED POLARIZATION AND AUTOPHOTOELECTRET STATE IN

POLYPYROMELLITIMIDE FILM

B. L. Sharma, J. K. Quamara and P. K. C. Pillai*

Department of Applied Physics
Regional Engineering College
Kurukshetra-132119, India

The photoelectret state in Kapton® polypyromellitimide has been investigated as a function of
polarizing temperature (30° to 250°C), field 85-214
kV/cm), time (5-25 minutes) and illumination intensity
(5000 lux). The optimum values of polarizing temperature and field were obtained for photo-electret charge.
With the help of these optimized parameters, the effect
of intensity of illumination on photo-polarization was
studied using visible light and ultraviolet radiations.
Experiments were conducted with heat-treated as well as
as-received Kapton samples. An increase in photoelectret charge with polarizing field and temperature
followed by a saturation has been observed which is a
characteristic property of photoelectrets. Increase in
polarizing temperature results in an increase in photopolarization but at high temperatures(above 200°C) the
photoelectret charge starts decreasing; this may be due
to an increase in thermal recombination of the charges.
The photoelectret charge is observed to increase with
the intensity of illumination followed by a saturation
which is in accordance with the reciprocity law. Ultraviolet radiation of 254 nm wavelength was found to be
more efficient for photo-polarization than visible
light. Heat treatment causes a reduction in photoelectret charge which may be due to the decrease in the
number of traps due to high temperature aging. The
nature of dark depolarization characteristics shows the

---

* Department of Physics, Indian Institute of
  Technology, Hauz-Khas, New Delhi-110016, India.

presence of shallow and deep trap levels in Kapton. The total volume trap density was estimated to be 3×10^{14} traps/cm^3. The autophotoelectret state, i.e., obtaining of polarization by illumination alone (without applying any electric field) has also been investigated in Kapton. The photoelectromotive force is seen to increase with temperature as well as with illumination intensity. The conducting glass-Kapton-Aluminum combination gives a photoelectromotive force of 1.15 volts at a temperature of 200°C.

INTRODUCTION

The photoelectret state in a material results from the simultaneous application of electric field and exciting electromagnetic radiation[1]. The carriers generated by the radiations undergo a directional trapping under the influence of the electric field and a polarization is set up which persists even after the removal of the electric field and illumination[2]. In some materials this persistent polarization can be produced by the application of illumination alone. This is known as the autophotoelectret state, first observed by Andrichin[3] in As_2S_3. A study of this persistent internal polarization gives a useful insight into the electronic processes taking place in the material and at the metal-polymer interface. It can thus be helpful in the selection of proper materials for various optoelectronic processes such as electrophotography and solar energy conservation.

Kapton-H polypyromellitimide film is known to possess good photoconductive properties[4,5,6] which make it a potentially useful material for electrophotography. This material is also reported to exhibit the photoelectret as well as the autophotoelectret state[7].

This article presents the results of a study of photo-induced charge trapping in Kapton under various conditions, viz, different polarizing fields, times, temperatures and intensities of illumination.

EXPERIMENTAL PROCEDURE

Kapton® samples of size 2.5 cm x 2.0 cm cut from a sheet of 2 mil nominal thickness were vacuum aluminized on one side for use in this investigation. The aluminum coating serves as the back electrode. The front electrode, made of tin oxide coated conducting glass, was pressed onto the sample held in a sample holder. The assembly is kept in a measurement cell (Figure 1) whose temperature may either be kept constant or varied in accordance with a preset program using an Indotherm 457 precision Temperature Controller/Programmer.

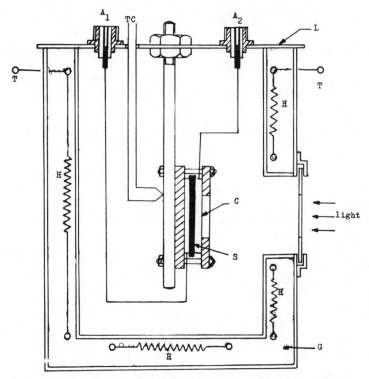

Figure 1. The measurement cell designed for photoconductivity,
photoelectret and thermoelectret measurements. A_1, A_2 – UHF
connectors; C – conducting glass/semi-transparent quartz electrode;
G – glass-wool insulation; H, H – heating elements (connected in
parallel); S – sample; T, T – heater terminals; TC – thermocouple.

The photopolarization with visible light is obtained using a
100 watt tungsten filament lamp and with ultraviolet by light from
a high pressure mercury discharge lamp with the outer (glass)
envelope removed. The intensity of illumination was varied by
changing the distance between the lamp and the sample.

After photopolarization, the sample was left in the dark for a
short time (15 seconds). It is then short circuited through an
electrometer (Keithley Model 610C) and the dark depolarization
current I_{dd} was recorded as it decayed with time using an x-y/t
recorder (Anica Model 2200A3). The photoelectret charge Q was
calculated from the I_{dd} vs time curves by integration (Q = ∫ I.dt).
In some cases, however, the magnitude of the initial I_{dd} was taken
to be proportional to the photoelectret charge Q. There is a direct
correspondence between the two and this fact has been utilized by
many other workers as well[1].

The autophotoelectret effect has been studied under illumina-

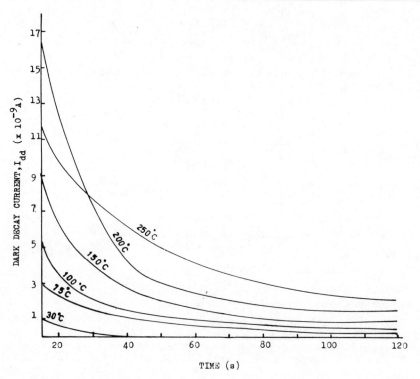

Figure 2. Typical dark decay current characteristics for Kapton photoelectrets prepared at different temperatures at E_p = 85.6 kV/cm, t_p = 5 min.

tion with visible light only. The photo currents/electromotive-forces were plotted with the x-y/t recorder.

RESULTS AND DISCUSSIONS

The variation of photoelectret charge with different parameters is shown in Figures 2 to 16. Typical I_{dd} vs time curves are shown in Figure 2 and in Figures 9 to 12, while Figure 5 shows a simple model proposed to explain the photoelectret state in Kapton. Figures 14, 15 and 16 show the autophotoelectret state in Kapton.

1. Polarizing Field Dependence

Thr role of the polarizing field in the photoelectret is to cause spatial separation of the trapped carriers[8]. With blocking contacts, the photoelectret charge Q, which is proportional to the dark depolarization current I_{dd}, is expected to saturate with the polarizing voltage E_p, because of the counteraction of the increasing internal field due to the trapped carriers. In the case of ohmic contacts, Q should increase linearly with E_p until carrier injection starts, resulting in homocharge formation and, hence, in the reduction of the initial $I_{dd}(I_o)$. The tin-oxide coated conducting glass (NESA)-Kapton-Aluminum structure is considered to be ohmic [10].

Representative dark decay curves for Kapton are shown in Figure 2. The initial dark depolarization current (I_o) is found to increase almost linearly with the polarizing voltage (Figure 3). In the range of field in which the investigations have been carried out I_o (and therefore Q) keeps on increasing with E_p. No reduction of I_o is noticed for values of E_p up to 214 kV/cm. However, saturation in I_o is noticed at the high temperature of 250°C for values of $E_p > 128.4$ kV/cm. The increase in I_o with electric field may be due to the increase in the photoconductivity in the sample with increase in the field[5].

The first three curves of Figure 3, namely, those drawn at temperatures of 30, 75 and 100°C show a distinctly abrupt rise in I_o with E_p around 150 kV/cm. This corresponds closely to the knee of the current voltage characteristics[7]. Similar correspondence has been reported by Pillai et al.[17] for Poly(vinyl carbazole). The abrupt increase in Q is obviously due to field assisted detrapping of carriers.

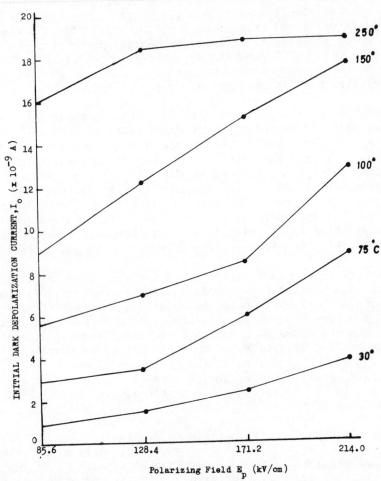

Figure 3. Dependence of initial depolarization current on polarizing field for Kapton photoelectrets prepared at different
temperatures and t_p = 5 min.

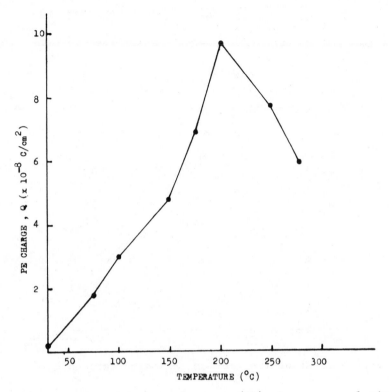

Figure 4. Dependence of photoelectret (PE) charge on polarizing temperature. E_p = 85.6kV/cm, t_p = 5 min.

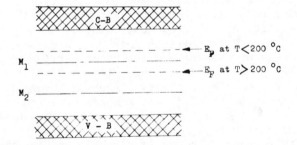

Figure 5. Energy band model to explain the photoelectret state in Kapton-H film. E_F are Fermi levels and M_1, M_2 are trap levels; 200°C is the transition temp.

The saturation in I_o noticed in I_o vs E_p curves at higher temperatures may be due to the establishment of an equilibrium between the process of field assisted detrapping on the one hand and the temperature activated recombination on the other.

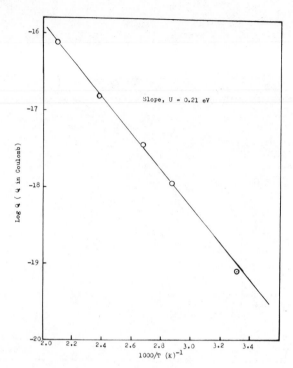

Figure 6. Arrhenius plot of photoelectret charge Q.

2.   The Effect of Temperature on PE Charge

In general, one might expect the photoelectret (PE) charge Q to decrease with polarizing temperature because of the higher de-trapping probability at elevated temperatures (being proportional to exp (U/kT), where U is the trap depth). The existence of a competing process of temperature activated carrier generation, however, causes the actual dependence of Q on T_p to be different from that expected on simple theory. Figure 2 shows the decay curves of I_{dd} with time at different temperatures and in Figure 4 is shown a dependence of Q on T_p as computed from the decay curves. Q is found to increase with T_p up to 200°C and thereafter it starts decreasing with further increase in temperature. The temperature activated trapping and detrapping are obviously balanced at the temperature of 200°C after which the temperature activated detrapping become more dominant.

The model represented in Figure 5 shows the Fermi level moving towards the valence band edge as the temperature increases. At low temperatures the photoelectret state is the result of levels M_1 and M_1. But as the material is heated the Fermi level crosses the M_1 levels converting them into recombination centers thereby decreasing Q.

494

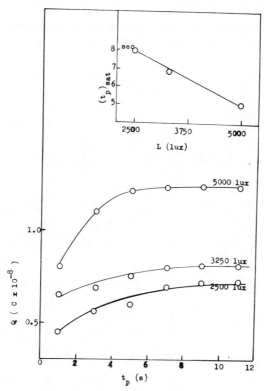

Figure 7. Photoelectret charge (Q) vs polarizing time (t_p) at different intensities of visible light. Inset: $(t_p)_{saturation}$ vs L showing verification of the reciprocity law.

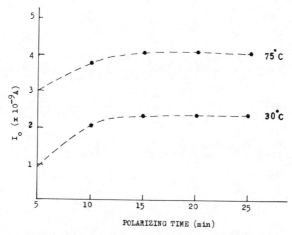

Figure 8. Initial dark depolarization current (I_o) vs polarizing time for 50 μm Kapton films (E_p = 85.6 kV/cm).

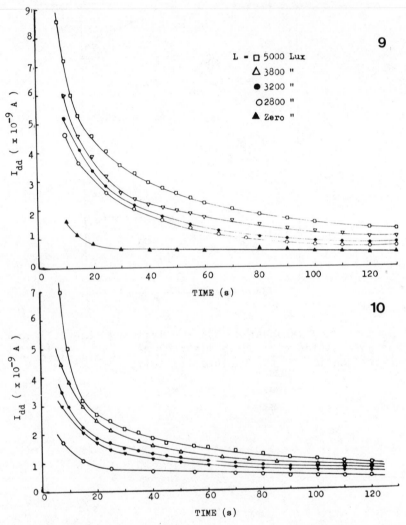

Dark depolarization characteristics of Kapton photoelectrets at various intensities of white light.
Figure 9. As-received samples of 50 μm film.
Figure 10. Heat-treated samples.

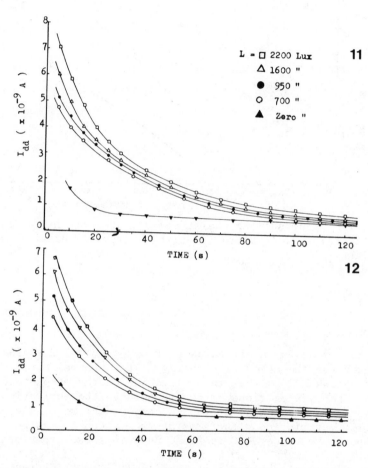

Dark depolarization characteristics of Kapton photoelectrets at
different intensities of UV light from high pressure mercury
discharge.
Figure 11. As-received samples of 50 μm film.
Figure 12. Heat-treated samples.

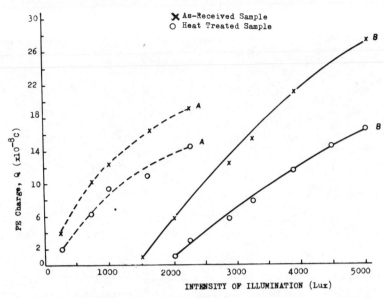

Figure 13. Diagram showing the effect of intensity of illumination (L) on PE charge Q. A: Ultraviolet light from HPM discharge; B: white light.

Figure 6 shows an Arrhenius plot of Q vs 1000/T giving a value of 0.21 eV for the activation energy of the photoelectret charge. This is somewhat less than the activation energy of photocurrent in Kapton[7]. The lack of close agreement between the activation energy of photo-current and that of the PE charge appears to suggest that the mobility variation with temperature may not be insignificant in the temperature range 30°C to 200°C used in these experiments.

3.   The Effect of Photopolarizing Time (t_p) on PE Charge Q

The dependence of Q on t_p at various levels of intensity and temperatures is shown in Figures 7 & 8. The photo-current in a photoconductor increases with time and reaches a saturation characterized by a response time which, in turn, is related to the recombination and trapping mechanism of charge carriers. Saturation occurs when equilibrium is established between the trapping and the recombination processes. It, therefore, follows that the photo-electret charge due to the trapped carriers should increase with time up to the response time and then saturate. However, due to the presence of deep traps, which by definition are not in thermal equilibrium with the transport band, the time required for photo-

498

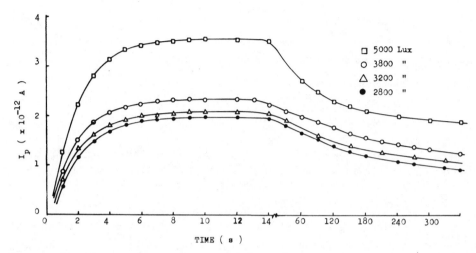

Figure 14. Photocurrent generated by Kapton autophotoelectrets at different intensities of white light (T_p = 200°C).

electret charge saturation could be more than the response time. It might be mentioned that the response time for the photocurrents as well as the saturation time of the PE charge should decrease with the intensity of illumination according to the equation:

$$N_t = \beta k L \tau n \qquad \qquad \dots \dots (1)$$

where N_t is the density of photogenerated carriers at time t, β the quantum yield i.e. number of electron hole pairs produced by a single photon, k the optical absorption coefficient, L the number of incident photons, and τn the life time of electrons.

This equation forms the basis of the reciprocity law[12] for photoelectrets which states that the space charge formed during depolarization on exposing a photoelectret to light is dependent on the exposure L x t_p and not on the individual values of light intensity (L) and exposure time (t_p)[13]. The inset in Figure 7 shows that Kapton obeys the reciprocity law.

The increase in photoelectret charge with duration of illumination (Figures 7 & 8) followed by saturation after a certain time is also a characteristic property of photoelectrets which is

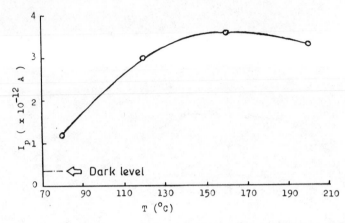

Figure 15. Diagram showing the current generated by the Kapton autophotoelectret at different temperatures (L = 5000 Lux).

observed in the present case[11,14,15]. The saturation may be due to the establishment of a dynamic equilibirum between the number of charge carriers trapped at discrete levels per unit time and the number liberated from those levels to the conduction band by illumination.

4.   The Effect of Illumination Intensity

Since the trapping probability is proportional to the product of trap density and the density of free carriers, charge carrier trapping becomes more effective at higher illumination levels[11,16,17]. But in the case of photoconductors where the Fermi level could sweep through the trap levels towards the transport band edges on illumination, significant changes in the density of trapping centers can result due to the conversion of the trapping sites into recombination centers[15,18,19]. Due to this reason the photoelectret charge can saturate or even decrease with the intensity[20] of radiation incident on the photoelectret.

The I_{dd} vs time curves for Kapton (both as-received as well as heat-treated) photoelectrets prepared at various illumination intensities of visible and ultraviolet (unfiltered) light are shown in Figures 9-12. The PE charge obtained by the time integration of I_{dd} vs t curves is seen, initially, to increase with the intensity of light and then approach saturation (Figure 13). The dark polarization is seen to be less than photopolarization even for low levels of illumination intensity. Also, the dark decay of the depolarization current is faster in the case of photopolarized

500

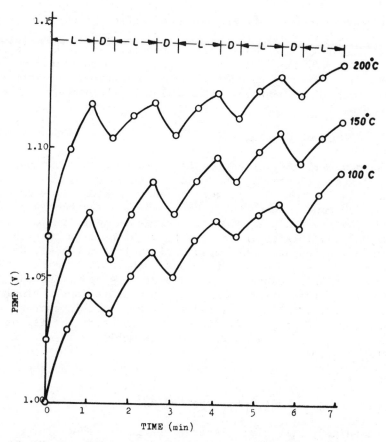

Figure 16. The autophotoelectret state in Kapton-H film: the variation of the photoelectromotive force (PEMF) with time at different temperatures (L = 5000 Lux).

samples as compared to that of the dark polarized samples. This definitely shows the conversion of some deep traps into recombination centers upon illumination and supports the model proposed already in the text (Section 2).

5.   The Effect of Wavelength

The effect of wavelength of illumination on the formation of photoelectrets has been studied by Kallman and Rosenberg[11]. Their observations on anthracene revealed that visible light was less effective in forming the photoelectret state than ultraviolet light. Our results (Figure 13) also support this observation. The UV radiations exhibit a better efficiency as expected from the results of photoconduction also where these radiations were found to be about three times as effective as the visible light in photo-

generation of carriers[5,6]. The enhanced effect of UV illumination
may be due to the excitation of $\pi \rightarrow \pi^*$ transitions in aromatic groups
of polyimide which make more free charge carriers available for
trapping. The slow decay of polarization in photoelectrets prepared
with UV radiations may be due to the larger number of carriers
trapped in the deep trapping sites owing to the higher energy of
the photons.

6.   The Effect of Crystallinity

The crystallinity of Kapton is known to be $\approx 5\%$ in the 'as-
received' state. Annealing increases it to about 9%. The data
presented in Figures 10 & 12 were taken on Kapton films annealed at
200°C for 20 hours. Compared to the results of as-received samples
(Figures 9 & 11) the photoelectret charge in heat-treated samples
is always less. This is in conformity with the view that the PE
state is a manifestation of imperfections in a material. Heat
treatment reduces these imperfections[22] thereby decreasing the
density of trapping centers and hence decreasing Q. The CdS does
not exhibit photoelectret behavior[23] when it is in the single
crystal form. Crystallinity affects the PE state in another way by
increasing carrier mobility. Since the saturation density of
trapped charge N_s decreases with carrier mobility μ in accordance
with the equation[9,24]

$$N_s = \frac{J_s M}{J_s + LN_o \ eE\mu \ exp \ [-U/kT]} \qquad \qquad .... \ (2)$$

Accordingly, the photoelectret charge decreases further with
increasing crystallinity. Here M is the trap density, N_o the density of states, L the illumination intensity and U the activation
energy of the trapping level.

7.   The Autophotoelectret State in Kapton

The photoelectret state which we have been discussing so far
had been the result of a simultaneous application of the electric
field and illumination to a photoconducting material. Adirovich, in
1970, showed on the basis of theoretical consideration that it
should be possible to obtain the photoelectret state in a material
by the action of illumination alone. The field in this case would,
according to him, be the internal photovoltaic field produced in
the material by the illumination. This of course presumes that the
material possesses the appropriate band structure to support the
photoelectret state. Such a state called the 'autophotoelectret
state' was experimentally realized for the first time by Andreichin[3]
in As_2S_3. Use was made here of the contact potential photovoltaic
effect known to exist in vitreous semiconductors. Similar behavior
has been reported by Ieda et al.[25] in low density polyethylene.

In Kapton-H the internal field or the photoelectromotive force is supplied by the contact potential difference formed across the aluminum-polymer-NESA conducting glass system[27-30]. When the sample is illuminated through one of the electrodes (SnO_2) a photoelectromotive force (pemf) is observed in the electrometer which rises exponentially (Figure 14) and then eventually reaches saturation. The saturation value of the pemf depends on the intensity of the illumination as shown in this diagram. It also depends on the temperature of the sample as seen from Figure 15. The decrease in pemf at higher temperatures may be due to an increase in recombination rate brought about by the increase in temperature. The maximum pemf observed in the present case is about 1.15V at 200°C (Figure 16) for NESA-Kapton-Al system.

The pemf decreases very little if the sample is darkened before the saturation value is reached. The voltage decreases very little and indeed continues to grow after illumination is resumed along a curve which is simply displaced (Figure 16). The persistence of pemf obtained in the dark represents the autophotoelectret effect. In short circuit the charges accumulated recombine through the external circuit. It appears, however, that some residual polarization remains but it is very small and difficult to measure as the field acting in the autophotoelectret state is 10^2-10^3 times weaker than that in the externally polarized electrets.

SUMMARY AND CONCLUSIONS

The results of a study of the photoelectret state, including the autophotoelectret effect, in Kapton-H polypyromellitimide film have been presented and discussed in this paper. The conclusions may be summarized as follows:

a)   The photoelectret charge Q is observed to increase continuously with voltage. It does not exhibit a maximum in the range of the electric field investigated (80 to 240 kV/cm). Saturation is reached under the combined action of field and temperature for values of $E_o \geq 125$ kV/cm and $T_p >$ 250°.

For lower values of T_p, a close correspondence is seen between the knee of the I-V characteristics (Figures 3 & 4) and the point at which the rate of increase of Q with E_p registers a sharp change.

b)   Q increases with temperature up to 200°C after which it starts decreasing. A simple model, which shows the Fermi level sweeping through a deep trap level as temperature is increased, has been proposed to explain this effect.

c)   The activation energy of Q is somewhat less than the average average value of 0.3 eV for photocurrent suggesting that the mobility variation with temperature may not be insignificant in the range of 30°-200°C.

d)   Q is seen to saturate with the photopolarization time t_p as well as the intensity of illumination L.

e)   Kapton obeys the reciprocity law. The polarizing time decreases with the increase in illumination intensity L.

f)   Q increases with intensity of illumination reaching a saturation at $L \approx 5000$ Lux.

g)   Ultraviolet radiations (unfiltered light from high pressure mercury discharge) are about 3 times as effective as the visible light in causing photopolarization in Kapton.

h)   Increasing the crystallinity of Kapton by annealing at 200°C for 24 hours decreases the density of deep traps and hence Q.

i)   Kapton shows the autophotoelectret effect. The photo-electromotive force increases with L and temperature (Figure 16).

The aforementioned facts lead one to conclude that Kapton polyimide film has good potential for applications as a photo-electret. What is more significant is the autophotoelectret effect in this material and because of this it may find useful application in optoelectronic devices.

ACKNOWLEDGEMENT

The Kapton samples used in this work were a gift from Du Pont (USA). The Ministry of Education, Government of India provided funds for this work.

REFERENCES

1. G. Nadzhakov. Chem. Rev., 204, 1865 (1937).
2. P. K. C. Pillai, S. K. Aggarwal and P. K. Nair, J. Poly. Sci. Polym Phys., Ed. 15, 279 (1977).
3. R. Andreichin, J. Electrostatics, 1, 217 (1975).
4. R. A. Barlett, G. A. Fulk, R. S. Lee and R. C. Weingart, IEEE Trans. Nucl. Sci., NS-22, 2273 (1975).
5. P. K. C. Pillai and B. L. Sharma, Polymer, 20, 1430 (1979).
6. B. L. Sharma and P. K. C. Pillai, Polymer, 23, 17 (1982).
7. B. L. Sharma, Ph.D. Thesis, I.I.T., New Delhi, 1982.

8. J. R. Freeman, H. P. Kallmann and M. Silver, Rev. Mod. Phys., 33, 553 (1961).
9. V. M. Fridkin, "Physics of the Electrophotographic Process", Focal Press, 1972.
10. R. P. Bhardwaj, J. K. Quamara, K. K. Nagpaul and B. L. Sharma, these proceedings, Vol. 1, pp. 521-536.
11. H. P. Kallmann and P. Rosenberg, Phys. Rev., 97, 1596 (1955).
12. P. K. C. Pillai and K. G. Balakrishnan, Il Nuovo Cimento, 15B, 284 (1973).
13. E. I. Adirovich, Sovt. Phys. Doklady, 6, 335 (1961).
14. V. M. Fridkin and I. S. Zheludev, "Photoelectrets and the Electrophotographic Process", D. van Nostrand, 1966.
15. P. K. C. Pillai andd R. C. Ahuja, Polymer, 17, 192 (1976).
16. P. K. C. Pillai and M. Goel, Physica Status Solidi, (A)6, 9 (1971).
17. P. K. C. Pillai, S. K. Aggarwal and P. K. Nair, J. Polym. Sci., 15, 379 (1977).
18. P. K. C. Pillai and K. G. Balakrishnan, Phys. Rev., B7, 3131 (1973).
19. P. K. C. Pillai and S. K. Arya, Solid State Electronics, 15, 1245 (1972).
20. K. Okamoto, S. Kusabayashi and H. Mikawa, Bull. Chem. Soc. Japan, 46, 1953 (1973).
21. E. Sacher, IEEE Trans. Electr. Insul., EI-14, 85 (1979).
22. T. Tanaka, S. Hirabayashi and K. Shibayama, J. Appl. Phys., 49, 784 (1978).
23. M. Moore and M. Silver, J. Chem. Phys., 33, 1671 (1960).
24. P. K. C. Pillai and K. G. Balakrishnan, Il Nuovo Cimento, 38, 225 (1971).
25. M. Ieda, Y. Takai and T. Mizutani, Memoirs of the Faculty of Engineering, Nagoya University, Nagoya, Japan, 29, 33 (1977).
26. P. K. C. Pillai and M. Mollah, J. Appl. Phys., 51, 2206 (1980).
27. V. I. Trubin, B. P. Usol'tev, V. V. Beskrovanoy and P. I. Khudayev, Akad. Nauk, USSR, 214, 813 (1974).
28. J. Van Turnhout, "Thermally Stimulated Discharge of Polymer Electrets", pp. 199- ,Elsevier, 1975.
29. Y. Sawa, D. C. Lee and M. Ieda, Jpn. J. Appl. Phys., 16, 359 (1977).
30. I. Thuizo, D. Bavancok, G. Vlasak and J. Doupovec, J. Non. Cryst. Solids, 24, 297 (1977).

ELECTRICAL PROPERTIES OF METAL-DOPED PYRIDINE-DERIVED POLYIMIDE FILMS

Eugene Khor and Larry T. Taylor[*]

Department of Chemistry
Virginia Polytechnic Institute and State
   University
Blacksburg, VA   24061

A series of polyimide films containing pyridine coordinating groups doped with cobalt or nickel ions has been examined. Using 3,3´,4,4´-benzophenone-tetracarboxylic acid dianhydride, films were prepared in combination with 4,4´-oxydianiline, 4,4´-dianiline sulfide, 3,3´-bis(aminopyridine)sulfide and 2,2´-dioxyphenelenebis(5,5´-aminopyridine). Room temperature volume resistivities have been found to decrease slightly with sulfur and pyridine containing diamines relative to diamines which do not contain these functionalities. Doping with cobalt and nickel additives results in further lowering of room temperature volume resistivity. Preliminary temperature profiling experiments have shown that volume resistivity decreases as T_g is approached.

INTRODUCTION

Modification of polymers is becoming an attractive field as it renders greater flexibility in the application of polymers.[1] One such endeavor is the improvement of electrical properties (by incorporation or doping with neutral or charged metallic species) of polymers. A number of polymer systems have been investigated in this regard and results have been varied.[2,3]

Our laboratory has been interested in the modification of polyimide properties by metal ion addition. To date the monomers used have been pyromellitic dianhydride, 3,3´,4,4´-benzophenone-tetracarboxylic acid dianhydride, 4,4´-oxydianiline and 3,3´-diaminobenzophenone. Interesting properties have been observed for a variety of dopants. The potential for enhanced electrical conductivity has been demonstrated with palladium doping.[4,5] Two palladium complexes produce films which show a decrease in electrical resistivity with specific dianhydride-diamine pairs relative to the undoped polymer. In the complex $Pd((CH_3)_2S)_2Cl_2$, the observed decrease has been attributed to a metallic surface produced on one side of the film during thermal imidization under normal atmospheric air conditions. This metallic surface employing x-ray photoelectron spectroscopy was found to be exclusively Pd(0) and is removable only with aqua regia. After etching, surface resistivity was the same as the undoped polyimide. The diminished resistivity, therefore, was probably only on the surface and no real lowering of volume resistivity was achieved. Lithium tetra-chloropalladate(II), Li_2PdCl_4, on the other hand, does not exhibit the above surface physical characteristics, although it showed a reduction in both electrical surface and volume resistivity. This held true for films cured in both atmospheric air and under nitrogen. X-ray photoelectron depth profiling revealed a very even distribution of palladium(0)/palladium(II) throughout the bulk of the polymer film. The presence of mixed oxidation states may facilitate a means whereby electrical charge can travel through the film.

This even distribution of metal through the bulk of the film was also observed in LiCl doped polyimides. An appreciable lowering of volume resistivity was observed with lithium chloride as the dopant. Due to the hygroscopic nature of the dopant, drying experiments were carried out to determine the role played by moisture. Results showed that after vacuum oven drying at 110°C, resistivities were similar to undoped films. Moisture up-take by the film was confirmed employing FTIR by the appearance of a broad absorption in the -OH stretching region.

The reason why some films exhibit appreciably lower resistivities and others do not has been left unanswered. This question has generated a need firstly to understand the manner of conduction and secondly to speculate on the effect of metal ion/atom distribution within the polymer bulk on electrical conduction. In an attempt to answer the first question, the study of electrical resistivity as a function of increasing temperature has been investigated to see if these films exhibit semiconductor behavior. In addressing the second question, the approach used has been the attempted "anchoring" of metal ions at regular intervals in the bulk of the polyimide, an occurrence not believed present in most films studied so far. The method employed to attain this regularity was by the introduction of pyridyl groups into the polyimide chain which then can act as a strong coordinating ligand towards metallic species. If coordination is achieved, the metal ions might be expected to distribute themselves evenly in the final film. The results of electrical resistivity measurements on films made with such monomers are reported here.

EXPERIMENTAL

Materials

3,3´,4,4´-Benzophenonetetracarboxylic acid dianhydride (BTDA) was obtained from Gulf Oil Chemicals Co. and dried before use. 4,4´-Oxydianiline (ODA) was donated by Mallinkrodt and purified by recrystallization and sublimation. 4,4´-Dianiline sulfide (DAS) and 3,3´-bis(aminopyridine)sulfide (BAPS) were supplied by NASA Langley Research Center. 2,2´-Dioxyphenelenebis(aminopyridine) (DOPBAP) was synthesized as per the original reference.[6] N,N-dimethylacetamide (DMAC) was obtained from Burdick and Jackson and used as received. Anhydrous cobalt(II) chloride was synthesized by drying the hexa-hydrate in a vacuum oven at 393K for 2 days and used directly. Nickel(II) chloride was used in the form of a DMAC solution by first dehydrating $NiCl_2 \cdot 6H_2O$ with 2,2´-dimethoxy-propane,[7] evaporation of any by-products under vacuum, and subsequent addition of DMAC to form the final solution.

Polymerization

Polymerization was achieved by first dissolving diamine in DMAC (15-20% w/w final solution) followed by dianhydride. The resulting solution was stirred at room temperature for 6h under nitrogen. Cobalt or nickel additives were then introduced and stirring continued for 1h under nitrogen. Solutions were kept in a refrigerator in the polyamic acid-metal "adduct" state until cast (diamine:dianhydride:metal ion; 4:4:1; mole ratio).

Film Preparation

Polyamic acid-metal "adduct" solutions were centrifuged at ~2000 rpm for 15 min and were poured onto glass plates. Solutions were spread using a doctor blade with a 12-18 mil gap, dependent on solution viscosity. Films were dried in static air at 333K for two hours. For ODA films, imidization was thermally achieved by heating in a forced air oven 1 h each at 373, 473 and 573K. All other diamine-derived polymers were imidized in a vacuum oven for 1 h each at 373 and 473K. Films were removed by soaking the glass plates in distilled water at room temperature. All films were air dried and cleaned with a methanol/ether mixture (1:1).

Characterization

Samples were sent to Galbraith Labs Inc., Knoxville, TN for metal analyses. Thermomechanical analyses (TMA) were performed on films in static air at a 5°C/min temperature program on an E.I. DuPont model 990 thermomechanical analyzer. Thermogravimetric analyses (TGA) were obtained on films at 2.5°C/min in static air. Electrical resistivity was measured following the standard ASTM method (D257-66) employing a Keithley voltage source and electrometer. The electrode assembly was modified to accommodate cartridge heaters and thermocouples and to facilitate measurements in vacuum. Heating and temperature monitoring were controlled manually or by a HP-85 computer. X-ray photoelectron spectra (XPS) were obtained using both a DuPont 650B spectrometer and a Physical Electronics 550 Auger/ESCA spectrometer.

RESULTS AND DISCUSSION

Cobalt and nickel doped films have been produced for all BTDA-diamine combinations (Structures I-V) except for BTDA-BAPS/Ni(II). All other combinations of dianhydride and diamine with cobalt and nickel dopants gave good quality, smooth flexible films. DMAC solvated cobalt(II) and nickel(II) chlorides were used as dopants in all cases. Undoped polymer films and nickel containing films were yellowish in color; while, the cobalt dopant produced greenish-colored films. ODA was the only diamine that produced good quality films when cured in air. When a BAPS film was cured under similar conditions, brittle pieces of film were obtained. This may have been caused by oxidation of the thioether bridge in BAPS, but the influence of curing environment on film quality has not been fully rationalized. Since previous curing of such pyridine-derived polyimides was achieved under vacuum,[6] the same procedure was adopted here which allowed good quality films to be obtained except in the BAPS/Ni(II) case.

510

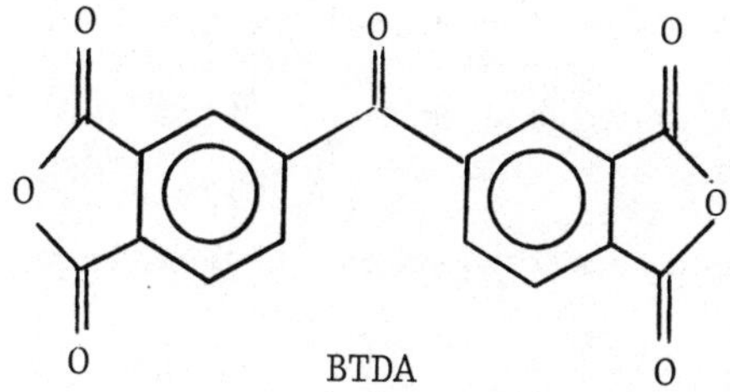

BTDA

STRUCTURE I

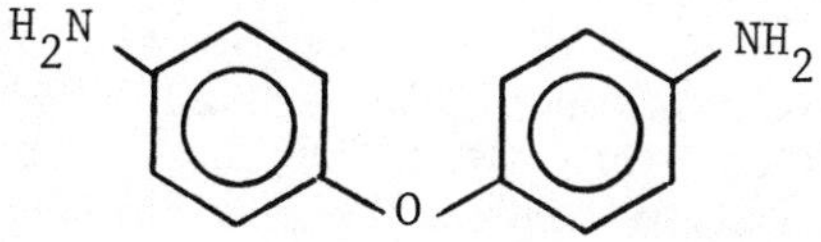

ODA

STRUCTURE II

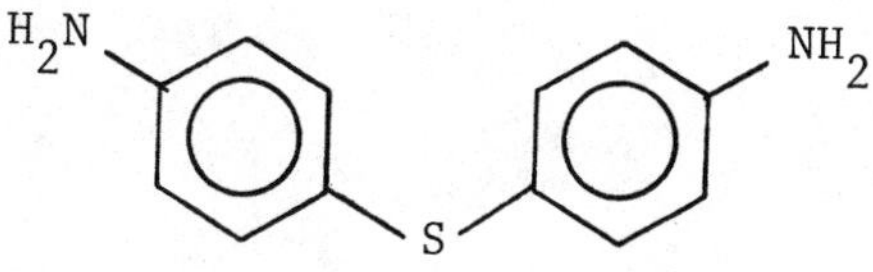

DAS

STRUCTURE III

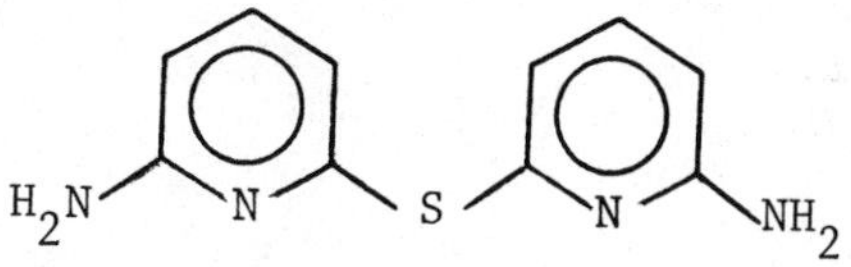

BAPS

STRUCTURE IV

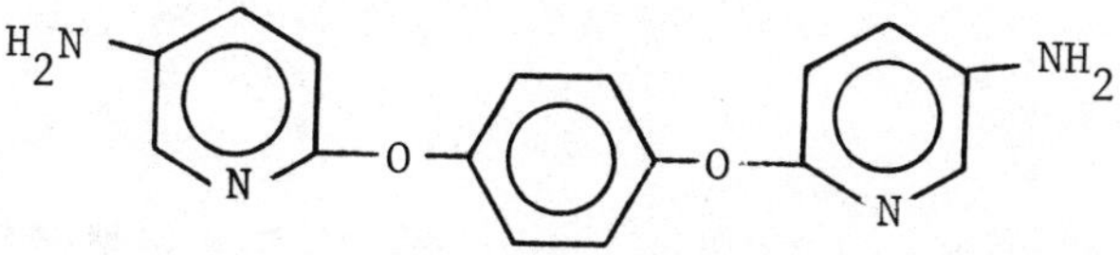

DOPBAP

STRUCTURE V

Thermogravimetric and thermomechanical data for the films are listed in Table I and Figures 1 and 2. The common trend of metal-doped polyimides is again observed.[5] Apparent glass transition (AGT) temperature increases slightly and polymer decomposition temperature (PDT) decreases with no observable dependence on metal concentration. In general nickel-doped films appear to have slightly better thermal stability than cobalt-doped films. It should be noted that in most of the doped and undoped films decomposition begins around 525K with 5% weight loss being reached about 675K. Nevertheless, films derived from pyridyl functionalized monomers doped with metals are just as stable thermally as those without pyridine.

Table I. Thermogravimetric and Thermomechanical Data for BTDA-Derived Polyimides.

Diamine[a]	AGT(K)[b]	PDT(K)[b]	% Metal
ODA	559	813	–
	608	751	1.51 Co[c]
	574	757	3.37 Ni[c]
DAS	555	851	–
	560	777	1.99 Co
	567	763	2.08 Ni
BAPS	530	785	–
	543	778	3.21 Co
	poor quality film		[d]
DOPBAP	617	803	–
	641	733	2.28 Co
	649	769	2.49 Ni

[a]ODA = 4,4´-oxydianiline, DAS = 4,4´-dianilinesulfide; BAPS = 3,3´-bis(aminopyridine)sulfide; DOPBAP = 2,2´-dioxyphenylene-bis(aminopyridine)

[b]Apparent Glass Transition Temperature = AGT; Polymer Decomposition Temperature = PDT

[c]% Cl = 2.09 for the cobalt film; % Cl = 1.64 for the nickel film

[d]Not Measured

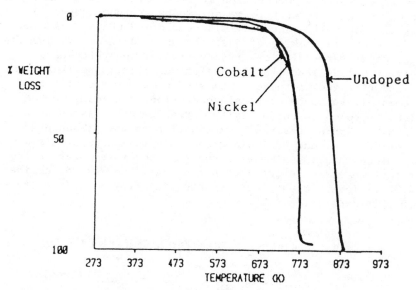

Figure 1.   Thermogravimetric Analyses of BTDA-ODA Films.

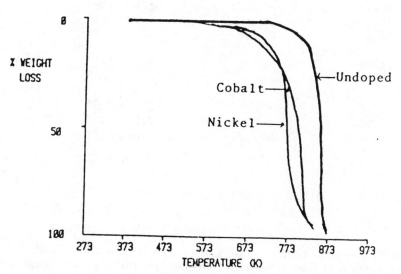

Figure 2.   Thermogravimetric Analyses of BTDA-DAS Films.

513

The main purpose of this work was to investigate (1) the effect of metal ion doping in polyimides that contain pyridyl groups on electrical conductivity and (2) to understand further the characteristics of conduction in these films. The results of electrical volume resistivity measurements at room temperature are reported in Table II. The polyimide employing ODA was used as the reference. In looking at just the undoped films, it appears that the other three diamines produce polyimides which have resistivity values less than the reference. It is unlikely that sulfur and nitrogen in these systems account for the slight enhancement in volume electrical conduction, according to the classical approach to conductivity for different classes of materials.[8] A more reasonable explanation is the presence of trace residual solvent or ionic groups due to incomplete imidization, since these polyimides were subjected to a lower curing temperature than the reference. A previous study has determined that 473K is insufficient for complete ring closure and/or drying of such polyimide films.[9]

Table II. Volume Resistivity Data for BTDA-Derived Polyimides.

Diamine[a]	Metal	Resistivity (Ω-cm)
ODA	–	6.4×10^{15}
ODA	Co	4.3×10^{14}
ODA	Ni	5.7×10^{12}
DAS	–	3.1×10^{14}
DAS	Co	5.6×10^{14}
DAS	Ni	1.5×10^{12}
BAPS	–	4.6×10^{14}
BAPS	Co	2.1×10^{14}
DOPBAP	–	2.9×10^{14}
DOPBAP	Co	2.0×10^{14}
DOPBAP	Ni	5.8×10^{14}

[a]See Table I for diamine abbreviations

In films with cobalt chloride as the dopant, the conductivity does not appear to be enhanced compared to its undoped counterpart. The ODA-derived film appears to be the only exception. XPS of the

ODA film (Table III) indicates a greater concentration of cobalt on the film surface exposed to air during curing. The surface in contact with glass during curing does not show any photoelectron peaks in the cobalt 2P region. This observation suggests a migration of cobalt to the air exposed surface during thermal imidization. This could account for the failure of this cobalt(II) additive to significantly lower the volume resistivity and could mean that a conducting channel through the polymer bulk has not been created as the metal (ion) concentration appears to prefer one surface. The absence of satellite structure in the x-ray photoelectron spectrum suggests that cobalt is present as Co(III) in the film. Although the binding energies for the $2P_{1/2}$, $2P_{3/2}$ photopeaks are lower than cobalt(II) chloride, it is not uncharacteristic for low-spin diamagnetic cobalt(III) compounds to have lower binding

Table III. XPS Data of BTDA-Derived Cobalt and Nickel Containing Polyimide Films.

Diamine[a]	Dopant	Film Side	Binding Energy (eV)			
			Main Peak		Satellite Peaks	
			$2P_{1/2}$	$2P_{3/2}$	$2P_{1/2}$	$2P_{3/2}$
ODA	$CoCl_2$	atm	794.4	778.6	–	–
		glass	–	–		
ODA	$NiCl_2$	atm	873.8	855.0	879.3	860.8
		glass	872.6	855.8	879.8	859.0
DAS	$NiCl_2$	atm	873.4	855.6	877.2	860.6
		glass	873.2	855.6	880.0	861.8
DOPBAP	$NiCl_2$	atm	–	–	–	–
		glass	872.6	855.0	876.0	861.0
–	$NiCl_2$	–	873.7	856.2	881.2	861.7
–	$CoCl_2$	–	798.2	781.7	803.9	792.1

[a]See Table I for diamine abbreviations

energies than high spin cobalt(II).[10] Furthermore the photopeak separation for cobalt in this film (15.7 eV) is 0.6 eV less than that for the cobalt(II) chloride additive. Based on these observations,[11] it is reasonable to think of the cobalt in this film as cobalt(III). The chemical environment of the cobalt(III) in the film is debatable. The cobalt additive may have undergone oxidation to its corresponding cobalt(III)-DMAC complex. Hydrolysis of the oxidized additive followed by dehydration is another possibility. Involvement of cobalt(III) with a heteroatom on the polyimide cannot be discounted with chloride ions serving as counter-ions. Such interaction should manifest itself by a binding energy shift for oxygen or nitrogen in the polyimide.

The nickel doped films exhibit by far the lowest resistivity in this work. As has been observed previously,[5] surface moisture can contribute to enhanced conductivity. The same situation probably occurs here, based on the observation that on heating the ODA and DAS films to about 363K in vacuum, the electrical resistivity increases to the undoped values (vide infra). In this instance, the DOPBAP film appears to be an exception having a resistivity value similar to its undoped counterpart at room temperature. XPS is used again to explain the behavior of this DOPBAP derived film. Whereas in the case of ODA and DAS, surface nickel is observed on both film sides, no nickel is observed in the DOPBAP film on the side exposed to vacuum. The presence of surface nickel on both sides appears to be crucial for conduction. This could be due to its interaction with surface moisture which subsequently generates a more convenient interface for electrons to access the film. The experimental data at the present time do not facilitate a further discussion of the distribution of metal in the polymer. This however, does not preclude that "anchoring" of metal ions has not been successful.

It is reasonable in considering the electrical conductivity of these films, that a first approach would be to treat them as semiconductors at best and as insulators at the very worst. The experiment to distinguish this behavior would be to measure electrical resistivity versus temperature. Preliminary results of DC conduction with a low constant applied potential of 100V for all eleven films in the relatively small temperature range 290 to 500K are presented here.

Figure 3 shows the volume resistivity-temperature plots for undoped and cobalt and nickel doped ODA-derived films. Except for the different initial resistivity, all three plots behave similarly. There appears to be an "induction" temperature above which resistivity decreases linearly with temperature according to the relationship:

$$\log \rho = \log \rho_0 - E/RT \qquad\qquad (1)$$

where ρ is the resistivity, E is the activation energy, R the gas constant, and T the absolute temperature.

Whether the "induction" temperature is a true behavior is debatable. Previous workers[12] have reported that conductivity is linear with temperature in the region where we have observed a constant resistivity. It is possible that the limit of current detection of 10^{-12} amperes in our electrometer, does not facilitate us to observe the linearity below 373K. If one looks at the data of Hanscombs et al.[12] (i.e. Figure 2b), one can see by plotting their data for our experimental conditions that at low fields, currents of less than 10^{-12} amperes must be measured.

The nickel doped film in this group shows a room temperature reading of 10^{12} ohm-cm which can be explained in terms of the interaction of moisture with surface nickel. Removal of this moisture by heating to 320K, essentially immobilizes the charge carriers and causes the resistivity value to increase to 10^{15} ohm-cm (e.g. the approximate value for the undoped polyimide). The cobalt-doped film does not exhibit surface moisture and therefore no change in resistivity is observed between 300K and 373K. From these temperature-resisitivity plots it is readily apparent that cobalt and nickel doping does not alter the volume electrical behavior of ODA-derived polyimides. It is interesting to note that the "on-set" of resistivity decrease begins at a lower temperature for the doped films relative to the undoped ODA film, an observation which might be correlated with the polymer decomposition temperature of each film (see Figure 1).

The DAS derived diamine exhibits similar behavior to the ODA series (Figure 4). Both the undoped and cobalt-doped films display a constant resistivity until above 425K when resistivity varies linearly with temperature. The nickel film again has the lower resistivity at the start but it quickly loses water to reach the maximum resistivity before again following the relationship of equation 1. Films derived from BAPS and DOPBAP follow a similar behavior as outlined for the ODA and DAS films.

While XPS measurements suggest a more uniform distribution of metal in these films, the effect of each dopant on polyimide resistivity is not great. Clearly other additives need to be examined and in higher dopant concentrations before more conductive polyimide films can be formed via this procedure. Whether "anchoring" of the metal in the polyimide has been achieved is, as yet, undetermined. Further work is underway to try to answer this question and address the validity of the low

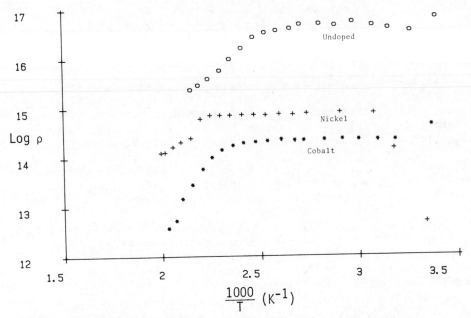

Figure 3. Resistivity Profiles of BTDA-ODA Films.

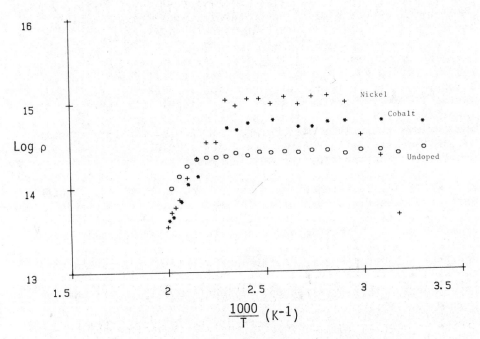

Figure 4. Resistivity Profiles of BTDA-DAS Films.

temperature-resistivity data presented here. The contribution of cobalt and nickel toward residual solvent[13] incorporation, the effect of doping on the activation energy for conduction and the resistivity-temperature profile on cooling from 500K is under investigation.

ACKNOWLEDGEMENTS

We thank the National Aeronautics and Space Administration (Grant NSG 1428) for sponsoring this research, Anne K. St. Clair for supplying the thermal data, Charles C. Johnson for continuing help in the development of the resistivity apparatus, and R. A. Pike (United Technologies) for the chloride analysis.

REFERENCES

1. C. E. Carraher, Jr. and M. Tsuda, Editors, "Modification of Polymers", ACS Symposium Series No. 121, American Chemical Society, Washington, D.C. 1980.
2. A. K. St. Clair and L. T. Taylor, J. Macromol. Sci. Chem., A16, 95 (1981).
3. R. B. Seymour, Editor; "Conductive Polymers", Plenum Press, NY (1981).
4. V. C. Carver, L. T. Taylor, T. A. Furtsch and A. K. St. Clair, J. Amer. Chem. Soc., 102, 876 (1980).
5. E. Khor and L. T. Taylor, Macromolecules, 15, 379 (1982).
6. K. Kurita and R. L. Williams, J. Polym. Sci., Polym. Chem. Ed., 12, 1809 (1974).
7. K. Starke, J. Inorg-Nucl. Chem., 11, 77 (1959).
8. D. A. Seanor, Editor, "Electrical Conduction in Polymers", p. 16, Academic Press, NY (1982).
9. V. L Bell, B. L.Stump, H. Gager, J. Polym. Sci., Polym. Chem. Ed., 14, 2275 (1976).
10. T. A. Carlson, "Photoelectron and Auger Spectroscopy", p. 251, Plenum Press, NY (1975).
11. C. V. Schenk, Doctoral Thesis, VPI & SU (1981).
12. a) J. R. Hanscomb and J. H. Calderwood, J. Phys. D., 6, 1093 (1973).
    b) E. Sacher, IEEE Trans. Electr. Insul., E1-14, 85 (1979).
13. E. Sacher, J. Phys. D., 7, L105 (1974).

METAL CONTACT EFFECT ON THE DIELECTRIC RELAXATION PROCESSES IN POLYPYROMELLITIMIDE FILM

R. P. Bhardwaj*, J. K. Quamara, K. K. Nagpaul* and
B. L. Sharma

Department of Applied Physics
Regional Engineering College
Kurukshetra-132 119, India

Experiments have been conducted to study the effect of electrode materials (Cu, Al, Pb & Ag) on various relaxation processes in Kapton® film using thermally stimulated discharge current technique. Thermoelectrets were prepared at 80°C(E_p=133.3 kV/cm) and 200°C(E_p=133.3 kV/cm and 280.6 kV/cm). Samples were depolarized from room temperature to 300°C at a linear heating rate of 2°C/min. TSDC spectra consist of three peaks, termed as β, α and ρ. The peak which appears around 108°C is seen to be unaffected by the nature of electrode material and therefore has been attributed to the dipole orientation polarization, associated with the residual reactive groups in polypyromellitimide. The α and ρ peaks appearing around 200°C and 250°C have been attributed to the transference of intrinsic space charge and to the surface charge injected from the electrodes, respectively. For both α and ρ peaks, the peak temperature and peak current is seen to be dependent on the work function of the electrode material. The values of activation energy and relaxation times were also calculated for these peaks. Further, analysis of these relaxation processes has been done by means of M_1-I-M_2 structure.

---

*Department of Physics, Kurukshetra University, Kurukshetra-132 119, India.

INTRODUCTION

The dipole relaxation process and the behavior of the trapped charges in polymers can be very well understood using the thermally stimulated discharge current (TSDC) technique[1]. Since various polarization processes which occur during electret formation are often superimposed upon one another, it is necessary for the correct interpretation of the TSDC data that all such possible mechanisms are fully analyzed. The study of the influence of electrode material in TSDC measurements can be very helpful in this respect, especially because the studies play an important role in distinguishing between dipolar and space charge polarization processes. The various metals used as contacting electrodes do not obviously affect dipolar polarization but these may lead to considerable modifications in formation and release of excess charges both of internal and external origin, owing to the difference in work function and blocking factor[2].

The thermally stimulated discharge current studies in Kapton thermoelectrets by the authors have revealed the existence of multiple polarization processes (dipole-orientation, space-charge motion, and charge-injection) in it. In the present communication these polarization mechanisms have been analyzed more critically by making use of electrodes of different metals.

The relaxation parameters were evaluated by using the concept developed by Perlman and Crewsell[4,5]. According to Perlman[5] the discharge current I(T) at an absolute temperature T is given by

$$ I(T) = A \exp \left\{ -(U/kT) - (\beta \tau_0)^{-1} \int_{T_0}^{T} \exp(U/kT) dT \right\} \qquad \ldots (1) $$

where U is the activation energy, A is a constant, β is the heating rate, k is Boltzmann constant and τ_0 is the pre-exponential factor of relaxation time.

The low temperature tail of Equation (1) is given by

$$ \ln I(T) = C - (U/kT) \qquad \ldots (2) $$

where C is a constant. Thus U can be obtained from the straight line plot of lnI versus 1/T in the range where the current rises initially. This is known as the initial-rise method of Garlick and Gibson[6].

If we differentiate Equation (1) and set $dI/dT = 0$, we obtain the condition for maximum current occurring at T_m

$$\tau_o = \frac{kT_m^2}{\beta U \exp(U/kT_m)} \qquad \cdots \cdots (3)$$

T_m being the temperature at which the maximum current is obtained. Time constant at any other temperature can be obtained by using the following equation:

$$\tau_T = \tau_o \exp(U/kT) \qquad \cdots \cdots (4)$$

EXPERIMENTAL

Kapton-H samples of 2 cm diameter were cut from a 3 mil thick sheet. Thermoelectrets using various electrodes, viz., Cu, Al, Pb, and Ag (work function $\emptyset_w$ = 4.47, 3.74, 4.02 and 4.28 eV, repectively) were prepared at 80 and 200°C. The polarizing fields 133.3 and 280.6 kV/cm were used at 200°C, while samples at 80°C were polarized at 133.3 kV/cm. These parameters were selected on the basis of a previous study[3] in order to obtain a complete view of all the relaxation processes in Kapton thermoelectrets (as occurrence of the charge-injection TSDC-peak requires high values of polarizing-field and -temperature, whereas the low polarizing temperature is favorable for the peak pertaining to dipole origin). The duration of application of the field was kept 1 hr. at elevated temperatures and 1/2 hr. at the time of cooling. After preparation of the sample it was kept short circuited for 10 minutes to remove the frictional and stray charges. The samples were then depolarized from room temperature to 300°C at a linear heating rate of 0.033°K sec^{-1}. The constant heating rate in the furnace was obtained by an Indotherm 457 precision temperature controller/programmer. The discharge current was measured with the help of a Keithley Model 610C electrometer. A schematic diagram for the formation of the thermoelectret and the measurement of TSDC is shown in Figure 1.

TSDC spectra were also obtained for Metal$_1$-Polymer-Metal$_2$ structure (M_1-P-M_2). Two combinations of electrode pairs (Cu-Al and Cu-Ag) were used for the polarization and depolarization of Kapton samples. Samples were polarized at 80°C (E_p = 133.3 kV/cm) and 200°C (E_p = 280.6 kV/cm). TSDC spectra were also obtained by reversing the polarity of the electrodes.

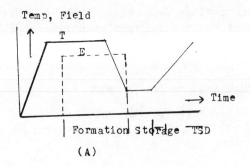

(A)

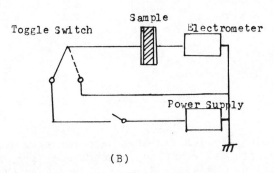

(B)

Figure 1. The schematic diagram for the formation of thermoelectret (A) and the measurement of thermally stimulated discharge current (B).

RESULTS

The TSDC spectra for samples polarized with different elec-trode materials (Ag, Pb, Al, Cu) at T_p = 80°C and E_p = 133.3 kV/cm are depicted in Figure 2. Two current maxima are observed, one around 108°C and the other around 200°C. These maxima have been termed as β and α peaks, respectively. The β-peak for different electrode materials overlaps in a very narrow region and therefore it has been shown by a single curve. This peak seems to be un-affected by the nature of the electrode material. A single peak (α) appears around 200°C in TSDC spectra of Kapton samples polarized at T_p = 200°C and E_p = 133.3 kV/cm with different electrode mater-ials (Figure 3). In addition to β- and α-peaks, as mentioned above, one more peak (termed as ρ-peak) is observed in thermally stimula-ted discharge current spectra of Kapton thermo-electrets prepared at T_p = 200°C and E_p = 280.6 kV/cm (Figure 4). The β-peak is absent for samples prepared at T_p = 200°C. It is observed that the peak-current (I_{max}) and peak-temperature (T_m) for both α and ρ peaks depend on the nature of the electrode material. The I_{max} for α-peak decreases with the increase in the work function of the electrode

524

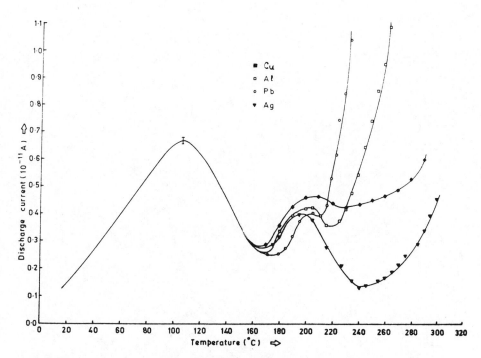

Figure 2. The TSDC spectra for samples polarized with different electrode materials at T_p = 80°C and E_p = 133.3 kV.cm. (■ Cu, □ Al, ○ Pb and ▼ Ag.)

material from aluminum ($\emptyset_m$ = 3.74 eV) to silver ($\emptyset_m$ = 4.28 eV). But with copper electrodes ($\emptyset_m$ = 4.47 eV) a sudden rise in I_{max} is observed (Figure 5-A). The nature of $\emptyset_m$ versus I_{max} curve for ρ-peak is opposite to that for α-peak (Figure 5-B). The variation in peak temperature (T_m) with $\emptyset_m$ for α and ρ-peaks has been shown in Figures 6-A, B. For both the peaks a decrease in T_m is observed with increase in $\emptyset_m$ from aluminum to silver electrode, but then it shows an increase for copper electrodes. The activation energy and other relaxation parameters for β, α and ρ peaks are shown in Table I.

Figure 7 and 8 represent the TSDC curves for Al-Kapton-Cu and Ag-Kapton-Cu structure, respectively. Samples were polarized at 80°C with E_p = 133.3 kV/cm. TSDC curves for samples polarized at these parameters but just changing the polarity of the electrodes have also been reported. Similarly, Figures 9 and 10 represent the

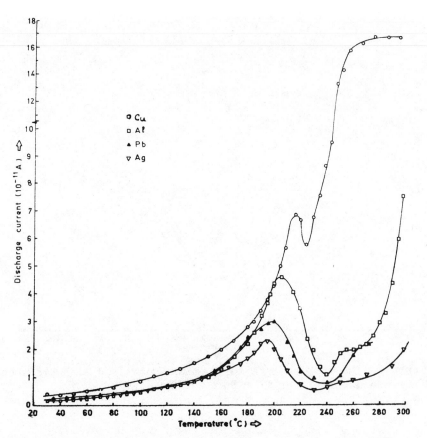

Figure 3. The TSDC spectra for samples polarized with different electrode materials at T_p = 200°C and E_p = 133.3 kV/cm. (○ Cu, □ Al, ▲ Pb and ▽ Ag).

TSDC curves for Kapton samples polarized in the same manner as mentioned above but with T_p = 200°C and E_p = 280.6 kV/cm. In all, three peaks P_1, P_2 and P_3 are observed in the TSDC spectra around 106°, 200° and 250°C respectively. The corresponding peaks in the TSDC spectra for samples polarized with reverse polarity (Al^- - Cu^+ and Ag^- - Cu^+) have been termed as P_1', P_2' and P_3'. The peaks P_1', P_1 are not observed for samples polarized at T_p =200°C. Similarly peaks P_3, P_3' are absent for T_p = 80°C. The magnitude of peak current (I_{max}) of P_2-peak is more than P_2', while the I_{max} of P_3 is less than the I_{max} of P_3'.

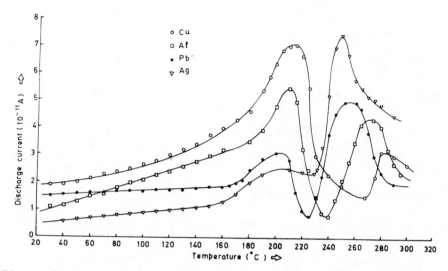

Figure 4. The TSDC spectra for samples polarized with different electrode materials at T_p = 200°C and E_p = 280.6 kV/cm. (○ Cu, □ Al, ● Pb and ▽ Ag).

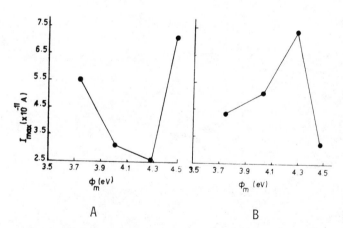

Figure 5. The peak current as a function of the electrode material work function (A) for α-peak (B) for ρ-peak.

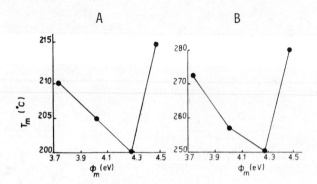

Figure 6. The peak temperature as a function of the electrode material work function (A) for α-peak (B) for ρ-peak.

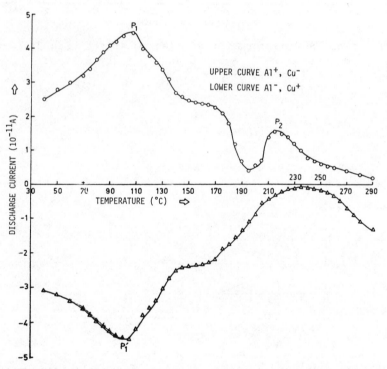

Figure 7. The TSDC-curves for Al-Kapton-Cu structure polarized at T_p = 80°C and E_p = 133.3 kV/cm.

Table I. Activation Energy, Relaxation Times at Room Temperature for Electrets Prepared using Electrodes of Different Metals.

$$(T_p = 200°C, \ E_p = 280.6 \ kV/cm)$$

Electrode metal	$\emptyset_m$ eV	U eV	τ_o sec	τ_{300} sec
		α-peak		
Cu	4.47	0.36	0.33	3.60×10^5
Ag	4.28	0.30	1.24	1.35×10^5
Pb	4.02	0.32	0.79	$1.86 \times 10 5$
Al	3.74	0.33	0.67	2.30×10^5
		ρ-peak		
Cu	4.47	1.59	1.66×10^{-12}	8.08×10^{14}
Ag	4.28	1.55	1.26×10^{-12}	0.30×10^{14}
Pb	4.02	1.53	1.38×10^{-12}	0.66×10^{14}
Al	3.74	1.56	2.03×10^{-12}	3.10×10^{14}

β-peak $(T_p = 80°C, \ E_p = 133.3 \ kV/cm)$

For all Electrodes	—	$0.12 \pm .01$	83.65	8.64×10^3

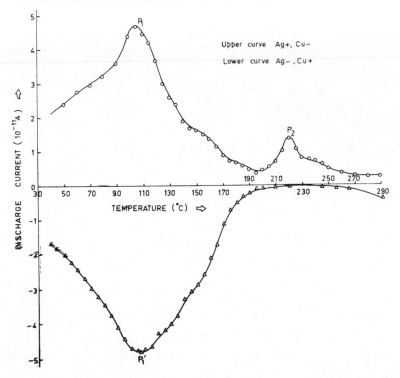

Figure 8. The TSDC-curves for Ag-Kapton-Cu structure polarized at $T_p = 80°C$ and $E_p = 133.3$ kV/cm.

DISCUSSION

(A) <u>TSDC Mechanism</u>:

According to the two-charge theory of Gross [8], the polarization in solid dielectrics subjected to an external electric field is due to two types of charges known as hetero- and homo-charge. The hetero-charge consists mainly of dipolar polarization and translational polarization, i.e., transference of intrinsic free charges (electrons or ions or both) towards the electrodes. On the other hand, the homo-charge is a space charge polarization of extrinsic origin which is obtained either by injection mechanisms from the electrodes or from Townsend-breakdown in the surrounding atmosphere [9].

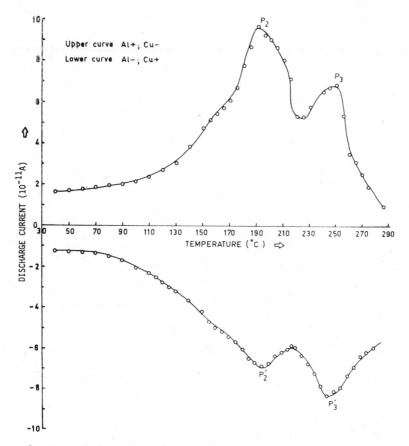

Figure 9. The TSDC-curves for Al-Kapton-Cu structure polarized at $T_p = 200°C$ and $E_p = 280.6$ kV/cm.

The TSDC spectra of Kapton thermo-electret samples reveal the existence of all the above mentioned polarization processes in it (depending on the polarizing parameters) in the form of peaks occurring at different temperatures. It is expected that during TSDC the current maxima for dipolar reorientation will occur at lower temperature than that due to the motion of excess charges. This is so because the former process requires only a rotational motion of molecular groups whereas the latter process involves a motion of molecular groups (ions) over macroscopic distances. The peak appearing around 108°C (termed as β-peak, Figure 2) has been attributed to the dipolar-relaxation process due to the cooperative motion of residual reactive groups in Kapton. The low activation energy (0.12 eV) associated with this peak further supports the involvement of the depolarization of the aligned dipoles of the main polymer chain. Moreover, the fact that this peak is not affected by the nature of electrode material (Figure 2) confirms

531

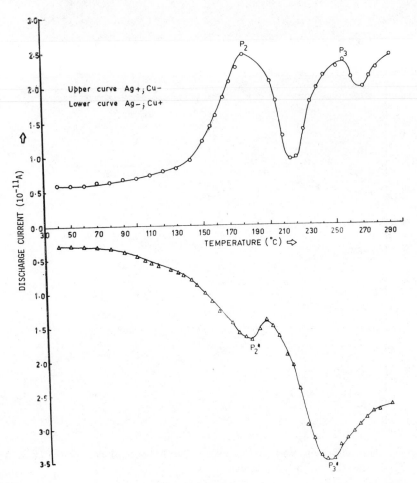

Figure 10. The TSDC-curves for Ag-Kapton-Cu structure polarized at $T_p = 200°C$ and $E_p = 280.6$ kV/cm.

this idea. The possible association of this peak with absorbed water of polypyromellitimide along with residual reactive groups has already been discussed by the authors[3]. The disappearance of this peak under high polarizing temperature (Figures 3 & 4) is plausible because at sufficiently high temperatures the specimen is expected to lose its absorbed water as well as residual side groups react almost completely[10].

The location (around 200°C) and the corresponding activation energy value (Table I) of the next peak (termed as α-peak) stand to infer its association with the intrinsic space-charge polarization. Under the influence of external electric field a free ion (some ions, electrons or holes, are supposed to be fairly free in comparison with their counterparts which are rigidly bound to their definite positions) in a certain dielectric can shift from one equilibrium position to another. If these ions happen to be trapped, some of them can stay in the sites slightly apart from their equilibrium situations. The polarization brought up in this manner will hold on even after the removal of the polarizing electric field at low temperature. The unsaturated carbonyl groups ($>c = 0$) present in Kapton can act, to a significant extent, as traps for charge-carriers, contributing to their transport and trapping. However, when the polarized specimen is heated, i.e., sufficient thermal energy is provided, these ions begin to get out of the traps to return to their original positions and the polarization due to them decays. The displacement current caused by this reduction of polarization gives rise to a peak in TSDC. The effect of polarizing-field and -temperature on this peak has already been discussed by the authors[3].

The peak which appears at 250°C and above, depending on the nature of electrode material, has been termed as ρ-peak (Figure 4). The origin of this peak may be attributed to the extrinsic space charge polarization (homocharge) due to the injection of charges from electrodes to the dielectric surface. The possible association of this relaxation process with the dielectric-electrode interface breakdown may also not be ruled out[9]. The absence of this peak at low polarizing fields is quite accountable in view of the fact that injection of surface charge from electrodes occurs at high fields only.**

(B) Electrode Effect:

The nature and type of contact of the electrodes with the dielectric play an important role in determining the polarization and depolarization processes in polymers[11,12]. As shown in Figure 11, a distinction can be made between neutral, injecting or ohmic, and blocking contacts. It depends primarily on whether the work function $\emptyset_p$ of the polymer is equal to, greater than or less than that of the metal $\emptyset_m$ with which it is in contact. When polymers are polarized under identical conditions with electrode metals of different work function (but less than the work function of the polymer used) the metal with higher work function injects a larger number of charge-carriers than the metal with lower work function[13]. The Kapton-H polyimide film is known to have a work function of 4.36 eV[13]. Consequently, it will form injecting contact with Al, Pb, Ag ($\emptyset_m$ = 3.74, 4.02 and 4.28 eV) and blocking contact with Cu ($\emptyset_m$ = 4.47 eV). The largest number of injected charge carriers is expected to occur from silver to Kapton and the least from aluminum to Kapton. With copper electrodes, no charge-injection would take place. Since the injection of charge carriers from electrodes to the polymer results in a reduction of the contribution of the hetero-

---

**   For an extensive discussion regarding the nature and origin of the multiple relaxations (β, α and ρ) in Kapton-H see Reference 16.

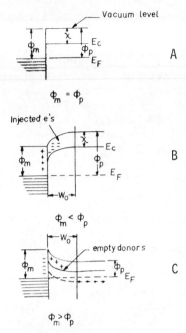

Figure 11. The energy-level diagram showing various types of metal-polymer contacts (A) neutral, (B) ohmic or injecting, and (C) blocking.

charges[14] (due to the charge carriers already present in the bulk of the material), so during the process of depolarization the peak current for α-peak would decrease from aluminum to silver electrodes. With copper electrodes the I_{max} for α-peak would be maximum since it does not inject charges to interact with the hetero-charge of the sample (Figure 5-A). A reverse relationship between $\emptyset_m$ and I_{max} for ρ-peak would be observed in comparison to that for α-peak. Since the ρ-peak is mainly due to injection of charges, the I_{max} for ρ-peak in case of silver electrodes will be maximum. The ρ-peak in case of copper electrodes is only due to Townsend break down[9]; hence I_{max} would be minimum in this case (Figure 5-B).

The dependence of peak temperature (T_m) for α- and ρ-peaks on the work function of the electrode material has been shown by many workers[2,10,15]. But, no proper explanation has been provided. It seems that the trap-levels present in the polymer play some role for such a behavior. The charge-carriers coming from the electrodes, initially, would reach at deep trap-levels. As more and more charges approach the dielectric surface, they would find a potential barrier due to the charges already present at deep levels and, therefore, would be trapped at surface (i.e., comparatively shallow) levels. Thus almost all of the charges injected from low work function electrode, being less in magnitude, would be trapped at deep levels; whereas the charges injected from high work function electrode, being more in magnitude, would be distributed from deep to surface levels. On heating the polarized sample, the charges from surface levels would be released earlier than the charges at deep levels. Hence a shift in T_m towards lower temperature may occur with increasing $\emptyset_m$ (Figure 6). However, with copper electrodes, the extrinsic polarization is minimum (only due to Townsend breakdown) and therefore all the charges captured by the dielectric surface may be assumed to lie at deep levels. Thus, during TSDC these charges would be released at high temperature (Figure 6).

The results of TSDC spectra of Kapton samples polarized with asymmetrical electrodes (M_1-P-M_2 structure) and with inversion in forming-field polarity can also be explained on the basis of injecting and blocking nature of the electrodes. The P_1 and P_1' peaks which are similarly located at 106°C and have polarity independent magnitudes are presumably of dipole-origin (Figures 7 & 8). The P_2 and P_3 peaks which show a change in the magnitude of peak current (I_{max}) on interchanging the polarity of the electrodes are resulted from the space-charge polarization and charge injected from electrodes, respectively (Figures 9 & 10). In both the combinations (Al-Cu and Ag-Cu), since copper is a blocking electrode, the change in peak-current would be mainly due to the charge-injection phenomenon from aluminum and silver electrodes. The suppression of peak P_2 in case of negative forming field applied to the aluminum and silver electrodes (Figures 9 & 10), can reasonably be ascribed to injected electronic carriers which reduce the contribution of hetero-charge, as has been discussed already. The larger magnitude of I_{max} for P_3'-peak than that for P_3-peak is quite obvious (Figures 9 & 10). Since in the latter case the silver and aluminum were given a positive polarity, thereby reducing the probability of electron-injection, the peak P_3 should arise due entirely to Townsend-breakdown; whereas the peak P_3' comprises contributions from Townsend-breakdown as well as charge-injection from electrodes.

ACKNOWLEDGEMENT

The Kapton samples used in this work were supplied by M/S Indmag Pvt. Ltd., New Delhi. The Ministry of Education, Government of India provided funds for this work.

REFERENCES

1. J. Van Turnhout, "Thermally Stimulated Discharge of Polymer Electrets", Elsevier, Amsterdam, 1975.

2. J. Vanderschueren and J. Gasiot, in "Thermally Stimulated Relaxation in Solids", P.Braunlich, Editor, p. 207, Springer-Verlag, Berlin, 1979.
3. J. K. Quamara, P. K. C. Pillai and B. L. Sharma, Acta Polymerica, 33, 205 (1982).
4. M. M. Perlman and R. A. Creswell, J. Appl. Phys., 42, 537 (1971).
5. M. M. Perlman, J. Appl. Phys., 42, 2645 (1971).
6. G. F. J. Garlick and A. F. Gibson, Proc. Phys. Soc., 60, 574 (1948).
7. American Institute of Physics Handbook, Second Edition, pp. 9-147, McGraw-Hill, New York, 1963.
8. B. Gross, J. Chem. Phys., 17, 866 (1949).
9. See, for example Ref. 2, page 138.
10. T. Tanaka, S. Hirabayshi and K. Shibayama, J. Appl. Phys., 49, 784 (1978).
11. B. Gross, in "Electrets", G. M. Sessler, Editor, p. 222, Springer-Verlag, Berlin, 1980.
12. A. Rose, "Concepts in Photoconductivity and Allied Problems", p. 129, John Wiley, New York, 1963.
13. D. K. Davies, Br. J. Appl. Phys., 2, 1533 (1969).
14. P. C. Mehendru, K. Jain and P. Mehendru, J. Phys., D-11, 1431 (1978).
15. P. C. Mehendru, S. Chand and K. Jain, Indian J. Pure Appl. Phys., 18, 183 (1980).
16. R. P. Bhardwaj, J. K. Quamara, K. K. Nagpaul and B. L. Sharma, Phys. Stat. Solidi (a), 77, 347 (1983).

ESCA ANALYSIS OF PMDA-ODA POLYIMIDE

P.L. Buchwalter and A.I. Baise

IBM Corporation
East Fishkill Facility
Hopewell Junction, NY 12533

PMDA-ODA polyimide gives rise to three major carbon 1s ESCA peaks: 285.0eV peak assigned to aromatic carbons not joined to nitrogen or oxygen, 286.1eV peak assigned to aromatic carbons joined to nitrogen or oxygen, and 288.8eV peak assigned to carbonyl carbons. The observed intensity of the 286.1eV peak is too high (about 2.5 fold) to be accounted for by the assignment given above. An alternative assignment is that the aromatic carbons in the PMDA ring are all rendered sufficiently electron deficient by the attached carbonyls that their carbon 1s peaks are shifted to the higher binding energy (286.1eV). This assignment has been confirmed by model compound studies and very good agreement with observed intensities was obtained.

INTRODUCTION

A great deal of work has been done to understand the chemistry of cured polyimide films.[1] Electron Spectroscopy for Chemical Analysis (ESCA) has proved to be an effective tool in elucidating the chemistry of these films.[2-4]

From earlier ESCA studies of pyromellitic dianhydride-oxydianiline (PMDA-ODA) films the following were found:[3,4]

a) two nitrogen 1s peaks: 400.8eV peak due to imide nitrogen and 399.3eV peak of unknown origin.

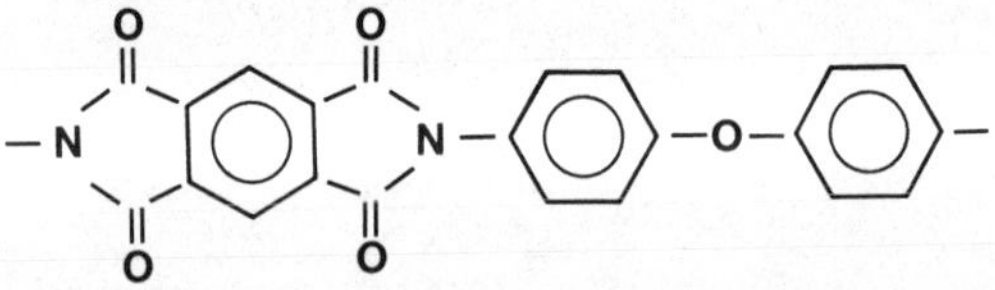

Figure 1. PMDA-ODA polyimide.

b)   carbonyl oxygen to ether oxygen ratio was about 74:26.
      Theoretically it should be 80:20 (Figure 1).

c)   Three carbon ls peaks: 285.0eV peak assigned to
      aromatic carbons not joined to oxygen or nitrogen,
      286.1eV peak assigned to aromatic carbons joined to
      oxygen or nitrogen, and 288.8eV peak assigned to
      carbonyl carbons.

    A more recent study,[2] employing ESCA and IR techniques,
has shown that the low binding energy nitrogen ls peak can tenta-
tively be assigned to isoimide[5] moieties (Figure 2). The
isoimide concentration was about 8% in cured PMDA-ODA polyimide
films.

    The observed carbonyl deficiency was then explained by the
presence of these moieties in the film. ESCA peak assignments,

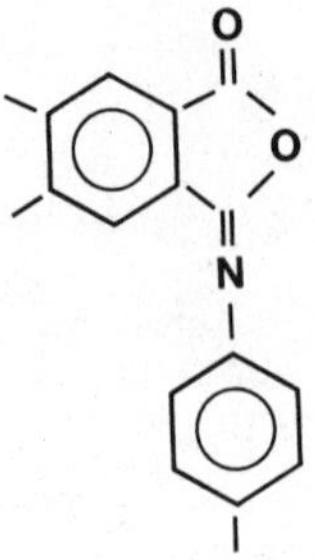

Figure 2. Isoimide structure.

the major carbon 1s peaks in particular, for cured PMDA-ODA polyimide will be discussed here. The intensity of the carbon 1s peak at 286.1eV is too high to be accounted for by the assignment given above. An alternative assignment for this peak based on model compound studies is proposed in this paper.

EXPERIMENTAL

PMDA-ODA polyamic acid ($\bar{M}_w \approx 30,000$) in N-methylpyrrolidone (NMP) was spin coated onto silicon wafers, baked at 85°C in air for 10 minutes, and then heated in a nitrogen atmosphere at temperatures up to 400°C. The following model compounds were synthesized in our laboratory:

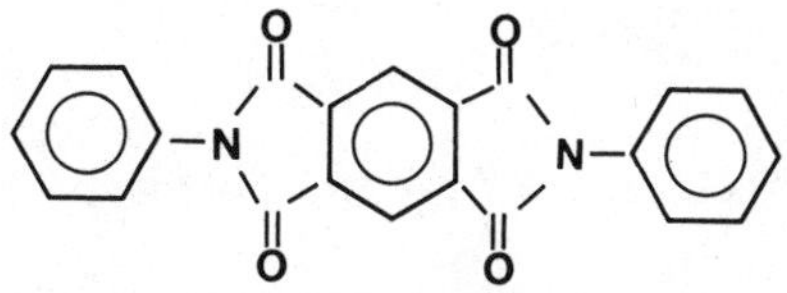

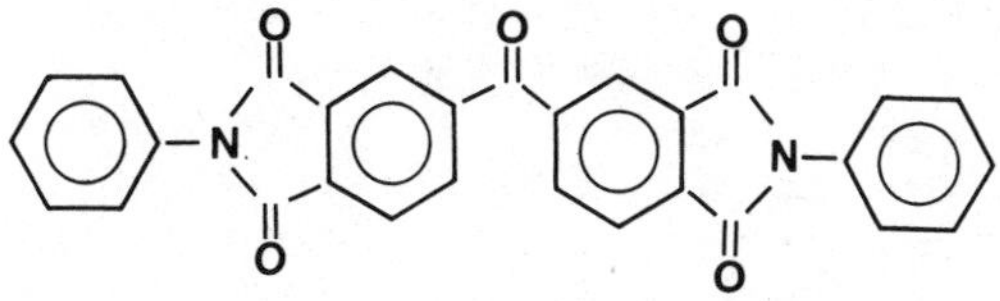

These compounds were made by first reacting the appropriate anhydride with aniline to give the corresponding amic acids. N-phenylphthalimide (model (2)) was obtained by subliming the amic acid precursor at 200°C in air to give needle-like crystals of the imide product (mp 209°C). The remaining two compounds were prepared by heating the amic acids to ~200°C in NMP, which led to precipitation of the corresponding imide. Analysis of model compounds (1) and (3): Model (1) calculated for $C_{22} H_{12} O_4 N_2$:C, 71.7; H, 3.3; N, 7.6. Found: C, 71.9; H, 3.5; N, 7.8. Model (3) calculated for $C_{29} H_{16} O_5 N_2$:C, 73.7; H, 3.4; N, 5.9. Found: C, 73.3; H, 3.5; N, 6.0.

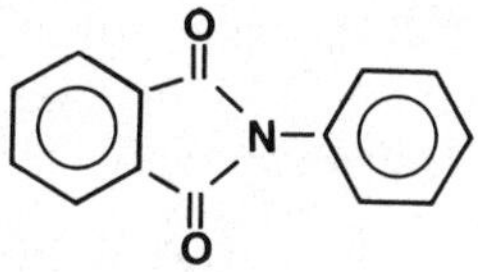

The ESCA spectra were recorded on a Hewlett-Packard 5950A spectrometer using a monochromatized AlK$_\alpha$ radiation source. The X-ray power was set at 800 W. The electrical insulating properties of the polyimide films and model compounds required the use of an electron emitting device (flood gun) to reduce shifts and to avoid distortion of spectra due to charging. The binding energies were referenced to the Au 4f doublet at 83.9eV and 87.7eV. Carbon ls binding energy of 285.0eV was used as an internal standard for correction of shifts due to charging.

The ESCA spectra were resolved into component peaks using a non-linear least squares curve fitting program.[6] The ESCA curves were synthesized by summing Gaussian peaks into a form identical to the experimental data. ESCA curve envelopes consisting of over-lapping peaks were resolved by fitting a minimum number of Gaussian peaks. The individual components were then plotted and relative areas determined.

RESULTS AND DISCUSSION

The elemental peak areas were corrected for photo-ionization cross-sections[7] and electron mean free paths.[8] Using corrections as delineated by I. Elliott et al.[9] or using photo-ionization cross-section only (without mean free path correction) gave slightly poorer correlation with the theoretical values for the model compounds (yet being within the limits of the accuracy of the technique). Elemental relative concentrations and theoretical values (calculated from the expected stoichiometry are shown in Table I. The $\pi \rightarrow \pi^*$ transition peak found in the carbon ls ESCA data has been taken into account in the curve fitting and in the calculation of element relative concentrations (Table I), but was not included in relative percentage peak areas (Tables II and III).

Table I. Elemental Relative Concentrations.

Sample	Theoretical Values			Experimental Values		
	C%	O%	N%	C%	O%	N%
Model (1)	78.6	14.3	7.1	79	13	7
Model (2)	82.4	11.8	5.9	83	11	6
Model (3)	80.6	13.9	5.6	81	14	5
PMDA-ODA	75.9	17.2	6.9	78	15	7

Table II gives the elemental peak positions (peak binding energies in eV) and relative peak areas for the samples.

540

Table II. Elemental Peak Positions and Relative Percentage Areas.

Sample	Carbon 1s			Oxygen 1s		Nitrogen 1s	
	285.0	286.1	288.8	532.1	533.8	399.3	400.8
Model (1)	46	37	17	100	--	--	100
Model (2)	79	7	14	95	5*	--	100
Model (3)	34	51	16	100	--	--	100
PMDA-ODA	39	46	15	76	24	8	92

*   As model (2) was sublimed from an amic acid, so
singly-bonded oxygen may be due to residual acid

The oxygen 1s peak at 532.1eV is due to carbonyl oxygen
whereas the peak at 533.8eV is caused by oxygen singly-bonded to
carbon.[4] The nitrogen 1s peak at 400.8eV is due to imide
species[2] whereas the peak at 399.3eV has been assigned to
isoimide moieties.   The carbon 1s data show the presence of
aromatic carbons at 285.0eV, -C-O- and -C-N- carbon species at
286.1eV and carbonyl carbons at 288.8eV. ($\pi \rightarrow \pi$*peak present at
about 292eV is not given in Table II.) On the basis of the data
presented in Tables I and II it can be said that the surface composition of the model compounds as well as PMDA-ODA film agrees
with the expected stoichiometry.

However, as indicated in the Introduction, the intensity of
the carbon 1s peak at 286.1eV (46%) is much higher than expected
(18.2%). Figures 3, 5 and 6 show an intense carbon 1s peak at
286.1eV, while in Figure 4 the carbon 1s peak at 286.1eV of the
N-phenylphthalimide (model 2) is much less intense. Table III
shows carbon 1s relative peak areas using two assignments for the
carbon peaks:

Assignment (I)   286.1eV peak due only to -C-O- and -C-N-
                     species.

Assignment (II)   286.1eV peak also caused by the carbons in
                     the PMDA aromatic ring.

The carbon 1s data for models (1) and (3) agree very well
with assignment (II), while N-phenylphthalimide (model (2))
results agree with the first assignment. A carbonyl group con-
jugated to an aromatic ring produces an electron deficiency in
the o- and p- positions, thereby rendering those carbon atoms in
the aromatic ring positive.

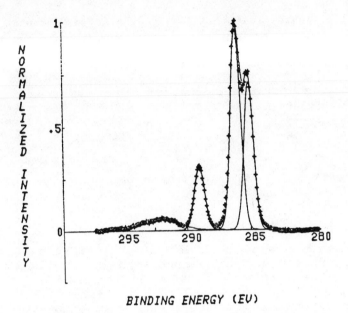

BINDING ENERGY (EV)

Figure 5. Model (3) Carbon 1s ESCA Spectrum.

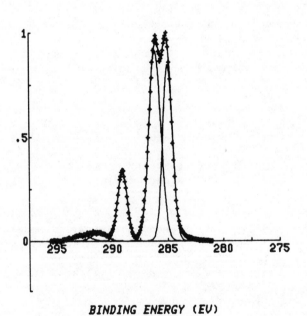

BINDING ENERGY (EV)

Figure 6. PMDA-ODA Polyimide Carbon 1s ESCA Spectrum.

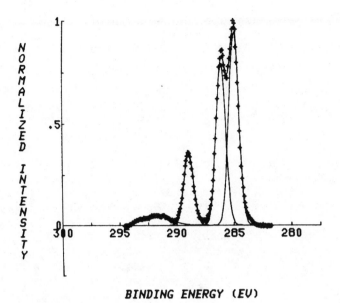

BINDING ENERGY (EV)

Figure 3.   Model (1) Carbon 1s ESCA Spectrum.

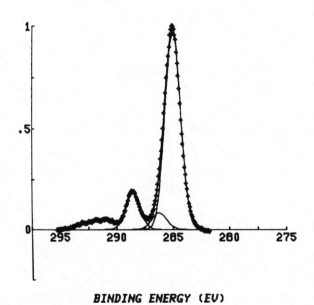

BINDING ENERGY (EV)

Figure 4.   Model (2) Carbon 1s ESCA Spectrum.

Table III.   Carbon 1s Peak Relative Percentage Areas.

Sample	Assignment (I)			Assignment (II)			Experimental		
	285.0	286.1	288.8	285.0	286.1	288.8	285.0	286.1	288.8
Model (1)	72.7	9.1	18.2	45.5	36.4	18.2	46	37	17
Model (2)	78.6	7.1	14.3	35.7	50.0	14.3	79	7	14
Model (3)	75.8	6.9	17.2	34.5	48.3	17.2	34	51	16
PMDA-ODA	63.6	18.2	18.2	36.4	45.4	18.2	39	46	15

The four carbonyl groups joined to the PMDA ring give rise to a sufficiently strong electron withdrawal from all the carbons in the PMDA ring that their core electron binding energies are shifted by $\simeq$1eV. Three carbonyl groups are also sufficient to cause this effect (the third model compound, Figure 5), but the two carbonyl groups in N-phenylphthalimide (model (2), Figure 4) are not sufficient to cause the shift of the carbon 1s electron binding energies in the phthalimide aromatic ring.

CONCLUSIONS

(1) Aromatic carbons in a PMDA ring give rise to a carbon 1s peak at 286.1eV.

(2) At least three carbonyl groups joined to a benzene ring are required to increase the binding energy of the aromatic carbons by $\simeq$1.0eV.

ACKNOWLEDGEMENTS

The authors express their gratitude to Dr. J. Greenblatt and Dr. H.R. Thomas for very useful discussions, and to P. Pacun for assistance in the experimental work.

REFERENCES

1. C.E. Sroog, J. Polym. Sci. Macromolec. Rev., 11, 161 (1976).
2. A.I. Baise and P.L. Buchwalter, unpublished results.
3. T. Wydeven, C.C. Johnson, M.A. Golub, M.S. Hsu, and N.R. Lerner, in "Plasma Polymerization," M. Shen and A.T. Bell, Editors, ACS Symposium Series No. 108, p. 299, American Chemical Society, Washington D.C., 1979.
4. H.J. Leary and D.S. Campbell, Surface and Interface Analysis, 1, 75 (1979).

5. F.P. Gay and C.E. Berr, J. Polym. Sci. Part A-1, 6, 1935
    (1968).
6. R.H. Lacombe, "Interactive Graphical Approach to the
    Analysis of ESCA Spectra", IBM Technical Report TR 22.2319.
    IBM Corporation, E. Fishkill Facility, Hopewell Junction,
    NY 12533.
7. J.H. Scofield, J. Electron Spectrosc. Relat. Phenom., 8, 129
    (1976).
8. C.R. Brundle, H. Hopster and J.D. Swalen, J. Chem. Phys.,
    70, 5190 (1979).
9. I. Elliott, C. Doyle, and J.D. Andrade, J. Electron
    Spectrosc. Relat. Phenom., 28, 303 (1983).

ULTRASONIC AND VISCOMETRIC INVESTIGATION OF POLYMER-POLYMER

COMPATIBILITY IN SOLUTIONS OF POLYAMIC ACID BLENDS

R.K. Gupta*, I.D. Gaba*, C.D. Pande*, Y.P. Singh† and
R.P. Singh†

*Defence Science Centre, Metcalfe House
Delhi - 110054, India
†Indian Institute of Technology
Kharagpur - 721302, India

The polymer-polymer compatibility of polyamic
acids having different chemical compositions has been
determined by ultrasonic and viscometric techniques.
The variation of ultrasonic velocity and viscosity with
compositions is indicative of limited compatibility and
strong polymer-polymer interactions in the blends of
structurally different polyamic acids. The compatibility data are expected to facilitate the development of
the polyimide blends possessing tailor-made properties.

INTRODUCTION

Recently, the polymer-polymer compatibility in polymeric blends has been investigated by some unconventional methods. Kuleznev et al.[1] have indicated that the melt viscosity behavior of polymer blends may reveal polymer-polymer compatibility in the blends. Hourston and Hughes[2] have applied the variation of sonic velocity with composition to reveal the polymer-polymer compatibility in the polymer blends in solid form. Singh et al.[3-6] have reported that the extent of polymer-polymer compatibility can be gauged by studying the variation of absolute viscosity and ultrasonic velocity wwith composition of blends in blend-solutions. The same trend is witnessed in the variation of ultrasonic velocity with composition of the blends in solid form. Suresh et al.[7] have observed a similar type of variation in the dielectric constant and ultrasonic absorption with composition in polyblends.

It has been observed by these workers that plots of both absolute viscosity and ultrasonic velocity vs. composition deviate from linearity according to the degree of compatibility of polymers in the solutions of polymer blends. The curves for compatible systems are linear while the curves are found to have S or inverted S shapes in case of incompatible blends. The behavior of the curves for semi-compatible blends falls in between the two extremes. The same behavior is reflected while the blends are studied in solid form by ultrasonic and dielectric techniques. It appears that the relaxational approach is also valid for studying polymer-polymer compatibility in polymer blends.

The molecular engineering of polyimide polymers can be achieved by changing the dianhydride/diamine components. It is sometimes preferable to achieve desired properties by blending two or more structurally different polyamic acids instead of synthesizing new polymers/copolymers. Since polyamic acids form an intermediate step in the formation of polyimide, the present paper reports the study of polymer-polymer compatibility in the blends of polyamic acids derived from 3,3', 4,4' - benzophenonetetracarboxylic dianhydride with 4,4" - diaminodiphenyl methane and 4,4"-diaminodiphenyl ether.

EXPERIMENTAL

Materials

3,3', 4,4' - Benzophenonetetracarboxylic dianhydride (BTDA) (TCI, Japan) was used after three crystallizations from freshly distilled acetic anhydride. Diaminodiphenyl ether (DDPE) and diaminodiphenyl methane (DDM) (TCI, Japan) were sublimed twice under vacuum. N,N - Dimethylacetamide (DMAC) (Koch Light, England) was kept over phosphorus pentoxide for 48 hours and distilled before use.

548

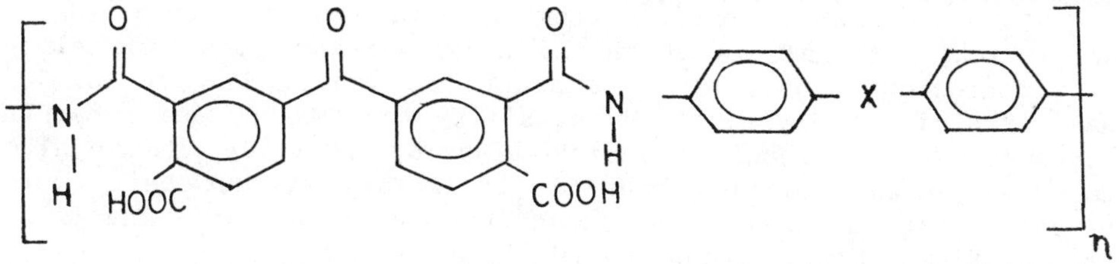

X = --CH$_2$-- for BTDA - DDM PAA and

--O-- for BTDA - DDPE PAA

Figure 1. General structure of polyamic acid.

Synthesis of Polyamic Acids

Polyamic acids (PAA) were synthesized from BTDA and DDPE as well as from BTDA and DDM using DMAC as solvent. Stoichiometric amounts of dianhydride and diamine were placed in a dry Erlenmeyer flask equipped with magnetic rotor and ground glass stopper. A measured amount of DMAC was then added to make 4% solution and the solution was kept stirred with cooling. The reaction was continued below 10°C until maximum viscosity was achieved. The structure of polyamic acids are represented by the general formula given in Figure 1.

Ultrasonic Measurements

The ultrasonic velocity measurements were carried out by interferometric technique using an ultrasonic interferometer supplied by Electronics Instruments and Controls, New Delhi, India. The accuracy of the ultrasonic velocity and the temperature stability were ±0.05% and ±0.05°C, respectively. The details of the measurements have been described earlier[8].

Viscosity Measurements

The viscosities of various blend solutions were determined at 25°C ±0.1°C using an Ostwald Viscometer which was calibrated using Standard Brookfield fluid.

549

RESULTS AND DISCUSSION

Polymer-polymer compatibility is, in general, obtained by sophisticated techniques of thermal, dynamic-mechanical and electron microscopic analyses[3-6]. As mentioned earlier Singh et al. have studied the degree of compatibility for PMMA-PS and PMMA-PVA systems. The ultrasonic velocity behavior of solid blends also conforms to the solution behavior to a large extent being more revealing as far as microheterogeneities are concerned. The findings of Singh et al.[5] are shown in Figures 2 and 3.

Ultrasonic velocity and viscosity measurements have been carried out in 4% solutions of blends of polyamic acids derived from BTDA-DDPE and BTDA-DDM. These are shown in Figures 4 and 5. The results of ultrasonic velocity measurements are plotted as ultrasonic velocity vs. composition of the blend. The general observation of the curve depicts its nonlinear nature showing continuous change in the ultrasonic velocity with composition. However, careful observation of the plot shows that it consists of various S and inverted S regions, viz., 0-40% (inverted S type), 10-60% (S type), 30-90% (inverted S type) and 50-90% (S type), of BTDA-DDM PAA respectively, though some parts of the curve are superimposed onto each other.

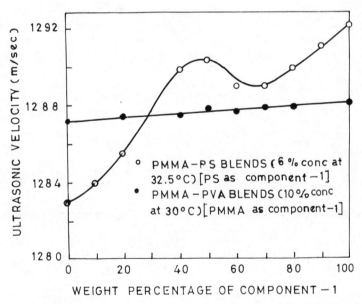

Figure 2. Ultrasonic velocity vs. composition of solutions of PMMA-PS (6% conc. at 32.5°C) and PMMA-PS (10% conc. at 30°C) blends.

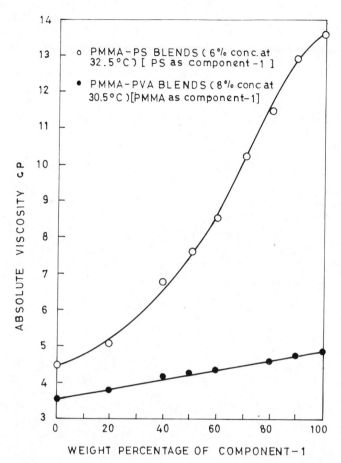

Figure 3. Absolute viscosity vs. composition of solutions of PMMA-PS (6% conc. at 32.5°C) and PMMA-PVA (8% conc. at 30.5°C) blends.

The overall nature of the curve shows continuous change in the morphology of the blends depicting several phase inversions. Hourston et al.[2] have observed that nonlinear variation of sonic velocity with composition of a blend represents incompatibility of the system. Similar behavior has also been observed by Singh et al.[3-6] in ultrasonic velocity. The continuous change in the ultrasonic velocity of the blend solutions may be attributed to the association and conformational change of macromolecules which is facilitated by the presence of solvent (DMAC). It has already been established that association of macromolecules is easier when they undergo decoiling to certain extent which is dependent upon the concentration of the polymer in the solvent and the presence of another polymer[10,11]. The solvent (DMAC) provides the environment for decoiling of the macromolecular chains and the blend is expected to depict true behavior at the molecular level. The association of the macromolecules may result in the formation of bigger size

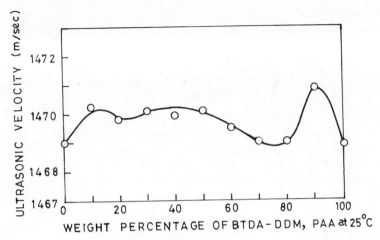

Figure 4. Ultrasonic velocity vs. composition of solution of blend of BTDA-DDPE PAA with BTDA-DDM PAA at 25°C.

species which break the polymer-polymer-solvent continuous matrix into discrete regions of a particular phase depending on its concentration in blends. The size of aggregating molecules may exceed the critical domain size of compatibility which is reflected in the ultrasonic velocity changes. The peak at 90% concentration of BTDA-DDM PAA in ultrasonic velocity is indicative of strong polymer-polymer interaction.

The absolute viscosity also shows the same type of variation with composition. It shows two inverted S type regions (0-30%) and 30-80%) with tendency to peaking at 90% of BTDA-DDM PAA. The average ultrasonic velocity appears to decrease as the absolute viscosity decreases. However, due to presence of various regions of incompatibility giving rise to various S type of curvatures which also superimposed upon each other, the variation in ultrasonic velocity is not pronounced as in the case of absolute viscosity.

CONCLUSION

These results indicate fair amount of correspondence between the two behaviors. It appears that there are two continuous regions of inhomogeneity or phase separation in the blends of BTDA-DDM PAA and BTDA-DDPE PAA. The study indicates that the polyamic acids in question have limited compatibility with each other in spite of similarity in chemical structure. The fact that the films of the blends are transparent at all compositions indicate that the refractive indices of both the polymers are of the same order.

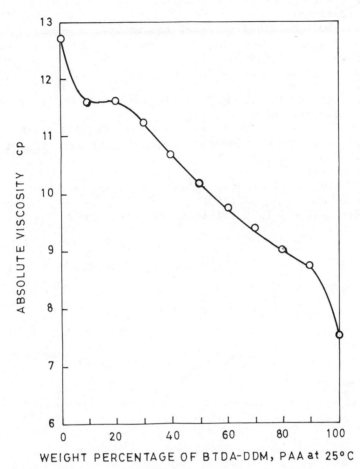

Figure 5. Absolute viscosity vs. composition of solution of blend of BTDA-DDPE PAA with BTDA-DDM PAA at 25°C.

ACKNOWLEDGEMENT

Authors express their sincere thanks to Dr. A. K. Sreedhar, Director, Defence Science Centre, Delhi for constant encouragement and for permission to carry out the work. Thanks are also due to Dr. S. R. Chatterjee, Head, Special Materials Division, Defence Science Centre, Delhi, for providing facilities and useful suggestions.

REFERENCES

1. V. N. Kuleznev, O. L. Melinkova and V. D. Klykova, Eur. Polym. J., 14, 455 (1978).
2. D. J. Hourston and I. D. Hughes, Polymer, 19, 1181 (1978).

3. Y. P. Singh, S. Das, S. Maiti and R. P. Singh, in "Proc. of International Conf. and Exhibition of Ultrasonics (ICEU)", p. 242 (1980), National Physical Laboratory, New Delhi, India.
4. Y. P. Singh, S. Das, S. Maiti and R. P. Singh, J. Pure Appl. Ultrason., $\underline{3}$, 1 (1981).
5. Y. P. Singh and R. P. Singh, Eur. Polym. J., $\underline{19}$, 535 (1983).
6. Y. P. Singh and R. P. Singh, Eur. Polym. J., $\underline{19}$, 529 (1983).
7. R. Suresh, Y. P. Singh, G. D. Nigam and R. P. Singh, Communicated to Eur. Polym. J.
8. G. V. Reddy and R. P. Singh, Acustica, $\underline{46}$, 342 (1980).
9. S. Krause, "Polymer-Polymer Compatibility" in Polymer Blends, Vol. 1, D. R. Paul and S. Newman, Editors, Academic Press, New York (1978).
10. S. Manabe, R. Murakami, M. Takanayagi and S. Uemura, Int. J. Polym. Mater., $\underline{1}$, 47 (1971).
11. R. Bachdahl and L. E. Nielsen, J. Polym. Sci., $\underline{15}$, 1 (1955).

EFFECT OF NEUTRON AND PROTON IRRADIATION ON SOME PROPERTIES OF KAPTON*

V. Shrinet, U. K. Chaturvedi, S. K. Agrawal,
V. N. Rai** and A. K. Nigam
Ion Implantation Laboratory, Physics Department
Banaras Hindu University
Varanasi 221 005, India

Photoacoustic, dielectric and surface-morpho-
logical properties of Kapton* were studied after exposing sheets (25 μm thick) separately to fast reactor
neutrons (E>0.1 MeV) and to 250 keV protons up to doses
1.2×10^{18} (maximum) and 7.0×10^{15} cm^{-2}, respectively. The
photoacoustic spectrum (PAS) of the maximum neutron
dose irradiated sample shows the red-shift of the
absorption peak from 450 nm to 470 nm, whereas in the
proton irradiated (7×10^{15} cm^{-2}) sample, a new flattened
absorption peak appears at $\lambda \simeq 515$ nm. There is a net
enhancement in the D.C. dielectric constant of about
13% after the maximum neutron dose. It is attributed to
the enhanced water absorptivity of the material due to
neutron induced radiation damage.

The scanning electron micrographs of the n and H$^+$
irradiated Kapton surfaces show swelled rough surface
and the scattered blisters of size $\simeq 10$ μm, respectively. The scratches on the surface of the polymer film
made during its extrusion, become "bold" and "decorated" due to H$^+$ implantation. The blistering mechanism
is correlated with the internal gases (CO and H$_2$)
released due to n and H$^+$ induced radiation damage. The
G value (i.e. the number of molecules released per 100
eV absorbed energy) calculations for radiation-released
CO and H$_2$ confirm this theory.

---

*   Trade name of DuPont.
**   Laser and Spectroscopy Laboratory, Physics Department,
     B.H.U., Varanasi 221 005 (India).

INTRODUCTION

Teflon* (Polytetrafluoroethylene), Mylar* (Polyethylene tere-phthalate) and Kapton* (Polypyromellitimide) are three heat re-sistant thermally insulating polymers which have attracted a great deal of attention, in the last few decades, from scientists and ma-terial engineers in different technologies, especially in nuclear and satellite-fabrication technology[1,2]. All these polymers are avail-able in thin sheet form (thickness > 10 μm) though Teflon is also available in the form of rods and thick plates. However, Teflon is much weaker in mechanical properties as compared to the other two and is also very prone to irradiation. A comparison of some of their useful properties is shown in Table I. Kapton has sufficient Ultimate Tensile Strength (UTS) and its working temperature ranges from liquid Helium temperature to 400°C in continuous use. In con-trast to Teflon and Mylar, it is one of the most radiation-resistant polymers. Thus, the properties of Kapton are very appeal-ing and useful.

Satellites in deep space face considerable thermal, vacuum and radiation threat, which, in the long run, degrade many properties of their fabricating materials, especially that of the outermost layer which faces the maximum exposure. This causes partial or complete failure of the satellite. The flux and type of these space radia-tions depends on the altitude of the satellite as listed in Table II. As the active life of the geosynchronized and other satellites is being increased day by day, and is expected to go well beyond 30 years in the near future[3,4], even the proton dose received by the outermost layer of the satellite will be as high as $\sim 10^{17}$ cm^{-2} (E ≥ 100 keV). This dose is accompanied by a high flux of electrons, U.V. and comparatively small flux of X-rays and γ-rays. This dose seems to be much more than the tolerance limit of most of the semi-conductors and insulators. The problem of radiation damage will be more crucial when nuclear fired rockets will come of age as they will have neutron flux several orders of magnitude higher than that of protons. The shielding of the satellite from these neutrons will be practically impossible as it will increase the payload consider-ably. Thus, these nuclear-fired carrier rockets, operating only for very limited time, may expose the whole satellite to a considerable dose of neutrons.

Das and Kaminsky[5] have shown that the effect of individual irradiation on matter is not the same as when the material is ex-posed to two or more radiations simultaneously. They have shown that the implantation of D$^+$ (0.5 C cm^{-2}) or He$^+$ (0.1 C cm^{-2}) on polycrystalline Ni does not show blistering if implanted one by one in any sequence, but a severe blistering occurs when both ion species hit the target simultaneously with their individual flux and dose. Thus for witnessing actual effects of space radiations on

Table I. Some Important Properties of Teflon*, Mylar* and Kapton* (at the ambient) (References are given in parentheses as superscript).

Serial Number	Property	Unit/ Condition	Teflon*	Mylar*	Kapton*
1.	Chemical Name		Poly(tetrafluoroethylene)	Poly(ethyleneterephthalate)	Poly(pyromellitimide)
2.	Dielectric constant; ε, at	60 Hz 1 kHz 1 MHz	$2.0^{(1)}$ $2.0^{(1)}$ $2.0^{(1)}$	$3.16^{(2)}$ $3.23^{(3)}$ $2.93^{(2)}$	$4.12^{(2)}$ $3.50^{(4)}$, $3.34^{(5)}$ $3.96^{(2)}$
3.	Loss tangent: $\tan \delta$, at	60 Hz 1 kHz 1 MHz	$0.0002^{(1)}$ $0.0002^{(1)}$ $0.0002^{(1)}$	$0.002^{(2)}$ $0.004^{(3)}$ $0.014^{(2)}$	$0.003^{(2)}$ $0.0025^{(4)}$ $0.011^{(2)}$
4.	Dielectric Breakdown Voltage (D.C.)	$V\ cm^{-1}$	$4 \times 10^{5\,(3)}$	$4 \times 10^{6\,(3)}$	$1.2 \times 10^{5\,(2)}$ $2.76 \times 10^{6\,(4)}$ (60 Hz)
5.	Volume Resistivity	$\Omega\ cm$	$10^{18\,(1)}$	$10^{23\,(2)}$, $10^{18\,(11)}$	$1 \times 10^{17\,(4)}$
6.	Density	$gm\ cm^{-3}$	$2.10\text{--}2.30^{(6)}$	$1.39^{(2)}$	$1.42^{(4)}$, $1.47^{(2)}$
7.	Refractive Index (Vis. Range)		$1.30\text{--}1.40^{(1)}$	$1.60^{(7)}$	$1.70\text{--}1.80^{(8)}$
8.	Tensile Strength (UTS)	$10^3\ psi$	$2.0\text{--}4.5^{(1)}$	$17\text{--}25^{(2)}$	$17.5^{(4)}$ (Along Machine Direction)

(continued)

Table I (continued).

Serial Number	Property	Unit/Condition	Teflon*	Mylar*	Kapton*
9.	Continuous Working Temperature Range	°K	4 to 530[1]	203[11] to 430[2]	4[8] to 600[2]
10.	Radiation Tolerance up to	Rads	10^{6}[9]	1×10^{7}[8]	5×10^{8}[10,8,4]

*Trade name of E. I. duPont de Nemours & Co., Inc.

REFERENCES

1. Robert C. Weast (Ed.); CRC Handbook of Chemistry and Physics; 58th Edition (1977-78).
2. U. K. Chaturvedi, S. K. Agrawal, N. Rajan and A. K. Nigam: ISRO Project Report Grant, 10/3/27 (1980 (references therein).
3. L. I. Maissel and R. Glang (Eds.), Handbook of Thin Film Technology, McGraw-Hill Book Company (N.Y.-1970); 19-24.
4. Kapton Polyimide Film: Summary of Properties (duPont publication): Data for 25 μm thick Kapton Type H film.
5. Reference 2, page 32 (unpublished).
6. J. A. Brydson, Plastic Materials; 3rd Edn., Newness-Butterworths (London-1975), p. 305.
7. A. Cunningham and I. M. Ward, Polymer, 15, 749 (1974).
8. E. T. Arokawa, et al., J. Appl. Phys., 52(5), 3579 (1981).
9. C. H. S. Dupuy (Ed.), Radiation Damage Process in Materials, Noordhoff, (London-1975), p. 247.
10. Varian/Vacuum Catalog No. VAC-24108B, 578 (Section 7) (1981).
11. Technical information of Mylar: Summary of Properties, (duPont Bulletin M-11).

any material, it should be either tested in actual flights or in ThermoVacRad type simulation chambers. However, investigation of the individual radiation effects on materials can provide important clues for these studies[6].

Although proton represents one of the main space radiations, no study has been made so far of the effect of proton implantation on any of these polymers, at least to the authors knowledge, except our first report of blistering in 250 keV implanted Mylar[7]. The statement that the effect of radiations on polymers is the same if the same amount of energy is deposited irrespective of the type of radiation passing through it[8,9] is somewhat misleading and should be taken carefully. If the radiation is neutral (electromagnetic waves, neutrons and other light neutral particles), this concept may be true to some extent[10]. However, in case of charged particles, the situation may be different in polymers which are electrical as well as thermal insulators. In these materials, unlike metals and metallic alloys, the electrostatic charge and heat "carried by" the ions does not get quenched immediately and this gives rise to local electrostatic and thermal effects[7]. These two phenomena can make ion implantation effects in insulators more complex as compared to metals and metallic alloys. The radiation-induced chemical reactions in polymers (crosslinking, scission, etc.)[9] render the situation more complex and distinct than metals and alloys.

It is noteworthy that nowadays polymers are included in nuclear materials as well, especially as insulating materials in Nuclear Fusion Reactors[2,11]. Glass Fabric Filled (GFF) polyimide is a very promising material for superconducting-magnet coil insulation in such radiation-environment[2]. However, such material can be expected to get radiation doses beyond their tolerance limit in the life span of Nuclear Fusion Reactors[12]. Polymeric materials are also being used in many ways in the containment building of the Nuclear Reactors, and in the worst case (e.g. liquid coolant failure accident) such materials will get a radiation dose as high as 5×10^8 Rads. It is desired that the properties of these materials should not be degraded even after such a high dose of irradiation. Hence qualitative as well as quantitative study of radiation damage of these so called "radiation resistant" polymers like Kapton at such high dose level is becoming very important.

The present study was made keeping in view these two major uses of these polymers. We have already reported[7] severe blistering in thin aluminum (750Å) coated Mylar due to 250 keV H^+ implantation up to a dose of 7×10^{15} H^+ cm^{-2}. In this paper, we report the same phenomenon in Kapton also, but not to that extent. The penetration depth of $\simeq$ 100 keV protons in these polymers is very limited ($\simeq 1$ μm). Therefore in order to attain the homogeneous radiation damage over the whole thickness of the sample, it was also irradiated by reactor neutrons.

Table IIa. Altitudewise Classification of Space Radiations.

Serial Number	Name of the Orbit	Altitude & Type of the Orbit	Major Space Ionizing Radiations
1.	Near earth Orbit (Orbiting Solar Observatory, OSO-H, etc.)	560 km x 327 km 33° Inclination Highly Elliptical	Solar U. V.
2.	Geosynchronized Orbit (ATS-1, etc.)	35,800 km Circular	Protons, Electrons and Solar U.V.
3.	Interplanetary Explorer (IMP-H and Explorer-47 etc.)	2,35,639 km x 2,01,599 km 17° Inclination Spin axis ⊥r to the Sun Elliptical	Solar U. V. Solar Flare Products (mainly directional protons, etc.)

Reference: Physics of Thin Films, Vol. 10 (Academic Press, New York, 1978), Chapter II (Edited by G. Hass and W. R. Hunter.)

Table IIb. Natural and Man-made Maximum Space Radiation Flux.

1.	Natural Sources	Low Energy	High Energy
	A. Solar U. V. $mW\ cm^{-2}sec^{-1}$	$\simeq 40$ 3 eV<E<125 eV	Soft and Hard X-rays of negligible flux
	B. Van-Allen Belts Protons $(cm^{-2}sec^{-1})$	$\sim 10^8$ 120 keV<E<4.5 MeV	$2\text{--}4 \times 10^4$ 30 MeV<E<700 MeV
	Electrons	$\sim 10^9$ E>10 keV	10^5 1.6 MeV<E<5 MeV
		10^8 110 keV<E<1.6 MeV	10^4 E>5 MeV
	C. Solar Flares Protons $(cm^{-2}sec^{-1},$ ~ 1000 sec.)	$\sim 10^9$ $E\simeq 1$ keV	$10^4 - 10^6$ E>30 MeV
2.	Artificial Sources		
	A. Starfish Exp. Electrons (Total)	3×10^8 500 keV<E<5 MeV	3×10^6 E>5 MeV
	B. Nuclear Fired Rockets Neutrons $(cm^{-2}sec^{-1})$	$\sim 10^{12}$ E>25 keV	$10^{12} - 10^{16}$ E>1 MeV
	Gamma-rays (Rads)		$10^5 - 10^7$

Reference: W. E. Price, IEEE, NS-12(6), 2 (1965).

EXPERIMENTAL

25 μm thick Kapton-H film was obtained from the Indian Space Research Organization (ISRO) as well as from E.I. duPont de Nemours & Co. Pieces of appropriate sizes were cut from this sheet and were cleaned in an ultrasonic cleaner with a commercial detergent, and then rinsed with deionized water several times. These samples were packed in a thin Al foil for irradiation with fast neutrons (E>0.1 MeV) at APSARA Reactor of Bhabha Atomic Research Centre (BARC), Bombay. The neutron flux was $1x10^{12}$ cm^{-2} sec^{-1}. The samples were irradiated up to doses of $3x10^{13}$, $6.1x10^{17}$ and $1.2x10^{18}$ cm^{-2}. The temperature of the samples during irradiation was below 45°C.

For H$^+$-implantation at 250 keV, our AN-400 Van-de-Graaff Accelerator was used. The magnetically analyzed mono-energetic beam was allowed to impinge on the Kapton sample mounted on a copper Faraday cup. The sample was coated with 750Å thick Al film on the front side (hereafter "front coat") and with 2500 Å thick Al film on the back side (hereafter "back-coat"). Coating was done by the vacuum thermal evaporation method, and thicknesses were measured by a quartz crystal film thickness monitor. The back-coat was deposited for proper thermal contact between the Faraday cup and the sample which was pasted on it by a conducting silver cement. The 750Å thick front-coat was side-connected with the Faraday cup and the latter was connected to a current integrator and a scaler. Besides measuring the number of protons impinging on it, the front-coat also helps to some extent in dissipating the heat generated in the sample due to implantation. Diffused, defocussed beam (dia ≃ 2mm) having flux density of $2x10^{13}$ H$^+$ cm^{-2} was used for the implantation up to a dose of $7x10^{15}$ H$^+$ cm^{-2}. It was experimentally confirmed that at this flux, even on prolonged implantation (>1 hr.), the temperature of the implanted area did not rise above 40°C. It is worth mentioning here that the 750Å thick Al film absorbs nearly 10 keV energy of the 260 keV protons[13]. So the effective implantation energy was 250 keV only. The schematic view of the Faraday cup assembly is shown in Figure 1. The other experimental arrangements and details were as furnished earlier[7].

For taking the photoacoustic spectrum, the photoacoustic spectrometer developed in the Laser and Spectroscopy Section of this Department was utilized using carbon-black as the photoacoustic material. In order to take the photoacoustic spectrum of uncoated, implanted sample, a larger sample area is needed. Therefore a 1 cm dia. defocussed beam-spot was obtained by increasing the distance between the switching magnet and the Faraday cup. For measuring the dielectric constant of n-irradiated samples, an effective area of 1 cm^2 was coated on both sides with thin (1500Å) Ag-film by the same vacuum thermal evaporation technique. The capacitances of such Ag-Kapton-Ag sandwiched capacitors were measured by TF-2700 Marconi Bridge at 1 kHz. The surface morphology of the neutron and proton irradiated samples were studied by using

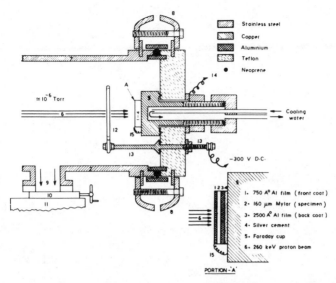

Figure 1. Faraday cup assembly for Proton Implantation. (7. standard beam pipe of 2' dia.; 8. fastening flange; 9. pumping port; 10. isolation valve; 11. ion pump; 12. electron suppression ring; 13. electrical feedthrough; 14. connection to current integrator; and 15. sample side-connection.)

Philips Scanning Electron Microscope PSEM-500 at 3 kV in the secondary electron mode. In the case of neutron irradiated sample, it was necessary to coat it with thin (500Å) Ag-film to avoid electron beam charging during the process of microscopy. However, in the case of H^+ implanted sample, the front-coat was sufficient to avoid any charging caused by the electron beam during the microscopy.

RESULTS

The variation of the dielectric constant $\bar{\varepsilon}$ at 1 kHz with the neutron irradiation dose is shown in Figure 2. The initial value of ε (3.34±1%) agrees with the value reported in the literature (3.50 for 25 μm thick Kapton H film)[14]. The dielectric constant increases slowly but steadily as the irradiation dose increases and finally it become 13% higher than its initial value. The photoacoustic spectra of the virgin, n and H^+ irradiated samples are shown in Figure 3. The neutron irradiated sample shows a shift in the absorption peak from 450 nm to 470 nm and development of an additional absorption peak at λ = 425 nm. The proton implanted sample develops one additional flattened absorption peak at $\lambda \cong$ 515nm. In general, the transmission decreases although it is greater than that of the virgin sample for $\lambda \leq$ 460 nm in both the cases and at $\lambda \geq$ 650 nm in the case of n-irradiated sample.

563

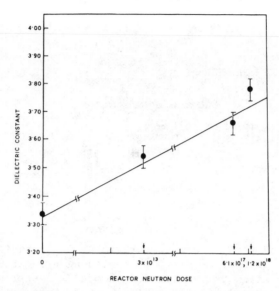

Figure 2. Variation of the dielectric constant (D.C.) with the Reactor neutron dose.

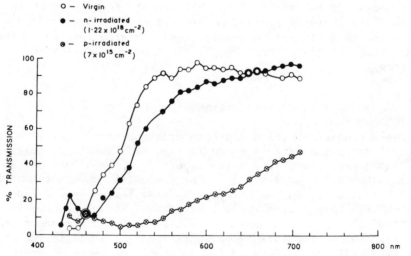

Figure 3. Photoacoustic spectra of virgin, n and H$^+$ irradiated Kapton.

The surface morphologies of the virgin, n-irradiated and H^+ implanted samples are shown in Figure 4. While the neutron irradiated sample shows in general a rough swollen surface in contrast to the flat, featureless surface of the virgin sample, the H^+ implanted surface clearly shows some scattered blisters of dia $\simeq 10$ μm. The scratches formed during extrusion of the film become slightly bold and decorated in case of neutron irradiated sample (Figure 4b). However, this effect is very much pronounced in the case of H^+ implanted sample (Figures 4d, f & h).

DISCUSSION

The increased dielectric constant due to neutron irradiation can be attributed to the increased water absorptivity and radicals produced due to radiation damage of Kapton. The dielectric constant of Kapton increases up to 30% when the film absorbs 2.9 wt.% of water vapor in it[14]. Due to radiation damage induced molecular weight reduction[15], the water absorptivity of Kapton will increase. The radicals produced due to bond breakage, e.g.

will increase the polarizability of the material which results in higher dielectric constant.

As far as the changes in the PAS after neutron and proton irradiation are concerned, these are also due to the radiation induced bond breakage and free radical formation. Though this radiation induced darkening of Kapton has been reported by many authors[2,3,12,14-16], yet no satisfactory explanation has been given by anyone. Nearly similar results have been obtained by Mirtich et al.[16] in 1 keV Ar^+ beam textured 8 μm thick Kapton after bombarding it at a flux rate of 1.8 mA cm^{-2}, for 60 min. Taking the tentative depth of 1 keV Ar^+ ions in Kapton to be $\leqslant 100$ Å, the dose comes out to be $\geqslant 10^{14}$ Rads. This dose is at least 3 orders of magnitude higher than the proton dose in our case (10^{11} Rads). This higher dose, coupled with the excessive thermal heating in the case of Mirtich et al.[16] (as the ion flux was nearly three orders of magnitude higher than that in our case) may be the cause of strong peaks developed at a slightly higher value of λ than in their cases.

When polymers are irradiated, many chemical reactions are possible which depend mainly on the chemical structure of the monomer and its chemical environment, and, of course, on the energy deposited by the energetic radiation. Kapton releases CO when it is heated[17,18] or irradiated[2,10,19] in addition to H_2. Two possible chemical schemes for the release of CO may be

565

(a)

(b)

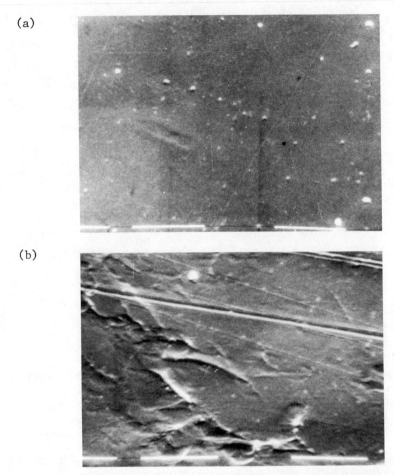

Figure 4. Scanning electron micrographs of the virgin (a), n-irradiated (1.2×10^{18}n cm^{-2} (b&c), and H$^+$ implanted (250 keV, 7×10^{15} cm^{-2} (d,e,f,g & h) Kapton. Micrographs "b" and "c" clearly show the surface roughness and swelling due to n-irradiation. Micrographs "d", "e" and "f" show scratch-line decoration (as compared to similar lines in "a") and surface swelling at several places. Micrographs "g" and "h" show some developed blisters. Marker 1 μm in "c", otherwise 10 μm. Micrographs "d", "e", "f" and "g" are taken at an inclined angle.

(c)

(d)

(e)

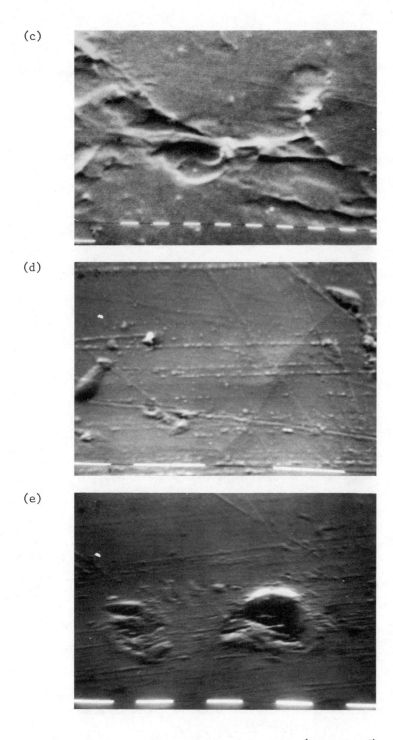

(continued)

(f)

(g)

(h)

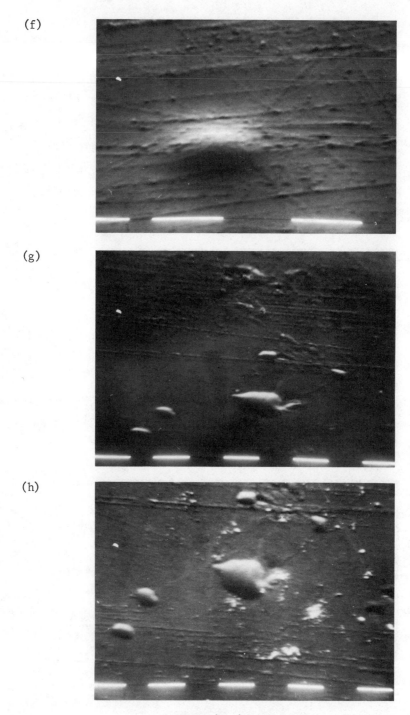

Fig. 4. (continued)

$$O = C \quad\quad C = N - \bigcirc - O - \bigcirc - O - \bigcirc \sim \quad + 2CO$$

$$+ 2CO$$

As dehydrogenation of irradiated aromatic structures is a very common phenomenon and has been reported in polystyrene[20] and also in polyimide[15], it may be one of the major sources of hydrogen gas in the irradiated Kapton. As much as 35.5% of the total hydrogen evolution from irradiated polystyrene is attributed to dehydrogenation of the phenyl sites[20]. However, in the case of polyimide, the dehydrogenation of the phenyl ring between the two imide rings is less likely because the three adjacent rings collectively form a very stable structure[15]. In spite of this, the intense irradiation breaks this intact ring and once it breaks, the rearrangement forming polyisoimide and releasing two CO molecules according to Equation (1) is possible. The need of two more H atoms of the imide-benzene ring is fulfilled by one of the post N phenyl sites, as shown in Equation (1). The imide-benzene ring is no longer in any intact ring-group and it can share its requirements with other similar adjacent rings.

The dehydrogenation of the phenyl sites in Kapton recovers partially afterwards[15]. This can be attributed to the large diffusibility of hydrogen in Kapton. The vacant phenyl sites will capture an hydrogen atom as it comes close to it. This ultimately nullifies the effect of dehydrogenation. However, the same is not true for CO due to its low diffusibility as well as the rearrangement of the sites from where it has been liberated. The chemical scheme shown in Equation (2) is less probable as two four-membered rings are formed in place of the two imide rings.

It would be interesting here to calculate the amount of CO and H_2 released due to irradiation in both cases. The G values of these two gases can be calculated from the gas-release data of Coltman et al.[12] and comes out to be 0.065 for H_2 and 0.013 for CO. The dose 7×10^{15} cm^{-2} of 250 keV protons having 2 μm range in Kapton will release 3.45×10^{16} and 6.89×10^{15} H_2 and CO molecules respectively in the implanted zone if we take this G value into account while the implanted volume has 7.1×10^{18} H_2 and 6.0×10^{17} CO molecules. A fraction of these released molecules, especially that of CO, can coalesce and will develop sufficient pressure to form blisters. H_2

molecules, having a high mobility in Kapton, will contribute less as most of them will be captured at the vacant phenyl sites within the monomer. If all the released gas molecules (4.2×10^{16}) coalesce and are able to form blisters of dia $\simeq 10$ μm, taking UTS of Kapton to be 16.7 kg mm^{-2}, about 1.1×10^4 blisters will be formed in the 2 mm dia H$^+$ implanted zone. If the 2 μm thick skin of the blister is not plastically deformed up to its UTS, the number of blisters formed will still be large. However, such a high number of blisters are not observed in our H$^+$ implanted sample. There are only some tens of blisters of diameter even less than 10 μm in the whole implanted region. This is due to the fact that most of the hydrogen generated is again captured, and all the CO molecules released are unable to coalesce due to their low diffusivity. Local coalescing of CO along the stress channels can decorate them. If it is assumed that only CO is important in blistering, the number of expected blisters comes out to be $\sim 10^3$ which is closer to the observed number of blisters. These calculations are based on similar calculations made in case of Mylar by us[7] and on metallic alloy by Das and Kaminsky[21].

One of the major differences between the proton and neutron irradiation is the penetration depths of these two particles. Neutrons are more penetrating than protons so the radiation damage due to neutrons is distributed over the whole thickness of the polymer more or less homogeneously. Whereas in the case of a proton irradiated sample, it is concentrated at a depth of 2 μm only. This is one of the main reasons for not observing blistering on the surface of neutron irradiated sample. The gas molecules released homogeneously over the thickness of the film have less concentration density and thus are unable to cause distinct blisters. However, they coalesce to some extent and in general swell the surface of the polyimide film.

ACKNOWLEGEMENTS

The authors are grateful to Dr. Vakil Singh, Department of Metallurgical Engineering, I.T., B.H.U. for his help in the Scanning Electron Microscopy. This work was financially supported by DAE, Govt. of India (Project No. 34/4/81-G). One of the authors (S.K.A.) is grateful to CSIR for the award of Senior Research Fellowship.

REFERENCES

1. D. M. Tasca and D. M. Long, Editors, IEEE Annual Conf. Nuclear and Space Radiation Effects, NS-28(6), (1981).
2. R. R. Coltman, Jr. and C. E. Klabunde, J. Nucl. Material, 103-104, 717 (1981).

3. W. Beggs, N. Harvey, G. Brown, J. Thomasson and R. Kurland, IEEE, NS-26(1), 1511 (1979).
4. U. K. Chaturvedi, S. K. Agrawal and A. K. Nigam, Nucl. Inst. Methods, 189, 331 (1981).
5. S. K. Das and M. Kaminsky, Argonne National Laboratory Annual Review No. ANL-78-66, 193 (1978).
6. G. Hass and W. R. Hunter, in "Physics of Thin Films", G. Hass and M. H. Francombe, Editors, Vol. 10, pp. 127-166, Academic Press, New York, 1978.
7. V. Shrinet, U. K. Chaturvedi, S. K. Agrawal, V. Singh and A. K. Nigam, Nucl. Inst. Methods in Physical Sciences, 209-210, 1193 (1983).
8. J. R. Bilinski, IEEE, NS-10, 20 (1963).
9. F. J. Campbell, Radiat. Phys. Chem., 18(1-2), 109 (1981).
10. "Standard Classification System for Polymeric Materials for Service in Ionizing Radiation", ASTM Designation No. D2953, (1971).
11. R. C. Coltman, Jr. and C. A. Klabunde, J. Nucl. Materials (in press).
12. B. A. Engholm, in "Proc. IVth American Nuclear Society Topical Meeting on the Technology of Controlled Nuclear Fusion", Oct. 14-17, 1980.
13. H. H. Andersen and J. F. Ziegler, in "Hydrogen Stopping Power and Range in all Elements", (Vol. 3 of the "Stopping Power and Ranges of Ions in Matter"), Pergamon Press, New York (1977).
14. Kapton Polyimide Film: Summary of Properties (Published by DuPont, Inc.).
15. J. E. Ferl and E. R. Long, Jr., IEEE, NS-28(6), 4119 (1981).
16. M. J. Mirtich and J. S. Sovey, J. Vac. Sci. Technol., 15(2), 697 (1978).
17. A. Burger, E. Fitzer, M. Heym and B. Tewiesch, Carbon, 13, 149 (1975).
18. D. M. Sharman, J. Vac. Sci. Technol., 15(3), 1113 (1978).
19. U. K. Chaturvedi, S. K. Agrawal and A. K. Nigam, ISRO Report Grant 10/3/27 (unpublished work) (1980)(may be obtained from Indian Space Research Organization, ISRO H.Q., Bangalore, India).
20. A. Charlesby, "Atomic Radiations and Polymers", Pergamon Press, New York, 1960.
21. S. K. Das and M. Kaminsky, J. Appl. Phys., 44, 25 (1973).

AMINOSILANE – POLYIMIDE INTERACTIONS AND THEIR IMPLICATIONS IN ADHESION

J. Greenblatt, C. J. Araps and H. R. Anderson, Jr.

IBM Corporation

Hopewell Junction, NY 12533

By means of a two-component dilution technique we have evaluated a series of primary and tertiary aminosilanes as adhesion promoters for polyimides and native SiO_2 in order to study the effect of chain-scission during covalent bond formation between the adhesion promoters and the polyimide. A rapid leveling-off of the increase in adhesion with an increase in the concentration of the reactive component may indicate that the scission of the polymer chains at the interface is detrimental to adhesion. An attempt to use a secondary aminosilane to obtain bonding without scission was unsuccessful. The reaction was not quantitative, as judged by model compounds, and the diamide product decomposed at elevated temperatures yielding a thermally stable imide via chain scission.

INTRODUCTION

The mechanism by which 3-aminopropyltrimethoxysilane (1) (or its triethoxy analog (1)a – Union Carbide's A-1100) functions as an effective adhesion promoter for polyimides and SiO_2, and other mineral or metallic surfaces, can be predicated on the occurrence of chemical reactions within the interfacial region.[1] It has been demonstrated that the intermediate trihydroxysilane generated by hydrolysis of (1) (or A-1100), can either adsorb onto surface OH groups or oligomerize and then adsorb, through a dehydration process, depending on the concentration of the treating solution[2] and its age.[1] It has been assumed that the pendant aliphatic amino group is also capable of reacting chemically with the polyimide although the details of the reaction have only recently been elucidated.[3]

In the case of pyromellitic dianhydride-oxydianiline (PMDA-ODA) polyamic acid (2), reaction of the aliphatic amine of the surface-bound adhesion promoter (1) with (2) results in preferential incorporation of the aliphatic amine at the expense of the aromatic amine of an ODA unit. Each such event results in chain scission at the point of attachment[3] (Figure 1).

This mechanism was arrived at by studying model compounds,[3] and confirmed by our studies of the reaction between 3-aminopropyltriethoxysilane (1)a and the polyamic acid (2) in bulk (vide infra). The net result of this mechanism is that each polymer chain will be attached to the substrate at a single point. Interaction of the already attached chain with a second molecule of (1) will result in chain scission and, therefore, attachment of two shorter segments of the original chain.

By changing and controlling the number of interactions between the surface bound adhesion promoter (1) and the polyamic acid (2), one may gain a quantitative insight into the contribution of the formation of covalent bonds and of the concomitant chain scission to the adhesion.

It is probable that a linear increase in the number of covalent bonds between a polymeric matrix and a substrate will result in a linear increase in adhesion. Indeed this was shown in an elegant way by Ahagon and Gent[4a] for a polybutadiene matrix and a vinylic silane adhesion promoter and recently by other workers using a similar approach[4b]. The occurrence of chain scission, however, may result in:

1. Decrease in adhesion due to reduced cohesive strength near the interface as a result of reduction in chain length, or

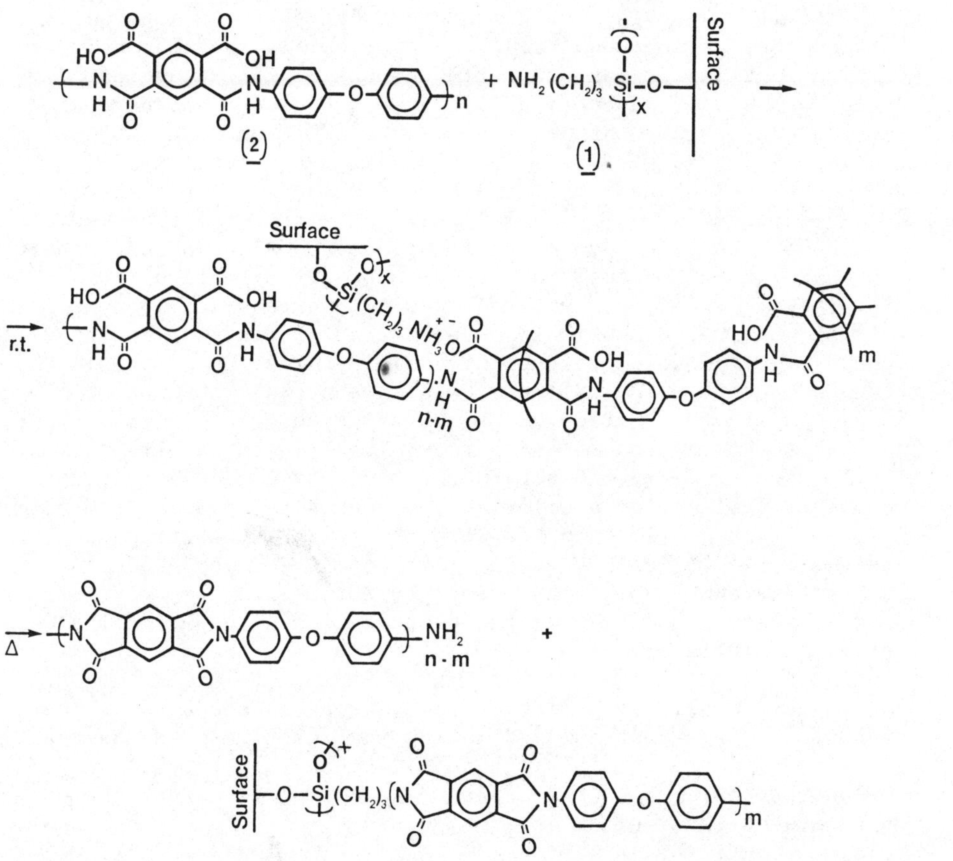

Figure 1. The mechanism of interaction of PMDA-ODA polyamic acid (2) with a primary aminosilane adhesion promoter (1).

2. Increase in adhesion due to better dissipation of stresses (lower Tg) due to the reduction in molecular weight.

A linear increase in the number of covalent bonds between the polyimide chains and the substrate accompanied by polyimide chain scission will result in a negative deviation from linearity if the scission is detrimental to adhesion, a positive deviation if the scission is beneficial, or no deviation if the scission does not affect the adhesion. For a recent review on the linear increase of the adhesion with linear increase in the concentration of adhesion promoters at the interface and deviations thereof, see reference 5.

We lack an obvious method of insuring that the interaction between the surface bound adhesion promoter (1) and the polymer is varying in a linear manner. From the scarce literature data, which includes a single adsorption isotherm on E-Glass[6] and several[7] reports on single point adsorption or desorption data on Pyrex[7], silica gel[8a,b] and E-Glass,[8b] native SiO$_2$[10] and metallic substrates,[10] it seems that in the normal concentration range of (1) in water (0.01 - 1%), a multilayered structure is formed on flat substrates.[8b] Indeed, a very elaborate vapor-phase technique was used to obtain monolayer coverage on a flat native SiO$_2$ substrate.[9] It is apparent that the trihydroxysilane polymerizes to yield a three-dimensional network that precludes deposition of a monolayer. A monolayer may, in principle, form from a monosilanol compound, or when the interactions between the trihydroxy silanol OH groups and the surface OH groups are much stronger than those between the trihydroxysilanols themselves (which is hard to imagine in the case of SiO$_2$). Since there is no way of depositing a monolayer of (1) from a solution, the deposited polyimide chains interact with a film of unknown thickness and, therefore, an unknown number of amino groups. As such it is impossible to control the number of interactions by controlling the concentration of the treating solution or the contact time since we will always obtain a multilayer.

To avoid all these difficulties, we have chosen to use Gent's method,[4a] by systematically diluting the reactive adhesion promoter (1) with an inert trialkoxysilane, at a constant total concentration level of the treating solution, to ensure the same multilayered coverage.

Thus, we have mixed the reactive component (1) with the "inert" N,N-dimethyl-3-amino-propyltrimethoxysilane (3), in five different ratios. It is expected that (3) will compete with (1) for the active sites of the substrate because of similar steric and electronic characteristics and, therefore, will adsorb on the substrate in a similar way. It will also form a siloxane copolymer with (1) due to their similarities. However, being a tertiary amine, (3) is not expected to incorporate into the polyimide structure. A second inert component that we evaluated was n-butyltrimethoxysilane (4), which is similar to (1) but its amino group is replaced by a methyl group.

A linear increase of the fraction of (1) in the treating solutions will result in a multilayered coverage with a linear increase in the number of covalent bonds between the promoter (1) and the polymer, and a concomitant linear increase in chain

scission. A fourth adhesion promoter studied by us was
N-methyl-3-aminopropyltrimethoxysilane (5). Having one of its
active hydrogens blocked by a methyl group, compound (5) was
expected to form a covalent amide bond to the polymer without
causing chain scission. This compound was compared to (1) in a
single adhesion experiment. The adhesion of the polyimide to the
different substrates was measured by a 90° peel test at room
temperature. The chemical reactions between the adhesion promoters
(1), (3) and (5) and model compounds were followed by IR.

EXPERIMENTAL

Adhesion Test

    Treating Solutions: The adhesion promoters (1), (3), (4) and
(5) were purchased from Petrarch Systems Inc., and were redistilled
prior to their use. We took IR and NMR spectra of the compounds to
ensure their identity. The IR spectra were taken on a Perkin Elmer
283 infrared spectrophotometer. The NMR spectra were taken on a
Varian EM360L spectrometer. Melting points were measured on a
Buchi 510 instrument and are uncorrected. The physical constants
of the silanes are listed below:

1. 3-aminopropyltrimethoxysilane (1): b.p. 30°C/0.65 mm Hg
    (104°C/30 mm Hg)[11]. IR (neat), cm^{-1}: 3360(w, NH_2), 3280(w,
    NH_2), 2940(vs, $-CH_2CH_2CH_2-$), 2840(vs, $-OMe$), 1600(w), 1460(m),
    1410(w), 1190(vs), 1080(vs). NMR ($CDCl_3$), δppm:
    0.60(2H,t,j=7Hz, $-CH_2Si$), 1.18(2H,s,$-NH_2$),
    1.53(2H,m,$-CH_2-CH_2-CH_2-$), 2.67(2H,t,j=7Hz,$-CH_2N$),
    3.57(9H,s,$-OMe$).

2. N,N-dimethyl-3-aminopropyltrimethoxysilane (3): b.p. 30°C/0.65
    mm Hg (106°C/30 mm Hg)[11]. IR (neat), cm^{-1}: 2940(vs,
    $-CH_2CH_2CH_2-$), 2840(vs, $-OMe$), 2820(vs, $-N-Me$), 2760(vs,
    $-N-Me$), 1460(s), 1410(w), 1375(w), 1360(w), 1260(m), 1190(vs),
    1080(vs), 920(w), 880(w), 835(vs), 780(s). NMR ($CDCl_3$), δppm:
    0.57(2H,t,j=7Hz,$-CH_2Si$), 1.50(2H,m,$-CH_2CH_2CH_2-$),
    2.17(8H,s+t,$-CH_2N(CH_3)_2$), 3.53(9H,s,$-OMe$).

3. n-butyltrimethoxysilane (4): b.p. 164°C (164.8°C/760 mm Hg)[12].
    IR (neat), cm^{-1}: 2950 (vs), 2920(vs), 2860(s), 2850(s),
    1460(m), 1200(s), 1080(vs), 900(s). NMR ($CDCl_3$), δppm:
    0.60(2H,t,j=7Hz,$-CH_2Si$), 0.87(3H,t,j=7Hz,$-CH_2CH_3$),
    1.33(4H,m,$-CH_2-CH_2CH_2CH_3$), 3.57(9H,s,$-OMe$).

4. N-methyl-3-aminopropyltrimethoxysilane (5): b.p. 30°C/0.65 mm
   Hg (106°C/30 mm Hg)[11]. IR (neat), cm^{-1}: 3920(w,NH),
   2930(vs,-CH$_2$CH$_2$CH$_2$-), 2830(vs, -OMe), 2780(s, N-Me), 1460(m),
   1180(s), 1080(vs), 830(s), 780(s). NMR (CDCl$_3$), δppm: .
   0.65(2H,t,j=7Hz,-CH$_2$Si), 1.01(1H,s,NH),
   1.62(2H,m,-CH$_2$CH$_2$CH$_2$-), 2.45(3H,s,-NCH$_3$),
   2.62(2H,t,j=7Hz,-CH$_2$N), 3.62(9H,s,-OMe).

   Solutions (6 x 10^{-3}M) of the four promoters in distilled
deionized water were prepared and mixed accordingly within one
hour, with the exception that the solution of promoter (4) was left
at room temperature for 24 hours prior to its mixing with (1) to
ensure complete hydrolysis.

   Substrate: Silicon wafers (57 mm diameter) were etched in 1:7
buffered HF for 0.5 hr., rinsed with water and treated immediately
with the treating solutions that were one hour old. The wafers
were flooded with the treating solutions for 30 seconds; then the
solutions were spun-off at 2000 rpm for 30 seconds. Areas (18 mm
wide) at the bottom of the wafers were masked by 3M's Scotch Tape
prior to flooding with the treating solutions and demasked after
the promoters were spun-off, and prior to polymer application. The
adhesive transferred from the tape provided enough residue at the
interface to ensure proper peel initiation.

   Polymer: Three layers of a PMDA-ODA type polyamic acid in
N-methylpyrrolidone were spun-on at 2000 rpm for 30 sec., with a
lag time of 3-5 minutes at 85°C between successive coatings. This
resulted in a uniform thickness of 23 μm. Five polyimide-coated
wafers were simultaneously cured for 1/2 hr. at 85°C, 150°C, 200°C,
300°C, and 400°C under nitrogen.

   Peel Test: The polymer samples were scribed into 1.59 mm
(1/16") wide stripes down to the silicon wafer and peeled off at
90°. The samples treated with the mixture of (1) and (3), as well
as the sample treated with (5), were peeled on a home-built 90°
peel tester, at a peel rate of 8.4 mm/min.[13] The 90° angle was
kept constant by mechanical coupling of the movement in the X and Z
directions. The samples treated with the mixture of (1) and (4)
were peeled using an Instron universal testing instrument Model
1122, having a load cell of 5N. The peel-rate was 5 mm/min. and
the 90° angle was kept 'constant' over the 2 cm peel path by means
of a 1 m long arm. At least three stripes per wafer were peeled.

Chemical Reactions

(A) Chemical reaction between 3-aminopropyltriethoxysilane ($\underline{1}$)a and polyamic acid ($\underline{2}$).

10 g (4 mmol) of 16% (w/w) polyamic acid ($\underline{2}$) in NMP (N-methylpyrrolidone) were mixed with 1.768 g (8 mmol) of the triethoxy analog of ($\underline{1}$) (A-1100). The gradual addition of ($\underline{1}$)a into ($\underline{2}$) resulted in a tremendous increase in viscosity and generated a large amount of heat. Overnight a white semi-solid separated from the reaction mixture; its IR spectrum was compatible with an ammonium carboxylate salt formed between the amino group of ($\underline{1}$)a and the carboxylic acids of ($\underline{2}$). New bands formed at 1560 cm^{-1} and 1335 cm^{-1} (carboxylate ions). Additionally, there was a decrease in absorbance in the region between 3500 cm^{-1} and 3000 cm^{-1} (-OH and -NH_2 groups) and a marked absorbance decrease in the 1720 cm^{-1} band of the amic acid carboxyl. A big absorbance increase in the 1070 - 1100 cm^{-1} region (Si-O bonds) indicated the presence of ($\underline{1}$)a.

We heated the mixture at 200°C in a stream of nitrogen for four hours until solvent losses were undetectable. This resulted in a brown compound, which upon sublimation at 175°C at a reduced pressure, yielded a quantitative amount of oxydianiline (ODA) with an IR spectrum (Nujol mull) identical in all respects to a spectrum of a sublimed commercial (Aldrich) material. The residual compound had imide IR (Nujol mull) absorbances at 1760 cm^{-1} and 1720 cm^{-1} similar to those of the diimide ($\underline{6}$) formed in the reaction between ($\underline{1}$)a and the pyromellitic dianhydride (PMDA). (See Figure 2).

(B) Chemical Reaction Between ($\underline{1}$)a and PMDA.

1.1 g (5 mmol) of PMDA were dissolved in 20 ml of THF. A solution of 2.2 g (10 mmol) of the triethoxy analog of ($\underline{1}$) in 10 ml of THF was added dropwise at room temperature. Evaporation of the THF followed by heating of the resulting mixture for five minutes at 160°C resulted in the evolution of ethanol and water and the formation of a viscous transparent liquid having IR (neat) absorbances at 1765 cm^{-1} and 1720 cm^{-1} (imide bands) (no further purification was attempted). (See Figure 2).

(C) Reaction Between ($\underline{1}$)a and Phthalic Anhydride ($\underline{10}$)

A reaction sequence similar to B was used. The imide formed had the following IR (neat) bands: 1768 cm^{-1} and 1710 cm^{-1}.

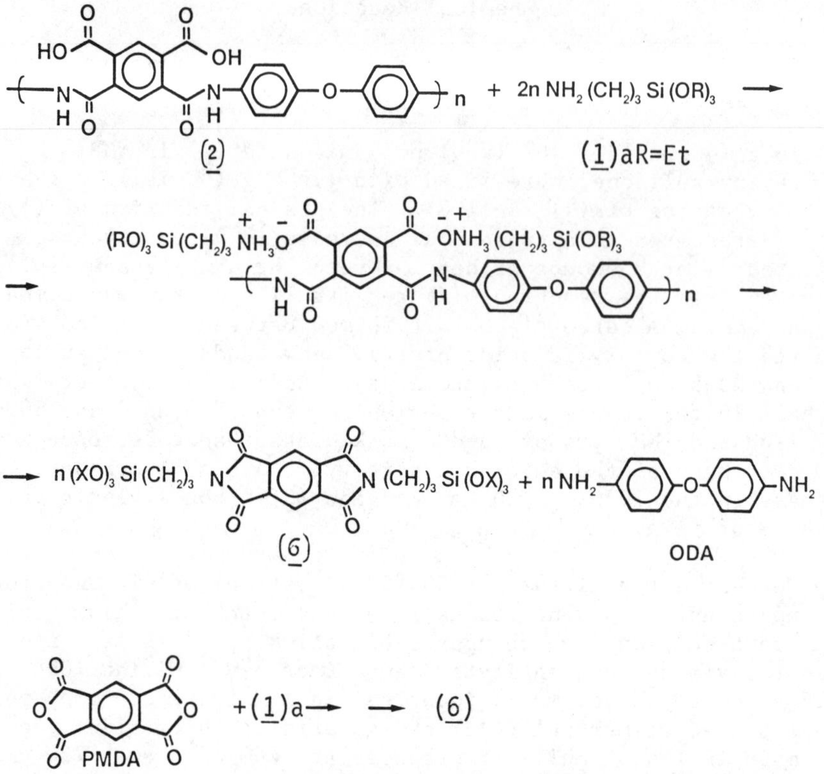

Figure 2. Reaction between 3-aminopropyltriethoxysilane (1)a and polyamic acid (2) and PMDA.

(D) Reactions Between 3-aminopropyltrimethoxysilane (1), N,N-dimethyl-3-aminopropyltrimethoxysilane (3), N-methyl-3-aminopropyltrimethoxysilane (5) and N-phenylphthalamic acid (8).

0.12 g (0.5 mmol) of (8) in 2 ml THF were mixed separately with 0.089 g (0.5 mmol) of (1), with 0.103 g (0.5 mmol) of (3) and with 0.096 g (0.5 mmol) of (5). In each case, after removing a few drops of the resulting reaction mixture for IR analysis, the solvent was blown off in a stream of nitrogen. The reaction vessels were covered with watch-glasses and were heated at 150°C for 1/2 hr. followed by 200°C for 1/2 hr. N-phenylphthalimide (11) (m.p. 206°C) sublimed out of the reaction mixtures of (3) and (5) in quantitative amounts and crystallized on the walls of the vessel and on the watch glass. Its IR (Nujol) spectrum had the

580

following bands: 1780 cm^{-1}, 1700 cm^{-1}, 1380 cm^{-1}, and 715 cm^{-1}. The IR spectrum was identical to that of N-phenylphthalimide obtained from the thermal reaction of (8) alone using the same technique. (Subliming (8) below its melting point of 173°C, even at low pressure, resulted in sublimation of (8) rather than of (11)). The IR (Nujol) absorbance peaks in this case were at 3320 cm^{-1}, 1718 cm^{-1}, 1640 cm^{-1}, 1600 cm^{-1}, 1390 cm^{-1}, and 690 cm^{-1}.

The THF samples withdrawn at the early stages of the reactions showed typical salt bands at 1550 - 1560 cm^{-1} and 1320 - 1340 cm^{-1} due to proton transfer from the carboxylic acid OH groups to the basic amine groups.

Heating the reaction sample of (1) and (8) at 85°C for two hours and then at 150°C for 1/2 hr. resulted in formation of a new imide, N-3-polysiloxypropylphthalimide (9). This imide persisted unchanged on thermal treatment of 200°C for one week. The imide bands of (9) were at 1765 cm^{-1} and 1710 cm^{-1}.

Contrary to the preceding reaction, the product of the reaction between (3) and (8) showed a decreased spectrum amplitude after heating at 150°C for 1/2 hr., and the spectrum showed the presence of (3) as judged by the Si-O bands at 1110 and 1020 cm^{-1}. The IR bands were located at 2930 cm^{-1}, 2810 cm^{-1}, 2750 cm^{-1}, 1455 cm^{-1}, 1110 cm^{-1}, and 1020 cm^{-1}. Prolonged (16.5 hours) heating at 200°C resulted in decomposition of the promoter with preferential loss of the two N-methyl groups as evidenced by the decrease of the 2750 cm^{-1} band.

A different result was obtained for the reaction mixture of (5) and (8). After 10 minutes at 127°C, some N-phenylphthalimide (11) formed, as judged by the appearance of three imide bands at 1778 cm^{-1} (w), 1735 cm^{-1} (s) and 1710 cm^{-1} (vs). Heating the sample at 200°C for 1/2 hr. resulted in sublimation of the N-phenylphthalimide (11); the overall intensity of the spectrum dropped, and new imide bands appeared at 1770 cm^{-1} (w) and 1710 cm^{-1} (m). Those were smaller in intensity than those of (11) that were obtained after 10 minutes at 120°C. A large band at 1630 cm^{-1} was formed as well. Heating for three hours at 200°C resulted in an increase in the intensity of the imide bands at 1770 and 1710 cm^{-1} as well as an increase of a 1670 cm^{-1} band with the 1630 cm^{-1} band losing some intensity. Seven hours heating at 200°C resulted in further increase in the 1770 cm^{-1}, 1710 cm^{-1} and 1660 cm^{-1} bands at the expense of the 1630 cm^{-1} band and the 2780 cm^{-1} band of the N-methyl group. At this stage, the IR spectrum was similar to that obtained for the reaction between (1) and (8), except for some bands at 3300 cm^{-1} possibly formed by decomposition of the unreacted silane (5) (similar bands appeared in the spectrum of the reaction mixture between (3) and (8) after 16.5 hours at 200°C).

Upon increasing the fraction of (1) in the treating solution that contained (3), we observed an increase in the adhesion (Figure 3). The overall response was flat because (3) turned out to be an excellent adhesion promoter for polyimide and native SiO_2. The absolute value of the normalized peel-force per strip width for (3) was 1.32 ± 0.03 kN/m as compared to 1.57 ± 0.02 kN/m for (1) or 0.56 ± 0.02 kN/m for bare SiO_2 (the last two values were obtained in the same experiment). Since (3) turned out to be an excellent adhesion promoter, possibly due to acid-base interactions[14] between the basic amine and the acidic polymer, the negative deviation from linearity upon increasing the concentration of (1) was quite small.

To eliminate the acid-base component of adhesion from the "inert" silane, n-butyltrimethoxysilane (4), was used in place of N,N-dimethyl-3-aminopropyltrimethoxysilane (3). The results were similar except that (4) alone did not function as an adhesion promoter. The peel-force value for pure (4) was zero. The peel-force for pure (1) was 1.30 ± 0.05 kN/m, slightly lower than in the previous experiment because we used a different peeling technique and a different polyimide batch. In this case we observed an increase in adhesion with increasing concentration of (1) with a distinct leveling-off after 50% of (1) had been added (see Figure 3). The initial increase in adhesion is due to covalent bonds as well as acid-base interactions between (1) and the polyimide. The leveling-off of the increase in adhesion with further increase in the concentration of the reactive adhesion promoter can be caused by two main reasons: (1) The inability of all of the amino groups in the interfacial region to react because of steric reasons or because of unavailability of enough reactive sites in the polymer, and thus once a threshold value is reached there will be no more covalent bonds formed between the polymer and the adhesion promoter despite an increase in the concentration of amino groups at the interface. (2) Deterioration of the mechanical properties of the polymer in the interfacial region, caused by the scission of the chains due to covalent bond formation.

Assuming that the uppermost layer of the promoter interacts with a 10Å deep polymeric layer, and the adhesion promoter is closely packed giving rise to a density of 5 amino groups per $100Å^2$ one can show by a crude calculation that the density of active sites of the polymer in the interfacial region almost equals the density of the amino groups. Thus, by taking into account the density of the polymer (1.42 g/cc) one can calculate the number of repeat units in a given volume to be $n = 1.42/382 = 3.72 \times 10^{-3}$ moles/cc. The number of moles of repeat units in a volume defined by a $100Å^2$ cross section and a depth of 10Å will be 3.72×10^{-24} moles per $1000Å^3$, or $6.02 \times 10^{23} \cdot 3.72 \times 10^{-24} = 2.24$ repeat units per $1000Å^3$. Since every repeat unit has two active sites, the

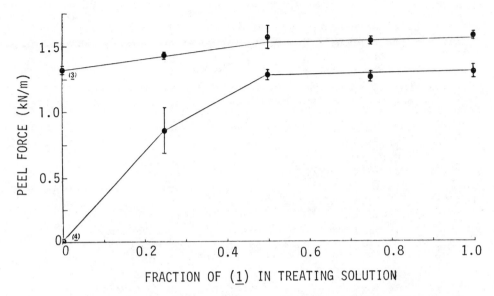

Figure 3. Adhesion of a PMDA-ODA polyimide to native SiO_2 as a function of the fraction of (1) in the treating solutions that contained N,N-dimethyl-3-aminopropyltrimethoxysilane (3) and n-butyltrimethoxysilane (4) at a total concentration of $6 \cdot 10^{-3}$M.

number of active sites in the "interfacial" region will be 4.5 or very similar to the number of amino groups (5 per 100Å^2).

The inability of all of the amino groups to react because of steric reasons is much harder to prove. If indeed this is the case in our system, to our knowledge, this will be the first time such an argument is used.

We prefer, therefore, the explanation that assumes deterioration of the mechanical properties of the polymer at the interface due to chain scission. Deterioration of mechanical properties of the polymer was claimed as the cause for deviations from linearity between the adhesion and the concentration of adhesion promoters at the interface, when the mode of application of the adhesion promoters was mixing them with the polymers.[5]

In another attempt to evaluate the detrimental contribution of polymeric chain scission, we evaluated the capacity of N-methyl-3-aminopropyltrimethoxysilane (5) to function as an adhesion promoter. The silane (5) has one active amine hydrogen blocked by a methyl group, such that the amine is capable of forming only an amide with the free carboxylic acid group of the polyamic acid. This could result in a structure such as (7) (see Figure 4).

Upon evaluation of (5) we found it to be, however, inferior to (1). This was true for as-received as well as for a freshly distilled material (which gave the same results). The peel test values were 1.07 ± 0.02 kN/m for (5) and 1.43 ± 0.04 kN/m for (1). Interestingly, the N,N-dimethyl analog (3), which is not expected to covalently bond to the polyimide gave adhesion values of 1.32 ± 0.03 kN/m as compared to 1.57 ± 0.02 kN/m for (1). The peel test values generated in this study are summarized in Table I.

Table I. Peel Strength* as a Function of the Composition of the Treating Solutions.

Fraction of (1) in the treating solution.

Diluent	0	0.25	0.50	0.75	1.0
(3)	1.32±0.03	1.43±0.03	1.56±0.09	1.53±0.02	1.57±0.02
(4)	0	0.86±0.16	1.28±0.04	1.26±0.04	1.30±0.05
(5)	1.07±0.02	--	--	--	1.43±0.04

* Peel force per unit width, kN/m.

In order to explain this behavior, we have checked the validity of the model proposed for the interaction between (5) and (2) (Figure 4), by reacting (1), (3), and (5) with N-phenylphthalamic acid (8) as a model for the polyamic acid (2). The chemical reactions between (8) and (1), (3) and (5) are represented schematically in Figure 5. Amino silane (1) reacted with amic acid (8) by replacing the aniline to form a new imide (9). This compound was similar to that obtained in the reaction between (1) and phthalic anhydride (10) (see Figure 5). The N,N-disubstituted amino silane (3) failed to react with (8), which merely imidized upon heating. The resultant N-phenylphthalimide

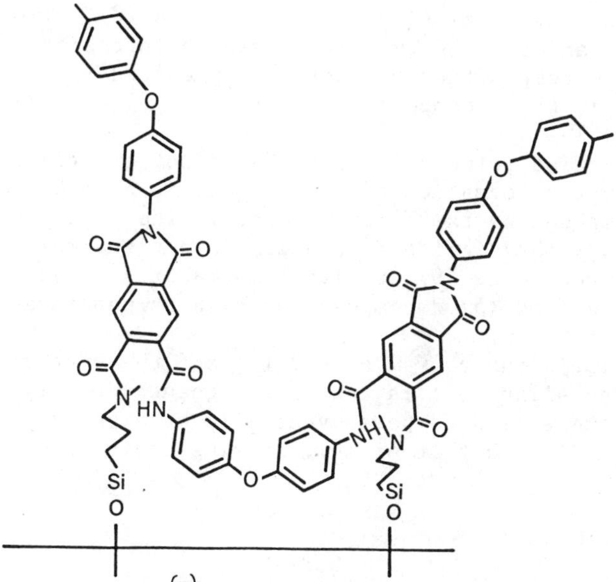

Figure 4. A proposed model for the interfacial interaction between PMDA-ODA polyimide and a secondary aminosilane (5).

(11), readily sublimed from the product mixture, leaving behind the polysiloxane condensation product of (3), which slowly decomposed at 200°C. The N-methylaminosilane (5), behaved in a different way. When reacted in the bulk in a covered beaker, most of the N-phenylphthalimide (11) sublimed out of the reaction mixture after heating 1/2 hr. at 200°C. On performing the reaction on a KBr crystal at a lower temperature, after 10 minutes at 120°C, some N-phenylphthalimide (11) formed, as evidenced by an IR spectrum. Further heating at 200°C resulted in its disappearance (sublimation) and in a decrease in the intensity of the spectrum. A new imide was gradually forming, an imide that was stable at 200°C for one week. New imide bands at 1770 cm^{-1} and 1710 cm^{-1} were forming at the expense of two bands at 1630 cm^{-1} and 2780

cm^{-1}, which can be ascribed to a diamide (12); the first is an amide carbonyl band and the second is an N-methyl C-H band. The two imide bands at 1770 cm^{-1} and 1710 cm^{-1} are similar to those of (9) obtained in the reaction between (1) and (8). Two other bands at 1660 cm^{-1} and at 3300 cm^{-1} increased in intensity. They can be ascribed to decomposition products of unreacted (5) (similar bands were formed in the decomposition of (3)).

The reaction between (5) and (8) is not quantitative as evidenced by the formation of (11). This is probably due to steric hindrance imposed by the N-methyl group. The reaction product is most certainly the diamide (12) which is not thermally stable and converts to the imide (9) at high temperatures (Figure 5). The residual unreacted (5) decomposes at high temperatures.

By analogy, the polyamic acid (2) and the surface bound N-methylamino silane (5) may not react quantitatively. The diamide (7) that forms will decompose eventually at high temperatures to a structure similar to that of (6), causing chain scission of the polymer.

SUMMARY

In conclusion, we have shown that (1) is an excellent adhesion promoter for PMDA-ODA polyimide and native SiO_2. With the two-component competitive dilution technique we have evaluated a series of primary and tertiary amino- and hydrocarbon-terminated trialkoxy silanes in order to evaluate the specific contributions of covalent bond formation and the concomitant polyimide chain scission. The results indicate that:

1. Polyimide chain scission is detrimental to adhesion as evidenced by the attenuation of the linear increase in adhesion with increased concentration of the primary amino silane (1).

2. N,N-dimethyl-3-aminopropyltrimethoxysilane (3) is an adhesion promoter for polyimides. The exact mechanism of its interaction is obscure, however. At low temperatures one may envision acid-base interactions. At high temperatures, however, (3) decomposes and its decomposition products may interact with the polyimide.

3. N-methyl-3-aminopropyltrimethoxysilane (5) is also an adhesion promoter for polyimides. However, it was slightly worse than (1) because, under the conditions used, the interaction between (5) and the polyamic acid is not quantitative. The resultant diamide (7) is unstable at high temperatures and converts eventually to the aliphatic imide by losing its N-methyl group and eventually causing chain scission.

586

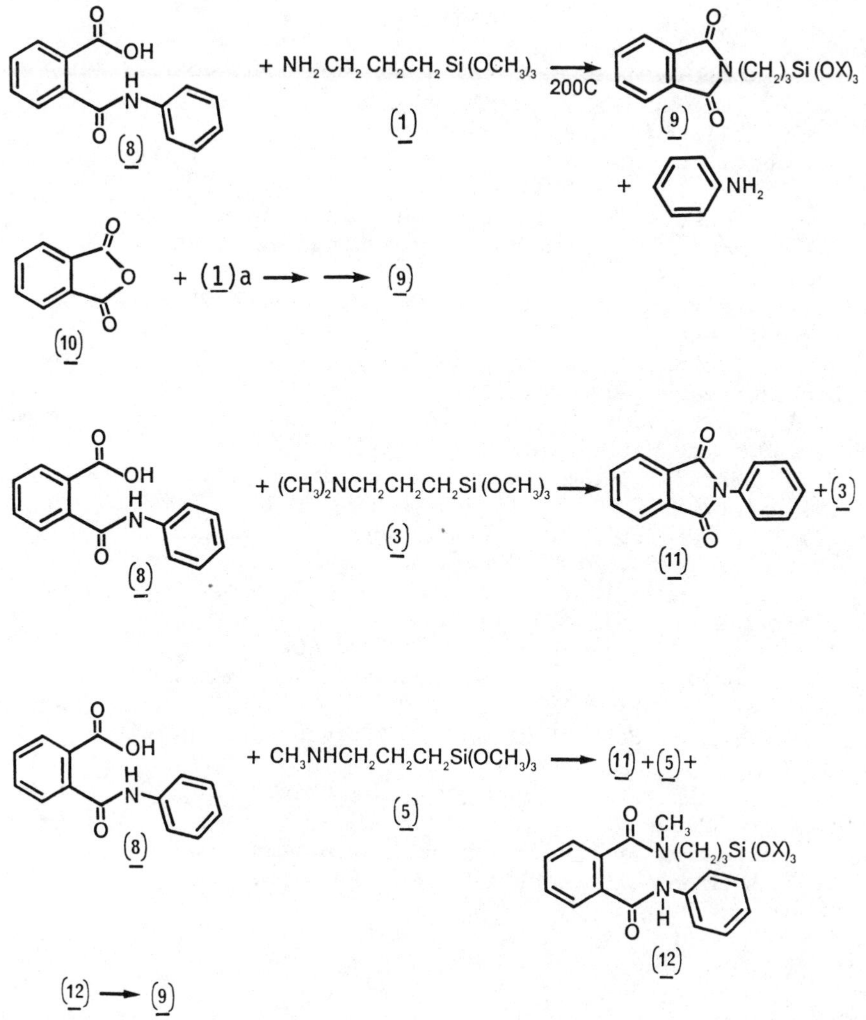

Figure 5. Reactions between primary (1), secondary (5) and tertiary aminosilanes (3) and N-phenylphthalamic acid (8).

4. It may be possible to use mixtures of (1) with inert silanes,
   as adhesion promoters, since it was shown above that a 1/1
   mixture of (1) and (4) was as good in promoting adhesion as
   pure (1). These inert silanes could provide for example
   higher thermal stability or additional modes of interaction
   such as acid-base or charge-transfer or interpenetrating
   networks to complement the chemical bond formation of the
   reactive component.

REFERENCES

1. For general reviews see: (a) M. R. Rosen, J. Coating Technol., 50, 70 (1978); (b) E. P. Plueddemann, J. Adhesion, 2, 184 (1970); (c: B. Arkles, Chemtech., 766, Dec. 1977.
2. H. Ishida, S. Naviroj, S. K. Tripathy, J. J. Fitzgerald and J. L. Koenig, J. Polym. Sci., Chem. Ed., 20, 701 (1982).
3. H. G. Linde, J. Polym. Sci., Chem. Ed., 20, 1031 (1982).
4. (a) A. Ahagon and A. N. Gent, J. Polym. Sci., Phys. Ed., 13, 1285 (1975).
   (b) P. Dreyfuss, Y. Eckstein, Q. S. Lien and H. H. Dollwet, Polymer Lett., 19, 427 (1981).
5. S. Wu, "Polymer Interface and Adhesion," pp. 420-422, Marcel Dekker, Inc., New York, 1982.
6. O. K. Johanson, F. O. Stark, G. E. Vogel and R. M. Fleischmann, J. Composite Materials, 1, 278 (1967).
7. M. E. Schrader, I. Lerner and F. J. D'Oria, Modern Plastics, 195, Sept. 1967.
8. (a) T. G. Waddell, D. E. Leyden and M. T. DeBello, J. Amer. Chem. Soc., 103, 5303 (1981).
   (b) C.-H. Chiang, H. Ishida and J. L. Koenig, J. Colloid. Interface Sci., 74, 396 (1980).
9. I. Haller, J. Amer. Chem. Soc., 100, 8050 (1978).
10. F. J. Boerio, L. H. Schoenlein and J. E. Greivenkamp, J. Appl. Polym. Sci., 22, 203 (1978).
11. J. L. Speier, C. A. Roth and J. W. Ryan, J. Org. Chem., 36, 3120 (1971).
12. A. Ya. Yakubovskaya, Zhur. Obshchei Khim., 25, 1124 (1955). CA. 50, 3218d.
13. H. Lee and R. H. Lacombe, unpublished results.
14. W. B. Jensen, Chemtech, 755, December 1982.

POLYMER COMPLEX FORMATION OF POLYAMIC ACIDS WITH AMINES

L. Minnema and J.M. van der Zande

Philips Research Laboratories
P.O.B. 80.000,5600 JA Eindhoven
The Netherlands

Upon the addition of amines or ammonia, to
solutions of different polyamic acids (PAA) it
was noticed that sometimes precipitates were
formed. This is particularly the case with the
PAA based on the condensation of pyromellitic
acid dianhydride with 4,4´-diaminodiphenylether,
(PAA(I)), and with primary amines or ammonia. It
is assumed that the addition of amines to polyamic acids causes the polymer to complex with itself; normally polymer complexation involves two
different polymers.

PAA(I) was neutralized with n-butylamine in
very dilute solution, and the reduced viscosity
was measured as a function of the degree of neutralization. The measurements provided an unusual
curve, demonstrating the contraction of polymer
coils at increasing degree of neutralization. It
is hypothesized that this contraction is due to
polymer complexation, implying that ladder-like
structures of cooperative hydrogen bonds are
formed. This assumption is furthermore supported
by the observation that a small temperature rise
is sufficient to cancel the polymer contraction,
indicating a cooperative effect.

An alternative explanation based on the different shielding of COO^- groups by primary, secondary and tertiary amines is unacceptable as in
these aprotic solutions the amines were found not
to be protonated. A second possible explanation

589

simply assuming that Θ-conditions are attained is not justified by the specific features observed ved in these systems.

Also with polymer complexation the well known "molecular aggregation" or "mesomorphous state" in polyimide films based on PAA(I) (e.g. Kapton H film) can be explained, at least partially, as well as the aggregation in polyamic acids upon concentrating and drying.

INTRODUCTION

While studying reactions between polyamic acids (PAA) and the well known coupling agent γ-aminopropyltriethoxysilane (APS), we observed that sometimes precipitates were formed. The system PAA-APS is complicated because of the reactivity of the siliconethoxy groups.

We therefore studied the simpler system PAA-amine or ammonia and again sometimes noticed precipitation to occur. In this paper we have tried to explain why in dilute solution PAA-amine macromolecules may contract, and why in concentrated solution precipitates are formed.

EXPERIMENTAL

Two commercial, IC quality, polyamic acid solutions were used, viz., Pyralin 2545, PAA(I), with 14.2% solids content; and Pyralin 2555, PAA(II), with 16.6% solids content (both products of Du Pont). The solids content was determined by TGA analysis. According to the manufacturers the viscosity of (I) is 0.9-1.3 Pa.s and that of (II) 1.2-1.6 Pa.s. The polyamic acid in solution (I) is the condensation product of pyromellitic acid dianhydride and 4,4´-diaminodiphenylether, in (II) it is the condensation product of benzophenonetetracarboxylic acid dianhydride and the same diamine. The solvent system in solutions (I) and (II) is N-methylpyrrolidone (NMP)-aromatic hydrocarbon (80:20). The amines used were triethylamine (TEA), diethylamine, n-butylamine (BuA) and ammonia.

FT-IR Analyses on PAA-amine systems were carried out with the Nicolet 7199 FT-IR spectrometer. All experiments were carried out with PAA(I) and the amines used were triethylamine, n-butylamine and ammonia. A polyamic acid solution was spun on a KBr-pellet and dried for 30 minutes in vacuo at room temperature. Then the surface was exposed to the relevant amine or ammonia for one minute, and the infrared measurement carried out. Measurements at 50°C were carried out in a thermostated cell, type CTC 250 Temperature Controller, of Beckman. We also

590

tried to carry out measurements on the same systems prepared in dilute solution. However when applied to a KBr-pellet these broke down on drying, so suitable comparisons between triethylamine and n-butylamine or ammonia could not be made. Subtraction FT-IR measurements were carried out on a 20% N-methylpyrrolidone (NMP) solution of o-carboxybenzanilide, C_6H_4 (COOH) $CONHC_6H_5$, (III), partially neutralized with n-butylamine; O-carboxybenzanilide was prepared according to Linde[1].

H[1]-NMR measurements were carried out on a 20% solution of (III) which was neutralized with n-butylamine. Deuterated dimethylsulfoxide, DMSO-d_6 was used as the solvent. The NMR apparatus was a Bruker WP 80 SY spectrometer.

The viscosity of dilute solutions was measured with a KPG-Ubbelohde Micro-Viscosimeter (Schott), capillary number 1c. The temperature was maintained at 25^0 C (30^0C-50^0C) ± 0.01^0C.

RESULTS AND DISCUSSION

When triethylamine is added to polyamic acid solutions (I) and (II), even in great excess of the COOH content of the PAA, a clear solution is obtained and precipitation never occurs. It is possible to cast clear films of such solutions. However when diethylamine, n-butylamine or ammonia are added to (I) in stoichiometric quantity, the mixture becomes turbid and a precipitate is formed. Usually these precipitates are very insoluble. The amount of insoluble products formed decreases in the order ammonia, primary amine, secondary amine. A routine X-ray analysis was carried out on these products and it was found that they were virtually non-crystalline. However no detailed investigation was made. With solution (II) a precipitate is formed only with ammonia, but not with primary and secondary amines.

So the following regularity is observed: (i) tertiary amines never give precipitates, and (ii) with ammonia, primary and secondary amines the tendency to form precipitates increases as the number of hydrogen atoms at the nitrogen atom increases.

This might be explained as a shielding effect: Primary and secondary amines are less insulated by alkylsubstituents than tertiary amines and therefore give stronger shielding of COO^- groups so that contraction of the polyelectrolyte coils in solution may take place, which under suitable circumstances may cause precipitation. These phenomena are well-known for polyelectrolytes in aqueous solutions[2]. In organic solvents the same kind of reasoning may hold.

However such an explanation is not compatible with the

results of the following experiments using o-carboxybenzanilide as a model compound instead of PAA.

We carried out H^1-NMR measurements on a 20 percent solution of o-carboxybenzanilide (III) in DMSO-d$_6$, which was neutralized with the stoichiometric amount of n-butylamine.

The α-amino methylene protons did not display the shift
corresponding to protonation of the amino group.

Also the OH protons of the carboxylic group failed to
disappear, in agreement with the previous observation.

On the other hand protons of the amino group immediately
showed a downfield shift, indicating that they are bound (possibly as a hydrogen bond, but this could not be decided by
this experiment).

Finally it was noticed that the NH proton of the amide
group of (III) upon the addition of the amine immediately displayed an upfield shift; apparently its intramolecular hydrogen bond[3] was weakened or broken down.

Admittedly these experiments were in DMSO-d$_6$. Therefore,
and because one should be careful to base conclusions on NMR-
measurements only, we carried out subtraction FT-IR measurements on a 20 percent solution of (III) in NMP, neutralized
with n-butylamine (25 mole percent).

The characteristic amino salt band did not appear, apparently the amine is not protonated.

However on drying a film of PAA-butylamine, salt
formation takes place rapidly as a function of time and
temperature, as we noticed from IR-measurements. So in
solution the amino group is not protonated. Hence there is no
reason to assume a shielding effect, as suggested above,
because ionization is absent.

We therefore propose a different explanation, assuming
the formation of polymer complexes due to hydrogen bonding
with the hydrogen atoms of the amines and ammonia, which are
bound to the carboxyl group of the polyamic acid.

Polymer complex formation normally involves two different
polymers. The most characteristic feature is the existence of
ladderlike structures with linear sequences of cooperative
bonds between the two polymers. The complexation may be due to
hydrogen bonds, as is the case with the system polymethacrylic
acid/poly N-methyl-2-pyrrolidone in water or dimethylformamide, DMF[4].

For general information on polymer-polymer complexation
reference is made to recent reviews[5,6,7].

It is usually difficult to characterize polymer-polymer
complexes. Often such products are very insoluble, and this
was also the case with our precipitates. In very dilute solution strong contraction of polymer coils can indicate complex
formation. But it may be difficult to prove the occurrence of
a cooperative effect, which is the characteristic feature of

polymer-polymer complex formation.

In the system PAA-amine hydrogen bonds may be formed e.g. with carbonyl groups from carboxylate groups (in the dry state, see below) (a), from amide groups of the polymer backbone (b), or from carboxyl groups in the case of non-stoichiometry (c), as follows:

(a)   (b)   (c)

The uncommon feature in this case is that the polymer must complex with itself, due to the fact that complementary binding sites are present in one and the same macromolecule. We noticed that by the addition in large excess of N-methylacetamide (NMAC) to the solution of (I) the formation of precipitates upon addition of n-butylamine or ammonia was prevented. It may mean that complexation now is not possible, hydrogen bonds being preferably formed with the hydrogen atoms of the amide group of NMAC.

We carried out viscosity measurements in dilute solution, the polymer concentration being below the homogeneous segment concentration, i.e. the concentration at which the polymer coils touch each other. The reduced viscosity $\eta_{red} = \eta_{spec}/c$ was measured with an Ubbelohde viscosimeter. Solution (I) was diluted with NMP to a concentration of 0.00251 g/ml. From this solution a number of other solutions were prepared by adding increasing amounts of n-butylamine and triethylamine, respectively, expressed in mole percent of amine with respect to the COOH stoichiometry of the PAA. The viscosity of these solutions was measured at 25° C (Figure 1).

Curve 1, for triethylamine (TEA), depicts the normal behaviour of a polyamic acid-amine solution, i.e. the viscosity increases upon the addition of the amine.

Curve 2, for n-butylamine (BuA), shows that at lower concentrations of the amine the viscosity increases. However at higher concentrations, the viscosity decreases, and at 100 or 150 percent stoichiometry of the amine, is considerably

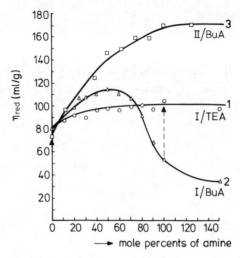

Figure 1. Reduced viscosity vs. mole percent of amine.

lower than the viscosity of solution (I), without amine. This is due to contraction of the polymer coils and it is assumed that cooperative hydrogen bonding takes place, when the degree of neutralization is sufficiently high to allow for the formation of ladder-like structures between chain sequences. From this experiment alone it is not possible to conclude whether this takes place intramolecularly or intermolecularly. In the latter case a strong increase of the apparent molecular weight may take place though the viscosity is low[4],[5],[6]. We also carried out viscosity measurement at 50°C on the PAA solution with the stoichiometric amount of n-butylamine, first measured at 25°C. Figure 1 shows that the viscosity increases strongly (single point as closed triangle). The viscosity of the PAA solution at 50°C is also indicated (closed triangle on the ordinate). Apparently the polymer complex breaks down. This agrees with the weakening of hydrogen bonds at the higher temperature, which is the normal behaviour of hydrogen bonds.

The existence of ladderlike structures of cooperative hydrogen bonds requires that these structures break down over a narrow temperature range (cf. melting). We therefore studied the viscosity of solution (I) with 150 percent stoichiometry of butylamine as a function of the temperature, between 20°C and 50°C. The polymer concentration again was 0.00251 g/ml. These measurements were carried out a year later than those just described. The viscosity of the PAA(I) solution had in-

594

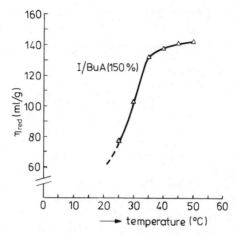

Figure 2. Reduced viscosity vs. temperature.

creased from η_{red}=79.2 ml/g to η_{red}=87.9 ml/g. In Figure 2
all values are also larger than those in Figure 1. Figure 2
shows that the polymer complex indeed breaks down in a narrow
temperature range. The viscosity change is about as large as
that expected in the corresponding case in Figure 1. A reliable measurement was not possible at 20°C, as solution (I) immediately became turbid upon the addition of the amine.

The experiments shown in Figure 1 were repeated with polyamic acid solution (II) (concentration 0.00255 g/ml), the
result for n-butylamine is shown in curve 3. Polymer contraction apparently does not take place, in agreement with the observation that in concentrated solution the addition of
n-butylamine to solution (II) does not cause precipitation.
The curve for the diluted solution (II) with triethylamine
nearly coincides with curve 1 and is not shown.
So in these dilute solutions the polymer based on pyromellitic
acid dianhydride and 4,4´-diaminodiphenylether may contract
but that based on benzophenonetetracarboxylic acid dianhydride
and the same diamine does not. The viscosity increase in curves 1 and 3 may be due to hydrogen bond formation and the molecular weight increase caused by the addition of the amine,
but not by ionization as was shown by our NMR- and FT-IR measurements described above.

Infrared analysis was carried out to obtain further support for the alleged hydrogen bonding mechanism. Unfortunately
this analysis was only possible on dry products.
It was not possible to carry out the analysis on the con-
tracted polymer generated in very dilute solution. Infrared

analysis, of course, cannot prove the existence of ladderlike structures, but it can prove that the "rungs" of the ladders consist of hydrogen bonds.

When ionization of the COOH group of an organic acid occurs, giving the COO^- group, resonance between the two C-O bonds is strong. As a result the characteristic carbonyl absorption splits up into two bands between 1600 cm^{-1} and 1550 cm^{-1} and between 1400 cm^{-1} and 1300 cm^{-1}, which correspond to the antisymmetrical and symmetrical stretching vibrations of the COO^- structure. Hydrogen bonding of the carboxylate group can lead to a small reduction in the frequency of the anti-symmetrical vibration[8].

For the anti-symmetrical vibration, ν_{as}, of the PAA(I) carboxylate salt with triethylamine we found a band at 1587 cm^{-1}. However for both the salts with n-butylamine and ammonia a band appeared at 1582 cm^{-1}. So a small but definite frequency reduction with these hydrogen containing compounds occurs.

For the polyamic acid of solution (I), the stretching vibration of the carbonyl group of the acid amide is located at 1661 cm^{-1} (case 1). When solution (I) is reacted with triethylamine it is located at 1666 cm^{-1} (case 2), with n-butylamine at 1662 cm^{-1} (case 3), and with ammonia at 1656 cm^{-1} (case 4). The measurements were carried out on samples entirely stripped of NMP: There was no NMP band at 1690 cm^{-1}.

This result is interpreted as follows: with triethylamine (case 2) the highest frequency is observed. It may be assumed that triethylamine cancels out the hydrogen bond between the carbonyl group of the amide and the carboxyl group. On the other hand the strongest hydrogen bonding takes place with ammonia (case 4) and relatively weak hydrogen bonding is found in the PAA(I) as such (due to the COOH group) and with n-butylamine.

Though the frequencies and their shifts could be measured accurately it is noticed that they are small. These measurements prove the existence of hydrogen bonds in the dry state. Hydrogen bonds also are highly probable in solution: As is well-known, already in solutions of polyamic acids as such, hydrogen bonds are responsible for their high viscosity (the viscosity of their esters is much lower).

It might be argued that the formation of precipitates in concentrated solution and the polymer contraction and precipitation in dilute solution simply mean that Θ-conditions are attained. However the following specific features favour the view of polymer complexation.
- The regularity with respect to the formation of precipitates observed in the range ammonia, primary amine, secondary and tertiary amine.
- The high insolubility of these products.

- The function of the amine causing chains or chain sequences to be linked.
- The behaviour in dilute solution and specifically the fact that the polymer contraction is cancelled by a small temperature rise, indicating a cooperative effect.

Evidence for the "Mesomorphous State" in Polyamic Acids and in Polyimides

In the context of this subject it is important to comment on the well-known "molecular aggregation" or "mesomorphous state" in polyamic acids upon concentrating and drying, and in polyimides, e.g. in film material[9,10,11]. In a recent study Isoda et al.[9,10] found that "molecular aggregation" of polyamic acid and polyimide cyclized at a low temperature gives amorphous structures. On the other hand, "molecular aggregation" of polyimide cyclized at high temperatures gives non-crystalline two-phase structures, as is also found in the well-known commercial polyimide film, Kapton H (Du Pont).

We suggest that at least the first type of aggregation represents an interesting case of polymer complexation. This conclusion is particularly based on Isoda´s Small-Angle X-ray Scattering measurements of polyamic acid and polyimide cyclized at low temperature[9], and on the observations summed up by Siderovich et al.[11]. On evaporating the solvent from a cast film, sequences of polymer chains may undergo complexation due to cooperative hydrogen bonding, now between carbonyl groups (of the COOH groups) and hydrogen atoms (of the acid amide groups of the polymer backbone). The relative position of complexed backbone sequences may be partially maintained upon imidization[9,10,11].

The idea of a hydrogen bonding mechanism is supported by the work of Ardashnikov et al[3]. They carried out an infrared study on o-carboxybenzanilide, both in different solvents and in the solid state. The existence of intramolecular hydrogen bonds in the solid state was clearly shown. In solution, these bonds are weaker, the extent depends on the solvent. The authors[3], among others, propose the following structure

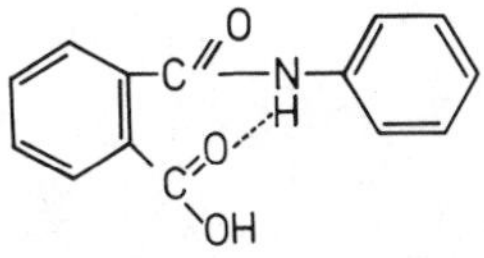

which favours cyclization to the imide (for other hydrogen
bonded structures, see original reference). In polyamic acids
such hydrogen bonds may be formed intra- or intermolecularly
and in a cooperative manner on evaporating the solvent.

Finally we notice that polymer complexation may be the
cause of the observed heterogeneity in polyamic acid solutions[12] (the same polymer as in solution (I) was studied) as
found by turbidity measurements. The explanation proposed by
the authors[12], based on inhomogeneous distribution of the
monomers in the reaction mixture, deserves reconsideration.

REFERENCES

1. H.G. Linde, J. Polym. Sci.: Polym. Chem. Ed. $\underline{20}$, 1031
   (1982).
2. F. Oosawa, "Polyelectrolytes", Marcel Dekker, New York,
   1971.
3. A.Ya. Ardashnikov, J.Ye. Kardash and A.N. Pravedinkov, Polymer Sci. U.S.S.R. $\underline{13}$, 2092 (1971).
4. D.W. Koetsier, G. Challa and Y.Y. Tan, Polymer $\underline{22}$, 1709
   (1981).
5. E.A. Bekturov and L.A. Bimendina, Adv. Polym. Sci. $\underline{41}$, 99
   (1981).
6. E. Tsuchida and K. Abe, Adv. Polym. Sci. $\underline{45}$, 2 (1982).
7. O. Olabisi, L.M. Robeson and M.T. Shaw, "Polymer- Polymer
   Miscibility", Ch. 4,5, Academic Press, New York, 1979.
8. L.J. Bellamy, "The Infra-red Spectra of Complex Molecules", pp. 198-199, Chapman and Hall, London, 1975.
9. S. Isoda, H. Shimada, M. Kochi and H. Kambe, J. Polym.
   Sci.: Polym. Phys. Ed. $\underline{19}$, 1293 (1981).
10. S. Isoda, M. Kochi and H. Kambe, J. Polym. Sci.: Polym.
    Phys. Ed. $\underline{20}$, 837 (1982).
11. A.V. Siderovich, Yu.G. Baklagina, V.P. Stadnik, A.Yu.
    Strunnikov and T.J. Zhukova, Polymer Sci. U.S.S.R. $\underline{23}$,
    1125 (1981).
12. M.M. Koton, Yu.V. Brestkin, V.V. Kudryavtsev, V.P.
    Sklizkova, L.M. Budanova, J.G. Silinskaya, S.Ya. Frenkel
    and O.S. Karchmarchik, Polymer Sci. U.S.S.R. $\underline{23}$, 94
    (1981).

ABOUT THE CONTRIBUTORS

Here are included biodata of only those authors who have
contributed to this volume. Biodata of contributors to Volume 2
are included in that volume.

S. K. Agrawal is with the Physics Department, Banaras Hindu
University, Varanasi, India where he received M.Sc. in Physics in
1975. His areas of interest include ion implantation, hydrogen
energy storage in Ti based alloys, neutron radiation damage of
materials, and space simulation. He is the author of 14 research
papers and shares one patent.

Herbert R. Anderson, Jr. is currently Manager of the Polymer
Science and Technology Group at IBM, East Fishkill, N.Y. He received
his Ph.D. degree in Physical Chemistry from Cornell University
and subsequently carried out postdoctoral research under the guidance of Prof. P. J. W. Debye at the same university. Before coming
to IBM, he was with the General Electric Co. and Phillips Petroleum
Co. In 1974 he was on a sabbatical leave from IBM to collaborate
with Prof. F. M. Fowkes in research leading to the experimental
evidence for acid-base relations in the adhesion of polymers to inorganic surfaces. He has contributed in a number of areas including
chemical kinetics, nuclear reactor technology, printing via offset,
tribology, and various applications of polymers to microelectronics. His work has led to a number of publications, $\sim$25 U.S.
patents, and a number of invited presentations. His work was recognized by the Mid-Hudson Section of the American Chemical Society by
selecting him to receive the Texaco Research Award in 1978. He has
served as Chairman of the Gordon Research Conference on the Science
of Adhesion and as Vice President and President of the Adhesion
Society.

Constance Joan Araps is presently Manager of the Polymer
Device Applications and Technology Department at the IBM Corporation in East Fishkill, N.Y. She received her Ph.D. degree in
Organic Chemistry from Princeton University in 1976. She has worked
in the areas of natural products synthesis, surface chemistry and
polymer science, and has publications and patents in these areas.

599

G. N. Babu is a Lecturer at the Indian Institute of Technology, Bombay, India. He received his Ph.D. degree in Polymer Science from the Indian Institute of Technology, New Delhi and has 48 publications to his credit.

Arnold I. Baise is with the IBM Corporation, Hopewell Junction, N.Y. He received his Ph.D. degree in Chemistry from the University of Wales, U.K.

A. Berger is currently Director of Research of PSI Venture Group at M&T Chemicals in Rahway, N.J. Prior to coming to M&T in 1977 he was in his own business (Bergston & Associates) in modification of polyimides via incorporation of aminofunctional-capped polysiloxanes. From 1960-1975 he was with the General Electric Company where he had more than 100 U.S. patents issued and was active in the program developing Ultem® . He received his Ph.D. in Chemistry in 1960 from the University of Kansas.

M. I. Bessonov is Head of the Polymer Mechanical Properties Laboratory of the Leningrad Institute of High Molecular Compounds, Leningrad, USSR. Has Doctor of Physics degree and has published one hundred papers on physical properties of polymers (including thermostable).

Ravi Prakash Bhardwaj is a Lecturer in Physics at S.G.R.R. Post Graduate College of Science at Dehra Dun, India. He received M.Sc. in Physics in 1971. In 1980, he was awarded a Teacher-Fellowship under Faculty Improvement Programme of Government of India to work at Kurukshetra University where he is associated with the research group headed by B. L. Sharma. His specific field of study is thermo- and photopolarization phenomena in polyimides.

R. F. Boehme is a Senior Associate Engineer at the IBM Thomas J. Watson Research Center, Yorktown Heights, N.Y. where he has been since 1981. He is currently studying EXAFS, and x-ray scattering for structural analysis of materials. He received his M.A. in Teaching Chemistry from Boston University in 1976 and was a Research Associate at the State University of New York from 1979-1981. He is the author of over seven papers.

P. L. Buchwalter is a Senior Associate Engineer at IBM Corporation, Hopewell Junction, N.Y., where she has been since 1980. She received M.S. in Chemical Engineering in 1976 from Helsinki University of Technology, Helsinki, Finland.

Harold D. Burks is a Materials Engineer at NASA Langley Research Center, Materials Division, Hampton, VA, where he has been since 1962. He was previously employed by the Air Force as a physicist at Tinker Air Force Base, OK. He received a B.S. in Physics in 1958 from Central State University, Edmond, OK. At Langley, he has been involved in a number of studies and is currently engaged in the development and characterization of experimental high temperature polymers.

G. S. Cargill, III is Project Manager for Microanalysis and Microscopy at the IBM Watson Research Center, Yorktown Heights, N.Y. where he has seen since 1975. He received his Ph.D. in Applied Physics in 1969 from Harvard University. He served as General Chairman for the Conference on Magnetism and Magnetic Materials in 1981. He was co-chairman of the Topical Conference on Atomic Scale Structure of Amorphous Solids held in 1978, and has been a member of the International Advisory Committee for the International Conference on Rapidly Quenched Metals since 1977. He has also served as   SRC Senior Visiting Fellow at the Cavendish Laboratory, Cambridge University, and as Overseas Fellow, Churchill College, Cambridge. He has published more than 55 papers in the fields of amorphous materials, magnetism, x-ray scattering and electron microscopy.

Robert D. Case is employed by the Armak Co., McCook, IL. He received his M.S. degree in Chemistry from Wright State University.

U. K. Chaturvedi is with the Physics Department, Banaras Hindu University, Varanasi, India where he received his M.Sc. in Physics in 1973. His areas of interest include space technology, thin films, accelerators, polymers, solar physics and solar energy, and has published more than 10 papers and shares one patent.

B. Chowdhury has since 1976 been a Senior Research Chemist for M&T Chemicals, Rahway, N.J. Prior to his current position, he was employed at the Bristol Laboratories, Syracuse, N.Y. (1970-1972) and Schering Corp., Bloomfield, N.J. (1972-1976). He received his Ph.D. in Materials Science from the California Western University. He has been involved in a number of areas including characterization of thermosets, thermoplastics and elastomers, and is currently studying vapor permeability and stressed viscoelastic field for possible elucidation of structure-property relationship due to diffusion boundaries of imperfect solids. He has authored a number of research publications.

Patricia M. Cotts is currently a Research Staff Member in the Polymer Science and Technology Department at IBM Research Laboratory in San Jose, CA. She received her Ph.D. degree in 1979 from Carnegie-Mellon University, Pittsburgh. Her interests include solution properties of unusual polymers such as rodlike and semiflexible polymers, polyelectrolytes, and water soluble polymers.

David R. Day is presently a Research Associate in the Department of Electrical Engineering at MIT. He received his Ph.D. degree in 1980 which was carrried out as a joint project both at Case

Western Reserve University and the University of Mainz, W. Germany. His research interests have included single crystal diacetylene polymerization, diacetylene monolayer polymerization, and monolayer structure analysis. His current research involves polymers for use in microelectronics and integrated-circuit dielectric property-sensors for chemical reaction monitoring.

Sue E. Delos is a Research Associate in the Chemistry Department at the College of William and Mary, Williamsburg, VA. She received her B.S. from the University of Michigan (1965) and her M.A. from the College of William and Mary (1979).

Carl E. Diener is a Staff Chemist with the IBM Corporation, Endicott, N.Y. where he has been employed for 19 years. Previously he held a position with the GAF Corporation, Binghamton, N.Y. He received his B.S. in Mathematics/Chemistry from Elmira College, and is the coholder of several patents in the field of photopolymerization.

P. J. Dynes is a member of the Technical Staff in the Science Center's Polymer Synthesis and Processing Group, Rockwell International, Thousand Oaks, CA which he joined in 1971. He received an M.S. in Chemistry in 1969. He has been involved in a number of research areas including hydrothermal aging of composites, and surface energy properties of graphite fibers. His current research includes the viscoelastic analysis of thermoset curing and the development of instrumental methodology for characterizing composite matrix resins. He has 32 publications to his credit.

Stephen A. Ezzell is currently a graduate student at VPI&SU, Blacksburg, VA. He graduated from St. Andrews College in 1981 and was employed at NASA-Langley Research Center during the summers of 1980 and 1981.

William A. Feld is Associate Professor of Chemistry at Wright State University, Dayton, OH. He received his Ph.D. degree in Chemistry from the University of Iowa and conducted postdoctoral research at Wright State University.

Paul D. Frayer is Senior Research Engineer in the Polymer Composites Group at Rogers Corporation, Rogers, CT. Prior to his current position, he had held a number of industrial positions and was Associate Professor, Polymer Science Department, University of Southern Mississippi (1974-1976). He received his Ph.D. in Macromolecular Science in 1970 from Case Western Reserve University and has a number of technical papers and three patents to his credit. He is the recipient of Tappi Honorable Mention Award for technical presentation, St. Louis Conference 1973; and Excellence in Teaching Award, USM, 1973-1974. His main interests are polymer processing, composites and surface modification of materials.

Michael Fryd is currently a Research Associate with the Fabrics and Finished Products Department of DuPont in Philadelphia, PA where he has been from 1963-1974, and 1980-present. During 1974-1980 he was Assistant Professor of Organizational Behavior, Rutgers University Graduate School of Business. He received his Ph.D. in Organic Chemistry from New York University, and Ph.D. in Group Dynamics from Temple University. He is the author of 24 patents and 4 papers on free radical reactivity, graft and block copolymers, nonaqueous dispersion polymerization and polyimide chemistry. He is a Fellow of the American Institute of Chemists and Sigma Xi.

Koji Fujisaki is with the Hitachi Research Laboratory, Hitachi-shi, Ibaraki-ken, Japan where he has been since 1961. He graduated from the Hitachi Technical High School in 1960. His interest is in the application of various polymers for electrical equipment and electronic devices.

I. D. Gaba is leading a group of polymer chemists at the Defence Science Centre, Delhi, India. He obtained his Ph.D. degree from Agra University and certificate in Characterization of Polymers from IIT, Kharagpur. For the last six years, he is engaged in the development of thermally stable polymers such as polybenzimidazoles and polyimides in different forms, and has developed a number of monomers and intermediates for the synthesis of these polymers. Presently, he is working on the fabrication and characterization of fiber-reinforced polyimide composites.

Romeo M. Galeos is with the Lockheed Advanced Composites Laboratory of Lockheed Missiles and Space Company, Sunnyvale, CA where he is now specializing in rheological testing. Prior to his current position, he had worked at Hexcel Corporation and Ford Aerospace Advance Materials Laboratory. He has been involved in the evaluation and analysis of polyimides and epoxy resin systems since 1971. He received his B.S. in Chemical Engineering from the University of San Carlos, Philippines in 1966.

Rochelle Ginsburg is currently a Senior Associate Chemist with IBM Corporation in Endicott, N.Y. She has an M.A. degree in Biological Sciences from SUNY/Binghamton. She has previously published in the fields of electron microscopy and bioorganic chemistry.

Jeremy Greenblatt is currently a postdoctoral Fellow with IBM Corporation, Hopewell Junction, NY which he joined in July 1981. He received his Ph.D. degree in Organic Chemistry in 1980 from the Hebrew University of Jerusalem, Israel and subsequently carried out postdoctoral research at Wayne State University. He was the recip-

603

ient of various honors and awards based on his scholastic excellence. He is interested in physical and synthetic organic chemistry, plasma deposition and surface characterization of polymers, adhesion of coatings, and use of polymers in microelectronics, and has published about 10 papers.

Ravi K. Gupta is a research and development Scientist at the Defence Science Centre, Delhi, India. He obtained his M.Sc. in Chemistry from Delhi University, India, and D.I.I.T. in High Polymer Technology from Indian Institute of Technology, Kharagpur, India. For the last four years, he has been working on the development and characterization of high temperature resistant polymers, especially polyimides for uses as coatings, films, and matrix resins.

C. L. Hamermesh is Manager of the Polymer Synthesis and Processing Group of the Science Center, Rockwell International, Thousand Oaks, CA. Since coming to Rockwell (formerly North American Aviation), he has been engaged in a broad range of programs involving polymers. Recent efforts have involved characterization of polyimide resins for graphite composites, evaluation of the effect of new cross-linking agents for PI's on the performance of graphite composites, and the curing efficiency of ethynyl terminated polymers. He received a Ph.D. in Chemistry from New York University in 1954. He is a member of the staff of the Extension School at UCLA where he teaches a course in polymers in engineering. He has 20 publications and a number of patents.

Frank W. Harris has recently accepted a position as Professor of Polymer Science at the University of Akron. He is also President of Daychem Laboratories, a company that carries out custom synthesis and scale-up work. Prior to moving to the University of Akron, he was Professor at Wright State University which he had joined in 1970. He received his Ph.D. degree in Chemistry from the University of Iowa in 1968 and then served two years as a Captain in the U.S. Army at Edgewood Arsenal, MD. He has been active in the ACS Division of Polymer Chemistry, chairing the Program Committee for the past six years. He has also been active in the Controlled Release Society and served as its first President. This year he received that Society's Founders Award. In 1982, he received the Dayton Engineering and Science Foundation's award for Outstanding Professional Achievement. He has published over 30 scientific papers, edited three books and chaired several symposia in the area of polymer science. In addition to developing polyimide systems for electronic and aerospace applications, he is currently pursuing projects on controlled release systems and membranes for advanced alkaline electrolyzers.

Lewis R. Iler is presently completing his M.S. degree in Chemical Engineering at North Carolina State University, Raleigh working in the area of interactions of penetrants in glassy polymer environments. He received his B.S. degree from the same institution, graduating Magna Cum Laude in May 1981. During his college years, he received three awards recognizing his undergraduate research projects.

Brian J. Jensen has been employed at the NASA Langley Research Center, Hampton, VA since 1980 and is currently working in the Advanced Materials Branch of the Materials Division on the synthesis and characterization of aerospace polymers. He is currently enrolled in the graduate program in Chemistry at the College of William and Mary.

Patricia K. Jue is currently a graduate student at Carnegie-Mellon University, Pittsburgh, PA. She received her B.S. from the College of William and Mary in 1981.

Eugene Khor is a Research Assistant at the Virginia Polytechnic Institute and State University, Blacksburg, VA where he began working toward his Ph.D. in 1979. His dissertation topic is on the synthesis and characterization of metal-doped polyimide films.

Noriyuki Kinjo is a Senior Researcher of Hitachi Research Laboratory, Hitachi-shi, Ibaraki-ken, Japan. He received his Dr. degree in Polymer Science from Hokkaido University, Sapporo, Hokkaido, Japan in 1974. He is a coauthor of Practical Polymers: An Introductory Text for Engineers (Japanese). His current research interests focus on the development of novel organic polymers and their application in electronic devices.

William J. Koros is in the Faculty of the Chemical Engineering Department at the North Carolina State University, Raleigh, which he joined in 1977. He received his Ph.D. degree in Chemical Engineering from the University of Texas at Austin. He is currently interested in polymer-fluid interactions and has authored/coauthored over 30 papers in the area of gas and vapor sorption and diffusion in polymers.

M. M. Koton is Head of the Polymer Synthesis Laboratory and the Director of Leningrad Institute of High Molecular Compounds, Leningrad, USSR. Has Doktor of Chemistry degree and is the Correspondent Member of the Academy of Sciences. Dr. Koton has published 400 papers in different areas of polymer chemistry.

David E. Kranbuehl is Professor of Chemistry at the College of William and Mary, Williamsburg, VA. He received his Ph.D. degree in Physical Chemistry from the University of Wisconsin in 1969 and was employed at the National Bureau of Standards - Polymer Division before coming to William and Mary in 1970.

605

V. V. Kudriavtzev is a senior member of the Polymer Synthesis Laboratory of Leningrad Institute of High Molecular Compounds, Leningrad, USSR. Is a Candidate of Chemistry and has published 70 papers on the synthesis and physicochemical properties of thermostable polymers.

N. P. Kuznetsov is a member of the Polymer Mechanical Properties Laboratory of the Leningrad Institute of High Molecular Compounds, Leningrad, USSR. Is a Candidate of Physics, and has published 40 papers on the physical properties of polymers (including thermostable).

L. A. Laius is a senior member of the Polymer Mechanical Properties Laboratory of the Leningrad Institute of High Molecular Compounds, Leningrad, USSR. Is a Candidate of Physics and has published 70 papers on the physical and physico-chemical properties of polymers (including thermostable).

David Landman is an Associate Scientist at Hysol Division, The Dexter Corporation, Pittsburg, CA, where he is responsible for development of high temperature service adhesives and resins. Prior to joining Hysol, he was employed by Milliken Chemicals where he was responsible for technical development of epoxy curing agents. He received his Ph.D. degree in Chemistry from the University of Sydney, Australia.

Abraham L. Landis is Senior Scientist, Materials Technology Department, Materials and Processes Laboratory, Hughes Aircraft Company where he has been since 1961. He has been involved in recent years in the synthesis and development of new high temperature polymers for composites and adhesives. In this capacity, he was one of the coinventors of the high temperature acetylene-terminated polyimide (HR-600 Series) polymers, and has also developed many other resins such as stannosiloxane elastomers, polyferrocenylene polymers, and polyphenylene resins. Prior to joining Hughes Aircraft Company, he was engaged in high temperature chemistry and thermodynamics at Atomics International, and high energy fuels at NASA-Lewis Flight Propulsion Center. He received his Ph.D. in Organic Chemistry in 1955 from the University of Kansas and has 8 journal publications and 10 patents to his credit. He is listed in American Men of Science.

Lynn H. Lanier is employed at Monsanto Chemical Company, Dekatur, AL. He received his M.S. degree in Chemistry from Wright State University.

T. T. Liao is a Staff Associate in the Science Center's Polymer Synthesis and Processing Group, Rockwell International, Thousand Oaks, CA where he has been for two years and has been involved in the development of instrumental techniques for char-

acterization of new composite matrix resin systems. He received an
M.S. in Chemical Engineering in 1980.

Milan G. Maximovich is a Materials and Processes Specialist
in the Advanced Composite Technology Group at Lockheed Missiles and
Space Company, Sunnyvale, CA which he joined in 1980. Prior to his
current position, he has held research and managerial positions in
a number of industries and was a Program Manager, Advanced Mater-
ials/Polymer Sciences, SRI International. He has extensive experience in the investigation, characterization, fabrication and
testing of advanced composites as well as a broad background in a
number of other areas. He did graduate studies in Polymer Chemistry
(1960-1963) at Stanford University and has numerous publications,
presentations and book chapters to his credit.

Lourens Minnema is a Senior Scientist with Philips Research
Laboratories, Eindhoven, The Netherlands where he has been since
1974. He has been working in the field of Polymer Chemistry and
Technology since 1953 and obtained a Ph.D. degree in 1957 from
Leiden University.

*Kashmiri Lal Mittal** is presently employed at the IBM Corpor-
ation at Hopewell Junction, N.Y. He received his M.Sc. (First Class
First) in 1966 from Indian Institute of Technology, New Delhi, and
Ph.D. in Colloid Chemistry in 1970 from the University of Southern
California. In the last ten years, he has organized and chaired a
number of very successful international symposia and in addition to
these volumes, he has edited eighteen more books as follows:
Adsorption at Interfaces, and Colloidal Dispersions and Micellar
Behavior (1975); Micellization, Solubilization, and Microemulsions,
Volumes 1 & 2 (1977); Adhesion Measurement of Thin Films, Thick
Films and Bulk Coatings (1978); Surface Contamination: Genesis,
Detection, and Control, Volumes 1 & 2 (1979); Solution Chemistry of
Surfactants, Volumes 1 & 2 (1979); Solution Behavior of Surfactants
- Theoretical and Applied Aspects, Volumes 1 & 2 (1982); and
Physicochemical Aspects of Polymer Surfaces, Volumes 1 & 2 (1983);
Adhesion Aspects of Polymeric Coatings, (1984); Surfactants in
Solution, Volumes 1, 2 & 3 (1983); and Adhesive Joints:
Formation, Characteristics and Testing, (1984). Also he is
Editor of the series, Treatise on Clean Surface Technology, a
multi-volume work in progress. In addition to these books he has
published more than 50 papers in the areas of surface and colloid
chemistry, adhesion, polymers, etc. He has given many invited talks
on the multifarious facets of surface science, particularly adhesion, on the invitation of various societies and organizations in
many countries all over the world, and is always a sought-after
speaker. He is a Fellow of the American Institute of Chemists and
Indian Chemical Society, is listed in American Men and Women of

---

* As the editor of this two-volume set.

Science, Who's Who in the East, Men of Achievement and many other reference works. He is or has been a member of the Editorial Boards of a number of scientific and technical journals. Currently, he is Vice-President of the India Chemists and Chemical Engineers Club.

Paul A. Moskowitz is presently working at the T. J. Watson Research Center, IBM Corporation, Yorktown Heights, N.Y. which he joined in 1977. He received his Ph.D. in 1971 in Physics from New York University. He was awarded an NSF Fellowship to work at the University of Grenoble, France, 1971-1973, followed by research work at the University of Mainz, West Germany, 1974, and the Joint Institute for Laboratory Astrophysics, Boulder, CO, 1975-1977. He has been involved in a number of research areas and is presently working on laser induced reactions, and superconductivity research.

K. K. Nagpaul is Professor and Head of the Physics Department at Kurukshetra University, India. During 1956-59, he worked in the High Energy Physics Group of Nobel-Laureate Prof. C. F. Powell of Bristol University (U.K.) where he obtained his Ph.D. degree. His current research interests include Fission Track Dating, a field in which he enjoys international standing. He has more than 100 research publications to his credit and has guided research of a dozen doctoral students.

Arthur B. Naselow is Manager, Materials Technology Department, Technology Development Laboratory, Hughes Aircraft Company. His activities at Hughes Aircraft have included development of processable polyimides (Thermiol 600P), colorless polyimides for solar cell and microelectronics applications, plasma cleaning processes for multilayer printing wiring boards and flexible and rigid/flex printed wiring boards. Before joining Hughes, he worked for the DuPont company for 17 years. At DuPont, he was instrumental in the development of "Pyralin" polyimide high temperature polymer systems, and his research efforts in the area of perfluorinated polymers led to the development of "Teflon S" coatings. He received his Ph.D. in Organic Chemistry from Temple University in 1963 and JD (Law) from San Fernando Valley College of Law in 1981. He has 2 journal publications and 9 patents in the high temperature polymer field, and is listed in American Men of Science.

M. Navarre is presently in charge of a group working on polymers and organic compounds characterization in the Chemical Laboratory of IBM Essonnes Plant, Corbeil-Essonnes, France. She has been with IBM France for 10 years and has been involved in many areas including resists and developers. She received her 3rd cycle Doctorat (Mineral Chemistry) from Jussieu University in Paris, France, and the subject of this doctorat was about photosensitive resists.

A. K. Nigam is presently Reader in Physics Department, Banaras Hindu University, Varanasi, India. He received his D.Phil. degree in 1969 from Saha Institute of Nuclear Physics, Calcutta. He was awarded Alexander von Humbold Fellowship (1974-1975) and UGC National Associateship (1976-1981). His areas of interests are radiation damage and ion implantation, Mössbauer spectroscopy, and nuclear spectroscopy and reaction. He is the author of about 70 research publications and one patent.

Steve O. Norris is employed by the Arco Company, Newton Square, PA. He received a B.S. degree in Chemistry from Wright State University and a Ph.D. degree also in Chemistry from the University of Iowa.

Shun-ichi Numata is with the Hitachi Research Laboratory, Hitashi-shi, Ibaraki-ken, Japan where he has been since 1970. He graduated from the Division of Applied Chemistry, Tohoku University, Sendai, Japan in 1970. His research interests are development of heat-resistant polymers, toughened thermoset resins, and polymers for semiconductor devices.

Sridhar M. Padaki is currently with Glidden/SCM Corporation in Strongville, OH. He received his M.S. degree in Chemistry from Wright State University and a Ph.D. degree also in Chemistry from Colorado State University.

C. D. Pande is a Senior Scientist in Special Materials Division of Defence Science Centre, Delhi, India. He received his Ph.D. degree from Agra University, India, and has been working on the development of urethane, isocyanurate and polyimide polymers, specialty coatings and foams. He has developed several resin formulations based on castor oil and aleuritic acid, a lac product, for production of coatings and foams. He is the author of a number of papers and patents in these areas.

Roscoe A. Pike has been with the United Technologies Research Center, E. Hartford, CT since 1967 as a Senior Materials Scientist in the Manufacturing Technology and Process Research Group. Prior to that, he was associated with Norton Company and the Silicones Division of Union Carbide He received his Ph.D. degree in Organic Chemistry from MIT in 1953. His patents and publications cover the areas of silicones, cutting fluids, high temperature resin based composites and adhesives, fuel additives and fibrillar carbon.

P. K. C. Pillai is Professor of Physics at the Indian Institute of Technology, Delhi, India where he is heading a large research group in electrets, electrophotography and polymer physics and has guided more than two dozen doctoral students. He received his Ph.D. degree in 1963, and as Colombo Plan Fellow in 1963-1964 he did advanced training and research at the Imperial College of

Science and Technology, London. During 1973-1975, he was a Senior Resident Research Associate at the Marshall Space Flight Center, USA. In 1980, he took an assignment with the Nigerian Government to head the Physics Department at the University of Sokoto for a period of two years. He is internationally known for his work in the field of electret and polymer physics and is the author of over 250 research publications and holder of a number of patents.

Jitendra Quamara has since December 1980 been working as a Research Associate in the Applied Physics Department of Regional Engineering College, Kurukshetra, India on electret and polymer physics. He received his Ph.D. degree in 1979 form Bhopal University, India for his work in the area of magnetoelectrets. He has 20 publications to his credit.

V. N. Rai is currently working for his Ph.D. degree in the Physics Department, Banaras Hindu University, Varanasi, India where he received his M.Sc. in Physics. His research interests include photoacoustic spectroscopy, optical properties and enhanced scattering of surfaces, and laser induced spectroscopy.

Bruce A. Reinhardt is employed by the Polymer Branch of the U.S. Air Force Materials Laboratory, Wright-Paterson Air Force Base, OH. He received his M.S. degree in Chemistry from Wright State University.

Robert K. Schellenberg received his B.S. from the College of William and Mary in 1980.

Stephen D. Senturia is Professor of Electrical Engineering at MIT which he joined after receiving his Ph.D. in Physics from the same institution in 1966. His research interests have included NMR and other physical studies in semiconductors. A parallel thread in his research has been the application of the methods of physical measurement to practical problems. These have included high-precision thermometry, oil-well logging, NMR instrumentation methods, and the automated reclamation of resources from urban solid waste. His most recent research development is the charge-flow transistor, a new solid state device that permits the fabrication of MOS integrated-circuit sensors for moisture measurement and chemical reaction monitoring. He is the coauthor of Electronic Circuits and Applications, a widely used introductory electronics text.

William Serfaty is Manager of Technical Marketing for the Ultem® Products Section, Plastics Business Operation, General Electric Company, Pittsfield, MA. He was previously employed by Eastman Kodak Company where he held several positions in research, sales and international marketing. He received his Ph.D. in Organic Chemistry from Purdue University and has published several scientific and technical articles. In 1982 he was the recipient of the

"Best Thermoplastics Materials and Foams Paper Award" of the SPE-ANTEC Conference.

B. L. *Sharma* is Professor of Physics and Chairman of the Department of Applied Physics at Regional Engineering College, Kurukshetra, India where he has been since 1968. He received his A.M. degree from Harvard University in 1963 where he was on a Gordon McKay Scholarship. From 1963-1965, he was with the General Dynamics/Electronics in Rochester and here he carried out the first known frequency modulation of a He-Ne laser using an intra-cavity electro-optical technique. In 1965 he joined the Central Scientific Instruments Organization in Chandigarh, India where he developed a number of electro-optical instruments including an Electronic Flame Photometer. His research interests include lasers, instrumentation, electrets, and electrophotography and has over 25 research publications. He holds a Ph.D. degree from Indian Institute of Technology, Delhi.

Vagish Shrinet is working toward his Ph.D. degree in the Physics Department, Banaras Hindu University, India, working on ion implantation in polymers. He had his graduate and postgraduate education from Gorakhpur University. His fields of interest include polymeric coatings, accelerators and solar energy, and has published several papers in these areas.

R. P. *Singh* is presently Assistant Professor involved in teaching and research in the fields of polymer physics and engineering at the Indian Institute of Technology, Kharagpur, India. Obtained   D.Phil. degree from Allahabad University, India in 1968 followed by postdoctoral research at the University of Glasgow and Liverpool (1968-1972). Was a Pool Officer at the University of Allahabad (1972-1975) and a Visiting Assistant Professor of Physics at the University of Baghdad (1975-1976). Dr. Singh has authored over 60 papers in scientific journals and conference proceedings.

Y. P. *Singh* is presently a graduate student in the Materials Science Centre, Indian Institute of Technology, Kharagpur, India working on multiphase polymeric systems. Had education at the University of Allahabad and is the author of six research papers.

Terry L. St. Clair is a Senior Scientist at NASA Langley Research Center, Hampton, VA where he has been since 1972. He was previously employed by E. I. duPont, Waynesboro, VA. He obtained a Ph.D. in Organic Chemistry from VPI&SU in 1972. His research at NASA has involved the development of high termperature resins for adhesive and matrix resin applications.

Sidney W. Street is New Product Manager with the U.S. Polymeric in Santa Ana, CA. Before coming to U.S. Polymeric in 1976, he had held managerial positions with a number of industrial concerns. He holds a B.S. degree in Chemical Engineering from Ohio State University and Business Management certificate from the Graduate School of Business at UCLA. He holds a patent and has had numerous technical papers published. Recently, he has been very active in polyimide synthesis and in the development of polyimide matrices for advanced composites. In addition to the development of V-378A, he has been a major contributor to the commercialization of PMR-15 and LARC 160 for NASA's CAST programs.

John R. Susko is currently a manager of a Materials Science Packaging organization at IBM Corporation, Endicott, N.Y. He received his B.S. degree from Elmira College. He has extensive experience in materials characterization and polymer science with numerous publications, particularly in the area of gas vapor transport phenomena in polymers. His patents are predominantly in polymer applications to electronic packaging.

V. M. Svetlichny is a member of the Polymer Synthesis Laboratory of Leningrad Institute of High Molecular Compounds, Leningrad, USSR. Is a Candidate of Chemistry and has published 30 papers on the synthesis and properties of thermostable polymers.

Larry T. Taylor is Professor of Chemistry at VPI&SU, Blacksburg, VA. He recieved his Ph.D. degree in Inorganic Chemistry in 1965 from Clemson University and spent two years (1965-1967) as Research Associate at Ohio State University. He was NASA-ASEE Faculty Fellow at Langley Research Center in 1976. He has co-authored about 100 papers in the areas of synthesis and characterization of metal-containing polymers, modification of polymer properties, small molecule reactivity with transition metal complexes, and analytical techniques for the speciation of complex synfuel mixtures.

Margarita Torres is currently working on an internship at the University of Chicago. She received a B.S. degree in Chemistry from Wright State University and an M.D. degree in medicine from the University of Cincinnati.

M. I. Tsapovetsky is a member of the Polymer Mechanical Properties Laboratory of the Leningrad Institute of High Molecular Compounds, Leningrad, USSR. Is Candidate of Physics and has published 10 papers in polyamic acid's cyclization.

Johan M. van der Zande is with the Philips Research Laboratories, Eindhoven, The Netherlands which he joined in 1979 after receiving his certificate in Chemical Engineering.

Sudarsanan Varaprath is currently conducting postdoctoral research at Iowa State University. He received an M.S. degree in

Chemistry from Wright State University and a Ph.D. degree also in Chemistry from Colorado State University.

Donald R. Vigliotti is with the IBM Corporation, Yorktown Heights, NY where he has been since 1968. At IBM he has worked mostly in the fields of microwave acoustics, laser micromachining, and plating and etching, and has coauthored several papers in these and other areas. Before joining IBM, he worked in the field of oceanography at the Hudson Laboratories (a Columbia University research laboratory) in Dobbs Ferry from 1957-1968. His primary job function was laboratory and field instrumentation, and he was responsible for the design and development of a precision electronic detonator for underground acoustic studies.

Willi Volksen is a research staff member in the Polymer Science and Technology department, IBM Corporation, San Jose Research Laboratory. He received his Ph.D. in Polymer Chemistry from the University of Lowell and spent one year as a post-doctoral fellow at Caltech and one year as a research chemist at the Jet Propulsion Laboratory before joining IBM in 1977. He is the author or coauthor of several patents and numerous publications in the area of polymer synthesis and characterization.

Robert J. von Gutfeld has been with the IBM Corporation, Yorktown Heights, N.Y. since 1960. Before joining IBM, he was with the Sperry Gyroscope Company (1957-1960) and worked on the study and design of electron beams for travelling wave tubes. His early work includes semiconducting switching speed phenomena and the study of heat pulse propagation in solids. Subsequent to these studies, investigations with dye lasers were undertaken, and this work resulted, inter alia, in the first nitrogen laser pumped dye laser, and a demonstration of a nanosecond write feasibility for write-read-erase optical memories using nitrogen pumped dye lasers to record information on chalcogenide films. His most recent work has been on laser enhanced plating and etching. He received his Ph.D. degree in 1965 from New York University and has authored/coauthored 46 technical publications. He is a Fellow of the American Physical Society.

E. F. Witucki is a member of the technical staff at the Rocketdyne Division of Rockwell International Corporation in Canoga Park, CA. He received an M.S. in 1953 from the University of Minnesota. He has considerable experience in synthetic chemistry and most recently he has been engaged in the synthesis of high energy binders of the poly-nitro-fluoro types. He has 20 publications and patents involving primarily the synthesis of new fluoro-nitro compounds.

David A. Yamaki is currently employed as an engineer with the Mobil Research and Development Corporation in Paulsboro, NJ. He was

613

employed at NASA Langley Research Center as a Research Assistant in the summer of 1980 when this work was done. In 1981, he received his B.S. in Chemical Engineering from MIT.

Dyi-Kang Yang is currently working toward his Ph.D. degree in Chemical Engineering at North Carolina State University, Raleigh, where his research is focused on the sorption and transport of water in Kapton® . He received his B.S. degree from the National Center University in Taiwan.

Philip R. Young has been employed with the NASA Langley Research Center, Hampton, VA since 1962 and is currently working the Advanced Materials Branch of the Materials Division on the characterization of aerospace polymers. He received his Ph.D. degree in Chemistry in 1976 from Virginia Polytechnic Institute and State University, Blacksburg, VA.

Robert Yui is currently working toward an M.S. in Chemical Engineering at North Carolina State University, Raleigh where he is interested in the characterization of the sorption and transport of gases in glassy polymers. He received his B.S. from the University of Illinois at Urbana-Champaign.

614

INDEX

Pages 1-614 appear in Volume 1
Pages 615-1182 appear in Volume 2

94
JB